Spatial Aspects of Development

Spatial Aspects of Development

Edited by

B. S. HOYLE

Lecturer in Geography
University of Southampton

JOHN WILEY & SONS
London · New York · Sydney · Toronto

Library of Congress catalog card number
73-2785

ISBN 0 471 41753 X

Made and printed in Great Britain by
The Garden City Press Limited
Letchworth, Hertfordshire SG6 1JS

Foreword

At a time when mankind is becoming increasingly aware of the limitations of his environment and when we must consider earnestly the ways and means at our disposal for the servicing and maintenance of our planet, a wide variety of specialised and interdisciplinary studies are being made in the field of development. In this context, contributions from geographers concerned with the understanding and eventual solution of development problems are particularly welcome. The spatial dimension is an important, basic framework within which considerations of developmental problems must be set; and contrasts between the problems of the less-developed countries and those of the advanced nations, both in environmental and socio-economic terms, require careful analysis. In these respects geographers who have lived and worked, often for many years, in the less-developed countries, have frequently made, and continue to make, important contributions.

The essays presented in this volume reflect both established traditions and new departures within the field of the geography of development. On the basis of original field work, methods of approach vary from synthetic description to analytical quantification and model-building. A basic objective throughout is to increase understanding of existing situations and to suggest new strategies which might be adopted in an effort to solve certain developmental problems. Varied ecological backgrounds, differences in history, society and culture, have led to rather widely varying perceptions on the part of governments, both in less-developed and in advanced countries, as to methods of attacking developmental problems. Geographical enquiry, while stressing this dual variety, may also help to illuminate interrelationships between contrasting areas of experience and activity.

It is important to stress repeatedly the significance to mankind of the less-developed world, perhaps especially the tropical areas where many ecosystems have a high biological productivity along with a great variety of animal and plant species. The trouble is that we still do not know enough about how such systems work to be able to manipulate and to tap their resources without destroying them. Conditions in tropical areas differ so markedly from those in temperate latitudes that we cannot just transfer our conventional agricultural methods there without substantial modifiction. Decomposition rates are 5–10 times higher than in, for instance, a tundra habitat; diseases are more aggressive and there is often no clear seasonal cycle. In this connection, the geographical approach spanning the wide spectrum from the physical environment to the spatial patterns of population, settlement, economy and political life is very helpful in the context of proposed action.

In comparison with the industrialised countries agricultural productivity in many tropical countries is very low. There is a lack of technical equipment, of fertilisers and pesticides. The infrastructure for communication, for marketing and for distribution is inadequate; although a more rapid industrialisation is important to trigger off the economic 'take-off', it seems that a main impulse must come from agricultural development. In many ways, the simple techniques of traditional tropical agriculture reflect social, political and religious situations and are also related to psychological attitudes amongst populations who would aim at reaching in a few generations a level of development that has taken the currently advanced countries many centuries to achieve.

Although we should not unquestioningly accept many figures on world food supplies, it is estimated that about 20 per cent of the population of the developing countries are hungry and underfed and 60 per cent suffer from various degrees of malnutrition. Protein deficiency amongst young children is a grave problem with severe consequences for their mental development. Out of the total number of the world's 950 million children, perhaps about 600 million are not fed properly. In order to change these conditions, the total food production in some developing countries must be quadrupled by the year 2000 and the production of animal foodstuffs must also be substantially increased. Although production figures during recent years underscore the success of the 'green revolution', its social and economic impacts are far from clear and more research is needed in the field. In addition, the 'green revolution' carried several considerable ecological risks. Borlaug, one of its architects, warned about population expansion as a 'frightening monster'. Solving the food problem alone is no solution. 'Miracle crops' can only buy us a few years of time. Much of the agricultural export surplus of the tropics has been used up by the growing population. Between 1920 and 1960 the population increase of the tropics and subtropics has been estimated to be in the order of some 800 million people; and this trend continues.

Attempts to solve the agricultural and nutritional crisis of the tropical world need a gigantic concerted effort of all mankind. Campaigns for literacy and education need large-scale assistance from developed nations. The building up of industries and communications, and the conservation of natural resources, need huge investments which can only be achieved by a more rational distribution of wealth between poor and rich countries. Within the framework of the Second Development Decade of the United Nations, the emphasis during the 1970s is changing from a too narrow-minded perspective of economic growth to a more comprehensive treatment of 'changing man'. It becomes apparent that agriculture, education, technological development, industrialisation and population and environmental control are closely linked with the future evolution of our world in which peace is the main condition for survival. This volume of essays on geography and development in the less-developed world may be considered as a contribution towards this goal. I hope that this book will be widely read in universities and colleges in many parts of the world where the problems of development are now beginning to receive the degree of attention they have long deserved.

W. Manshard
Director
Department of Environmental Sciences
and Natural Resources Research, UNESCO

Preface

The essays contained in this volume have been assembled in order to emphasise the relevance of geographical analysis to the increased understanding and eventual solution of development problems in the less-developed parts of the world. In recent years important changes have taken place both in the methods and application of geographical studies and in approaches to the problems of the less-developed countries. The growth of development studies and of area studies in universities and colleges in many parts of the world has reflected increased interest in, and concern for, the problems of development; and it has emphasised the importance of an interdisciplinary approach to the solution of development problems and the improvement of living conditions for most of mankind. Changes within geography, itself an interdisciplinary field of study, have involved the increased use of quantitative techniques, a broadening range of methods of analysis and growing recognition within other fields of study of the importance of the spatial dimension; for a variety of reasons these trends have been more clearly apparent within the relatively advanced nations rather than in the less-developed countries; and the role of modern geography, both actual and potential, within the sphere of development studies has been inadequately promoted and sometimes misunderstood. There is thus a need for a *rapprochement* between geographical techniques and development studies; the present volume is offered as a contribution towards this end.

The international panel of contributors to the volume have attempted to explain and illustrate various ways in which geographical approaches, methods and techniques may positively contribute towards an increased understanding of the problems faced by the less-developed countries and towards the attainment of higher levels of living. Each contributor has drawn upon his own research experience in the context of a particular problem, a specific area of reference and a selected technique or method of inquiry. No attempt has been made to provide a systematic coverage of the problems of the less-developed countries or of the available range of geographical techniques. The essays presented here should rather be seen as bringing together a series of illustrations and analyses of common problems; methods used range from synthetic description to analytical quantification and model-building; problems surveyed include both environmental hazards and social situations; and areas of reference are drawn from Latin America, Africa, the Middle East and South-East Asia.

The arrangement within the volume of the studies thus linked together by a common theme has followed a conventional pattern: essays dealing with ecological aspects of the environment and with the analysis, appraisal and application of land and water resources are followed by studies of rural and urban development strategies. A larger group of essays deals with agricultural patterns and potentials and with problems of industrialisation and transport development. The political dimension is not neglected and a final essay discusses the geography of development policy.

It is a pleasure to record my gratitude to the contributors for their full cooperation and to the many authorities and individuals in both the advanced and the less-developed countries without whose aid the research on which these essays are based could not have been accomplished. The maps and diagrams were drawn, with much skill and considerable patience, by Mr Alan S. Burn, assisted by Mr R. E. Smith, in the Cartographic Unit of the Department of Geography at Southampton University. I am further indebted to Professor W. A. Hance and Professor R. W. Steel for their helpful advice and to Professor J. H. Bird for the encouragement he has given to the study of less-developed countries at Southampton University.

B. S. HOYLE
Department of Geography
University of Southampton

Contributing Authors

Dr B. S. Hoyle	*University of Southampton*
Dr Anthony Young	*University of East Anglia*
Dr Gustavo Antonini	*University of Florida*
Dr Katherine Carter Ewel	*University of Florida*
Dr John J. Ewel	*University of Florida*
Dr L. W. Hanna	*University of Newcastle-upon-Tyne*
Dr Ian R. Manners	*University of Texas*
Newton Rose, B.S., M.S.	*New York University*
Michael Mortimore, M.A.	*Ahmadu Bello University*
Dr Douglas L. Johnson	*Clark University*
Dr D. R. F. Taylor	*Carleton University*
Dr Reuben K. Udo	*University of Ibadan*
Dr Stella Lowder	*University of Liverpool*
Dr David A. Preston	*University of Leeds*
Dr D. E. Short	*University of Hull*
Dr Alan Gilbert	*University College, London*
Dr J. P. Dickenson	*University of Liverpool*
José A. Smith, M. A.	*Kingston Polytechnic*
Dr B. J. Turton	*University of Keele*
Dr James Osborn	*Center for the Study of Democratic Institutions, Santa Barbara*

Contents

List of Figures

Chapter 1

Spatial Analysis and the Less-Developed Countries

B. S. Hoyle

Introduction

Development studies are part of a wide upsurge of interest, manifest in many parts of the world, in the problems of the less-developed countries. In recent years there has been a growing awareness of the widening gap in levels of development between advanced and less-developed countries, and between regions within the less-developed countries; of the variety of problems involved in the development process, and of the nature of development itself. There are many reasons for this increased interest and growing understanding. Although most Latin American LDCs achieved political independence during the first quarter of the nineteenth century, the mid-twentieth century has seen a more dramatic emergence of Asian and African states which collectively and sometimes individually now have an important influence in world affairs. Physically the intensifying fabric of contact between states old and new, tropical or temperate, has meant that more people are actively involved in development and are able to appreciate at first hand the problems that arise. Economically, it can be argued that the prosperity of the advanced nations is at least partially dependent upon the continued poverty of the less-developed countries; but also that it is in the long-term interests of world society as a whole that the development process should be encouraged. Morally, further basic issues are raised; many would argue that the advanced countries have a moral responsibility, quite distinct from political or economic expediency, to assist the progress of the less-developed countries, or at the very least not to place obstacles in their path Pearson, 1969; Myrdal, 1970; Seers and Joy, 1971; Streeten, 1972).

The growth of development studies has involved a certain measure of change and reorientation within established academic disciplines. The political process of decolonisation has been accompanied not only by a growing interest in the problems of poverty in less-developed countries but also by an increasing awareness of the need to transcend inter-disciplinary boundaries in order to attack those problems. The trend away from traditional subject areas towards a multidisciplinary approach is not of course confined to development studies; a similar approach underlies the recent growth of area studies (especially Latin American, African and South-East Asian studies) in universities in many countries and also characterises the growth of the environmental sciences. Economics and geography are two subject areas closely involved with development problems today, although in neither case could it be claimed that the study of less-developed countries is a central concern within the subject. Development economics has emerged—not without controversy—as one of the newer branches of the subject and claims to be concerned 'with the explanation of underdevelopment and with a discussion of measures employed or advocated to overcome it' (Elkan, 1973, p. 9). Many economists have contributed in the last twenty years towards the formulation of a substantial body of economic theory concerning the problems of development, although the central focus of the subject has tended to lie in growth in advanced areas rather than in the solution of underdevelopment. Whether it can now be claimed that there has emerged a distinctive geography of development is a question frequently discussed, especially in the context of the so-called 'new' geography (Gilbert, 1971). Interest on the part of geographers in the less-developed world is as old as the subject itself, but it may be argued that 'geography's typically idiographic approach to problem formulation and solution . . . has prevented the emergence of a theoretical analysis of the development process which is necessary to complement and reinforce the work being undertaken in other disciplines' (Gilbert, 1971, p. 123). The emergence of a 'new' geography in which the emphasis is upon a nomothetic approach, involving quantitative analysis and model building, provides an opportunity for geography to contribute more positively towards the understanding and eventual solution of development problems (Keeble, 1967). There are, however, several dangers inherent in a theoretical approach by geographers to development, and some of these are discussed below.

PATTERNS AND DYNAMICS OF DEVELOPMENT

Spatial variations in poverty

The less-developed world is so described partly because it exhibits a range of symptoms of underdevelopment (Clarke, 1971, p. 5; Hodder, 1973, pp. 1–7); partly because its known development potential vastly exceeds present levels of achievement; and partly because of the existence of obsolete production methods and inhibiting forms of social organisation which slow down rates of progress. Although poverty is obviously one of the broad common characteristics and preoccupations of the LDCs, the concept of spatial variety within poverty is an important qualification. When we consider the spatial differences in development levels on a global scale, it is clear that the world is not neatly divided into three parts, with the LDCs as a uniform base-level. Myint has written that 'the first thing to bear in mind about the underdeveloped countries is that beyond the broad common fact of poverty it is rarely safe to make generalisations about them without carefully specifying the *type* of underdeveloped country one is considering . . . Instead of a monolithic theory to cover all the underdeveloped countries it will be more fruitful to look for alternative theoretical models to suit different types of underdeveloped country' (Myint, 1967, p. 14).

Many less-developed countries have tended to remain so because their overall condition is dominated by one overwhelming problem—aridity, population pressure, or a dearth of exploitable resources. Countries comparable in one sense are often greatly contrasted in other ways: India and Mauritania have little in common apart from the fact that both produce iron ore in quantity; similarly, the economies of Brazil and Kenya depend considerably upon coffee, but otherwise these two countries are strongly contrasted. It is particularly important to distinguish those LDCs where overpopulation is the most serious restrictive factor in development terms —India is an obvious example, but Indonesia and Upper Volta may also be cited—from others such as Argentina or Zambia which do not yet suffer from serious overall population pressure. Characteristically uneven population distributions within many LDCs create 'islands' of severe population pressure, in relation to available resources and current levels of development, set in vast tracts of territory where low population densities reflect in part a scarcity of resources and a minimal level of development achievement.

The development spectrum

Knowledge of the combination of factors and problems in area gives us one method of analysing the variety that exists within the poverty of the less-developed world, both in terms of macro-regions, states and areas within states. An alternative and complementary method is to consider the overall level of development attained by individual countries, as measured by a wide range of criteria. The pattern which this produces, described as a 'development ladder' or 'development spectrum', shows how the countries of the world are extended through a kind of continuum which is perhaps the most basic of the dynamics of development. For this continuum does not imply any immutable rank order; nor, many would say, that countries towards the lower end of the spectrum are necessarily aiming at the socio-economic systems which now characterise those countries currently further up. The development continuum represents only the state of affairs at a given moment in time; relative positions may change dramatically, as the experience of Japan, Hong Kong, Kenya and Venezuela has shown. Interesting results have come from an attempt to quantify the development ladder and to relate these results to population and transport network data (Berry, 1960; Haggett, 1965, pp. 76–79).

The continuum from agriculture to industry

Discussions of development in terms of interaction and integration between different economic sectors have important spatial aspects and implications. Economists, geographers and other specialists have considered at some length the relationships that exist between the transport sector and other aspects of economic development (Fromm, 1965; O'Connor, 1965a and b; Owen, 1968; Hoyle, 1973). More broadly, the relationships between industry and agriculture have occupied a good deal of attention and represent another kind of continuum; the two sectors have sometimes been thought of as mutually exclusive and many have argued that industrial expansion offers the fastest road towards a higher level of development. A contrary view is that only increased and diversified agricultural production provides a sound basis for successful industrialisation at a later stage. Today, although there is a tendency to emphasise the undoubted importance of rural development strategies and the spatial diffusion of economic growth rather than the concentration of activity in urban industrial growth poles, there is widespread recognition of the essential interpendence of agriculture and industry advocated some years ago by Lewis (1955) and Nurske (1959). However, as Hodder has emphasised, 'the validity of the hypothesis that sustained industrial growth is closely linked with the expansion and improvement of agriculture must vary considerably from country to country' (Hodder, 1973, p. 162). The adoption of a multisectorial development programme involves the promotion of a two-way flow of materials, labour, capital and technology between the agricultural and industrial sectors. In the long term this may imply, for many LDCs, a gradual transition from an agriculture-dominated to an industrially-orientated economy. While this continuum is often conceived in economic sectoral terms, it is important to emphasise that there is no sharp distinction between these two forms of economic activity either in a theoretical economic context or in terms of geographical realities.

The vicious circle of poverty and the diffusion of innovation

The promotion of economic development sometimes seems to be frustrated by many factors which, taken together, operate as disincentives and create a 'vicious circle' of poverty and stagnation (Bauer, 1965); the poverty of a low-level subsistence economy produces little or no surplus for reinvestment and in the absence of external influences poverty is frequently perpetuated. This concept, which implies an almost unchanging traditional economy and an almost static spatial system, may have some relevance in the least developed areas of the world; but the circle is not unbreakable —the application of capital to a wide range of activities, the diffusion of innovation, the enlargement of markets and the stabilisation of prices can produce a form of balanced growth which induces a circular process on an altogether higher level and encourages cumulative advance (Nurske, 1953). The spatial implications of growth and change, under the influence of internal and external factors, involve trends fundamentally important in terms of the dynamics of development; the degree of inter-regional inequality in development within a country (Williamson, 1965) and the extent to which investment and growth are concentrated in poles of development are basic issues in economic and social planning. Much attention has been devoted in recent years to the spatial effects of the diffusion of innovations amongst population groups, geographical areas and economic sectors (Hagerstrand, 1952 and 1967; Rogers, 1962; Siddle, 1970). Development means change; and the geography of development is basically concerned with the spatial diffusion of innovation—the spread of new ideas, the adoption of new techniques and the establishment of new forms of production, all of which involve changes in human distributions and movements and in patterns of human activity. On a macro-scale this relates to the whole process of long-continued contact, dating back to precolonial times, between the LDCs and the western world; on a micro-scale today it involves, for example, experimentation with new agricultural techniques under the guidance of an agricultural extension service. In this way, fishing communities on the coast of south-western Nigeria have expanded their range of economic activity to include new food and cash crops; the geographical pattern of their economy has thus been greatly diversified and their way of life transformed (Ajaegbu, 1970).

Problems of approach

Attitudes and approaches towards the patterns and dynamics of development vary considerably between disciplines, but all share three common problems. The first is the importance of *objectivity*. There is a strong case for the view that the advanced countries of the world today have a moral responsibility, quite distinct from their economic interest, to help to narrow the widening gap between rich and poor nations; many who become involved in a variety of ways in the development process do so partly because they accept a share in this responsibility. Concern for the problems of development may arise from personal overseas experience, political involvement or religious conviction, as well as from academic interest. The value and validity of participation in the development process arising from non-academic causes is not, of course, in doubt; but the active promotion of development on a broad front requires objective analysis and planning which piecemeal participation cannot always provide. The ultimate value of the academic study of development lies in the fact that it is, or should be, relatively independent of political positions, but the application of the results of academic enquiry may, nevertheless, require an assessment of policy that is deeply influenced by political positions and by value judgements.

A second problem is that of *Eurocentricity*. Much research and writing on development problems today, both by Europeans, Americans and indigenous academics in developing countries, is undertaken by those whose higher education and training has taken place in European or American universities. In consequence, much discussion of development problems takes place in the context of a body of theory and a set of assumptions and value judgements derived from Euroamerican experience. It has been suggested that in recent years geography has tended 'to follow too closely the Eurocentric approach of economists' (Gilbert, 1971, p. 124) whose 'elegant models . . . fit very uneasily when applied outside Euroamerica' (Chapman, 1969, p. 338). The reasons for these trends clearly lie in the attractions of the nomothetic rather than the idiographic approach in modern geography; in the problem of inadequate and unreliable data, especially statistical data, on many development problems; and in the consequent difficulty experienced in attempting to formulate theories and models based upon a wide spectrum of developing countries. There are serious dangers implicit in the transfer of theory from one subject to another and from one area to another; the experience of advanced countries is not necessarily closely relevant in theoretical terms to the problems of the less-developed countries today; models devised by economists are often useful to geographers, but they contain a very different set of assumptions; and spatial variety within the less-developed world is such that the experience of a narrow range of LDCs should not be extrapolated into a general theory of development without very careful consideration of the implications.

Thirdly there is the problem of *scale of approach*. Gottman has defined the three scales of concern relevant to studies in human geography: 'First, the "human scale" dealing with the interests or behaviour of individuals or small groups such as the family unit; second, and especially in our time and in urban affairs, the mass-scale, which deals with people but in large numbers as collective masses; and, finally, the scale of administrative units, through which the other two scales operate and correlate' (Gottman, 1973, p. 88). The geographer's approach to development studies involves an awareness of these varying scales of human activity and of the interaction between them. It is equally important to analyse the spatial aspects of development at the macro-level—in terms of global situations, interstate relationships and patterns within states—and at the micro-level in terms of small communities, family groups and individuals who provide the touchstone of reality in development studies.

Geography and the less-developed countries

From the classical period of Ptolemy and Strabo to modern times geographers have always displayed a considerable interest in the less-developed parts of the world, although this interest has probably never been a central concern to the practitioners of the subject as a whole. There is a case for the view that Alexander von Humboldt was the first development geographer, in view of his strong emphasis upon the interrelationships between environmental and human phenomena in nineteenth-century Latin America. Since that time the attitudes of geographers towards less-developed countries have changed and grown, partly in response to changing world political and economic situations and partly in the context of the changing nature of geography itself.

It is possible to identify a series of phases in the evolution of geographical thought about the less-developed countries. An initial phase, illustrated by Humboldt for Latin America and by Livingstone and other explorers for Africa, was concerned primarily with a general search for knowledge, with the discovery of parts of the world little known to Europe, with the solution of age-old geographical conundrums such as the source of the Nile, and with an initial attempt to assess the extent of resources which Europe might find useful. The role of geographers in promoting African exploration, for example, is well documented; the support given by the Royal Geographical Society to explorers working in tropical Africa was a very important contribution to the increasing knowledge of the continent on the part of Europeans in the mid-nineteenth century (Bridges, 1963, 1973).

The process and form of colonisation, and the resultant spatial patterns, have stimulated a great deal of interest on the part of geographers. On the one hand, the establishment of colonial administrations in Asia and Africa provided an increasingly substantial flow of information on less-developed areas, and geographers and others have made good use of this material (Perham, 1965). Pioneer geographers working in tropical lands were able to assemble and interpret data on basic environmental and demographic problems—soils, vegetation, human distributions—about which very little was known before the Second World War. Clement Gillman, who worked in East Africa between 1905 and 1946, provides an example of such a pioneer (Hoyle, 1965). On the other hand, contemporary geographers have shown considerable interest in various aspects of colonial development: studies of patterns of agricultural development in the Kenya highlands, for example, rely heavily upon an understanding of the spatial processes of land apportionment and land use in colonial times (Odingo, 1971); similarly, an understanding of the patterns and problems of contemporary urbanisation in the less-developed countries has necessitated close attention to the process of urban growth as an expression of, and a medium of, colonial rule (McMaster, 1968; Mabogunje, 1968).

There is a sense in which geographers have always provided much of the background information necessary for any contribution to development planning, but it is only in the post-1946 period that close attention has been given by geographers and others to problems of development, particularly economic development, in a narrower context (Grigg, 1973). In recent years numerous works from many disciplines have attempted to explore the reasons for the widening gaps between richer and poorer states and regions, and to attempt to explain the resultant patterns and problems (Harvey, 1969). Numerous economists have drawn attention to the importance of regional variations in levels of prosperity and economic growth, and have devised models which describe and attempt to explain the concentration of development in a few selected areas (e.g. Myrdal, 1957; Hirschman, 1958). Geographers have shown increasing interest both in the broad sweep of patterns of economic development (Lacoste, 1965; Hodder, 1973. Mountjoy, 1969 and 1971; O'Connor, 1971; Odell and Preston, 1973); in problems of economic integration on a continental scale (e.g. Odell, 1967); in the analysis of patterns of economic development within groups of states (Safier, 1969); in the description and analysis of specific regional development programmes (Wikkramatileke, 1972); and in the application of geographical techniques to the understanding and solution of development problems in a wide variety of contexts. Current trends emphasise especially the role of geography in development planning (Hoy, 1968), ways in which geographical approaches and methods can make a positive contribution towards a fuller understanding of development situations, and the importance of the spatial dimension in association with contributions from other disciplines. In the context of interdisciplinary inquiry in development studies, the central focus of attention is upon problems and the selection and use of appropriate methods of investigation, including spatial analysis, rather than upon the comparatively confined approach of particular disciplines. As Hodder has emphasised, 'the scope of the enquiry must be defined only by the problem set' (Hodder, 1973, p. 246).

CONTEMPORARY PROBLEMS AND APPROACHES

Environmental problems and resource appraisal

When we examine contemporary situations in the less-developed world and consider the problems that affect the position of individual countries in the development continuum and the operation of the dynamics of development, we are inevitably faced with a formidable array of difficulties. Those arising from the physical environment are among the most intractable, and deserve emphasis since these problems have sometimes been undervalued by development economists. From a spatial viewpoint the importance of variety in environmental and ecological circumstances cannot be too strongly stressed; most less-developed countries provide a spectrum of environments including harsh areas where low rainfall, poor soils and other factors combine to offer only a very limited range of opportunities for human development (Grigg, 1970); but extending also to other areas, often of restricted size, where the combination of environmental conditions is much more favourable in human terms and the range of opportunities correspondingly greater (Hance, Kotshcar and Peterec, 1961). The extent to which these varied landscapes are inhabited and developed by man depends upon many factors—physical, economic, social, political, technological—but among these water supply is the most vital. A map showing the pattern of average annual rainfall within an underdeveloped country may suggest that much of the area is reasonably well-watered, but this is frequently not the case since it is the seasonal variation, the intensity and above all the unreliability of the rainfall that matter. In many LDCs areas receiving a well-distributed rainfall of 750 mm (about 30 inches) or more, conducive to high-level crop production, are relatively rare; water conservation is thus a vital aspect of life in many areas, and the complex ecological relationships between water supplies, soils, vegetation and man are often inadequately understood.

The concept of natural resources extends of course beyond landscapes, soils and water to include wildlife, forests and other flora, domesticated animals and minerals. The developmental role of the natural resource endowment of a state has been much discussed (United Nations, 1970) and was once thought largely to determine the pattern and level of growth, just as economists have sometimes argued that capital, with trade, is the only true instigator of development. Neither of these extreme views is now acceptable; modern opinion stresses that the fortuitous geographical distribution of natural resources certainly does not provide the only, and probably not even the principal, explanation for differences in levels of prosperity. However, resources may certainly confer an important advantage at a crucial stage in the development process, although they are not a precondition for development, which requires the spatial and temporal interaction of favourable factors in physical, economic, political and social terms. There is thus no causal relationship between resource endowment and economic development, but resources may help to create conditions in which other favourable factors may operate.

The demographic framework

Within varied and often hostile physical environments, the peoples of the less-developed world present complex patterns and problems (Clarke, 1971). The distribution of a structurally complex population presents in many areas a pattern of striking unevenness; areas of greater economic opportunity, where numerous factors combine to create a more favourable environment, emerge as zones of high population density and as 'islands' of economic advancement which provide magnets for migration over long distances (Prothero, 1964; Hance, 1970); in other areas where the range of economic opportunity is much narrower, lower population densities reflect a lack of prosperity which may result in seasonal hunger and which often encourages seasonal or permanent out-migration (Hunter, 1967). Population density and population distribution provide in many respects a most sensitive spatial parameter of development: for together these factors reflect the interaction of an extremely wide range of physical, political, economic and social influences in a context of human ecology and represent, often with striking clarity, inter-regional differences in levels of prosperity.

The migration of peoples within and between the countries of the less-developed world represents both a cause of, and a form of adjustment to, this pattern of inter-regional inequality in development and opportunity. Although many social and political factors affect migration flows, the economic motive is today uppermost in the minds of most migrants, who seek a better livelihood in the restricted areas where economic opportunity is greater (Mabogunje, 1971). Special problems are created by the fact that many such areas of attraction focus upon the towns, frequently in many countries upon one large capital cityport which is multifunctional and in terms of population size may be primate several times over (Hoyle, 1972). Below the level of the leading city many less-developed countries offer no true urban hierarchy and the continued rapid growth of the main urban centre inevitably exacerbates the gulf between urban and rural societies. More people are involved, however, in rural-rural than in rural-urban migration in the less-developed countries, although less attention has been devoted to their problems. In recognition of the ways in which overurbanisation can become a constraint upon economic development (Mabogunje, 1965), many development plans now devote increased attention to rural development strategies, the promotion of rural growth centres and the expansion of both the agricultural and industrial sectors in rural areas by means of intermediate technology (e.g. Republic of Kenya, 1969).

The dual economy

We are presented therefore with a dual set of contemporary problems of development arising from the environmental and the demographic conditions of the less-developed countries today. Within this framework, economic and social development is taking place within a context dominated economically and spatially by what is traditionally referred to as the dual economy. It is true that in many parts of the less-developed world subsistence systems prevail, characterised by low productivity and frequently by seasonal hunger, and by associated features which include high disease incidence and high mortality rates (Chakravarti, 1970). Subsistence production probably occupies over two-thirds of the adult male population of the less-developed countries and more than two-thirds of the cultivated land. Production for money and for export is still, in many LDCs, largely a matter of producing a fairly small, though increasing, range of crops for the world market, plus whatever minerals are commercially exploitable. The cash economy of an average LDC is thus still, basically, an inherited colonial-type system based upon the export of primary and semi-processed goods and the import of manufactured products. In many of the more advanced LDCs this basic pattern has changed significantly as employment in manufacturing and service industries has increased; more primary products are being processed before export, thus providing employment and reducing transport costs; industrialisation is actively promoted, partly by overseas investment, and to some extent expensive imports may thereby be reduced. As the cash economy grows, in area and significance, so more and more people are drawn more fully into it; and the distinction between the cash and subsistence sectors, which in some less-developed countries was never very precise (e.g. O'Connor, 1971b, p. 13) becomes less distinct both economically and geographically.

Transport and development

The transportation network is one of the key elements in the infrastructure required in the process of development from an agriculturally-based to an industrially-orientated society; and this demands priority attention from LDC governments. The inter-relationships between transport and development have occupied much attention both in advanced and less-developed countries, and constitute a field of considerable practical as well as theoretical importance. The interaction between the level and pattern of transport resources and the average level of living of the population of an area is a critical factor affecting economic and social progress and must be taken into account at all stages of national and regional development planning. The cost of overcoming the restrictive impact of distance directly affects progress in the less-developed countries and the promotion of increased mobility is everywhere a primary objective of development planning in the socio-economic sphere. Most of the less-developed countries have progressed beyond a stage when almost any transport improvement seemed bound to bring economic benefit, to one in which the main need is not just for more and better transport facilities but rather for a great degree of precision in the establishment of transport policies in the light of some appreciation of what transport can and cannot be expected to do and of what else must be done if transport investment returns are to be maximised. In the developing countries, as elsewhere, the elaboration of modern transport systems may relieve pressures and provide opportunities; but transport does not work miracles, and integration of transport planning with other sectors in economic and spatial terms is essential to development strategy. The more primitive transport modes, however,—head porterage, canoes, bicycles—account for a high proportion of goods movements in the less-developed countries; in view of the importance of transport at this level, a high priority of transport policies and of rural development strategies should be the improvement of road systems in rural areas. In spatial terms, the transport problems of the less-developed countries should not be viewed in isolation from one another or from the rest of the world system of which they form a part. Transport media provide a framework for the spatial integration of economic development; the power of transport to foster inter-regional and inter-state understanding should not be underestimated.

The political dimension

It is impossible to consider in depth any aspect of development today without reference to the political dimension. Interest in development problems stems in part from the relatively recent emergence to political independence of many of the countries of Africa and Asia, coupled with their desire for rapid economic progress. The colonial relationship was a fundamental fact of political and economic life in most of the countries now grouped as less-developed; this relationship was only temporary and in many areas lasted for less than a century; but it occurred at a time of rapid (especially technological) change in the advanced countries and initiated a period of rapid transformation in at least some aspects of the less-developed countries also. Today colonialism is generally abandoned and derided as an ideology and as a method of political and economic organisation, but in spite of its shortcomings colonialism in the less-developed countries helped to provide a basic fund of information which in recent years has been greatly expanded as interest in the problems of development has grown. In the post-colonial period political instability has held back the rate of economic growth in many countries, and 'the revolution of rising expectations' has involved both economic and political disillusionment. At the same time there has been a revival of interest in pre-colonial states and cultures; for example, Balandier has written that 'Independent Africa is renewing her links with the distant past, and is searching there for the roots of her modern personality and sometimes for an indication of her future boundaries; unless, contrariwise, she is analysing past situations the better to contest the conclusions to which they lead' (Balandier, 1957, p. 259). One of the dilemmas facing many less-developed countries, in Africa and elsewhere, is how to balance increasing knowledge of, and identification with, the traditions of the pre-colonial past and the urgent need to foster forward-looking pan-African political alliances and economic groupings. The reconciliation of past traditions and future hopes rests in some measure in the realm of political geography; for it turns on the broad issues of continuity and change in man–land relationships and more specifically on the preservation *pro tempore* of the interstate boundaries inherited from the colonial period. The study of traditional political forms in the less-developed countries highlights the fact that the modern political map is in many respects and in many areas a highly artificial innovation. The less-developed countries are in part a product of the processes of political interaction between the advanced and less-developed areas of the world over an extended time period; such interaction continues today, in different forms, and the spatial analysis of development requires an understanding of these processes past and present.

THE ROLE OF GEOGRAPHICAL RESEARCH

Within the spectrum of contemporary problems facing the less-developed countries it is possible to identify several major priority research areas to which geographical analysis has made, and is continuing to make, effective contributions (Dresch, 1967; Abler, Adams and Gould, 1971). In broad terms, geographical methodology is especially useful in the analysis of national-scale patterns of socio-economic structure (Horton, McConnell and Tirtha, 1970) and in the discussion of overall development strategies in spatial terms, especially in the context of the degree of polarisation or diffusion of development and the application of growth pole theory (Perroux, 1955; Jansen, 1970; Robinson and Salih, 1971; Siddle, 1971). Beyond this general level of approach the first priority area is the study of the physical environment in relation to man's activities, present and potential. In this context the use of land-systems analysis has proved a fruitful method of approach in the appraisal of land resources in less-developed countries (Young, 1968 a and b, 1969; Moss, 1968 and 1969; Vink, 1968; Ollier, 1970). Other geographers have directed attention towards the analysis of specific resources in relation to patterns of human development; the work of geographers within multidisciplinary research teams is important in this respect, as for example in the Bureau of Resource Assessment and Land Use Planning in Tanzania (BRALUP, 1970). The discussion of patterns of water use and the analysis of water resources is a vital element in development planning in any context, but especially in those less-developed countries characterised by low and unreliable rainfall over large areas (Kenworthy, 1964; Lebon, 1968). Geographers have also written extensively on the problems associated with the geomorphology, biogeography and ecology of less-developed countries; in all these spheres there is an increasing interest in the application of research results to practical problems of development (Dasmann, Milton and Freeman, 1973). While the importance of malnutrition and disease as a restrictive factor in development has long been recognised, only in relatively recent years has there been close cooperation between medical research workers and geographers concerned with the spatial analysis of disease distribution (Prothero, 1965; Hutt, 1967; Hall and Langlands, 1968).

The measurement and analysis of patterns of human distribution within and between both rural and urban areas is a second major preoccupation of geographers concerned with the problems of development. Here there is often a problem of unreliability or non-availability of data (Eighmy, 1968), but where statistical material is accessible the geographer can often still perform a useful role by mapping and analysing spatial distributions of populations; development planning is unlikely to prove successful unless basic aspects of the demographic context are fully explored and understood (Morgan and Shaffer, 1966; Caldwell and Okonjo, 1968). More extended treatments of the population geography of the less-developed world as a whole (Clarke, 1971), of individual states (Clarke, 1969) and of regions and districts within states (Langlands, 1971) are increasingly available and there is a growing awareness amongst all who are concerned with development of the importance of spatial variations in population pressure both at macro- and micro-scales (Mortimore, 1967; Steel, 1970).

Particular concern has been shown in many spheres of development studies for the problems of urban areas and geographers have made many contributions to the analysis of the urbanisation process in less-developed countries (Breese, 1966 and 1968; McGee, 1971) and to the understanding of the problems of particular urban centres (e.g. Dwyer, 1968; Mabogunje, 1968). The concentration of physical, economic and social problems in the rapidly growing cities of the developing world provides a fruitful field for interdisciplinary cooperation in research which is potentially of great practical value in a context of development planning. Concern has been repeatedly expressed in many quarters for the over-rapid expansion and 'ruralisation' of some of the larger cities of the less-developed world (Grigg, 1973, p. 35) and, on the other hand, for the lack of expansion and difficulty of promoting development in many smaller urban centres (Harrison Church, 1967 and 1972). The city-size distribution of many less-developed countries is not conducive to the spatial diffusion of development; most LDC governments aim to spread the benefits of development as widely as possible throughout all economic sectors and geographical areas under their control, yet in present circumstances what is socially and politically desirable is frequently not feasible economically. The spatial dimension is fundamental to economic planning in this context and, wherever possible, geographers should ensure that development planning teams are fully aware of the facts and implications of the demographic conditions within which they operate.

Within the sphere of economic development, there is now widespread agreement that the agricultural and industrial sectors are in many respects interdependent and that their simultaneous promotion can lead to numerous mutual benefits. The promotion of agricultural development requires, basically, an understanding of existing patterns and systems and of factors affecting the diffusion of agricultural innovation. In this context, the International Geographical Union has sponsored an attempt to establish a world-wide system of agricultural typology as a basis for planning change in this sector of less-developed economies (Kostrowicki and Tyszkiewicz, 1970); this work will complement the World Land Use Survey, in progress since 1950, also under IGU auspices. The role of agricultural geography in development studies has been widely discussed (Benneh, 1967; Coppock, 1969) and studies of land-tenure systems and land-settlement schemes have been made with reference to a wide variety of environmental conditions and socio-political and economic circumstances (Thomas and Whittington, 1969; Grigg, 1970 and 1973). At the most primitive end of the agricultural spectrum, recent studies have helped to correct some of the misconceptions associated with primitive forms of shifting cultivation (Maude, 1970); and increased attention has been devoted to the special problems of subsistence agricultural systems (McMaster, 1962; Oluwasanmi and Dema, 1966; Antonini, 1971; Blaikie, 1971; Denevan, 1971). Whilst such systems form the essential basis of agricultural activity in less-developed countries in terms of food production and employment (both in the wage and non-wage sectors), the spatial patterns of crop production for export have also received widespread attention in view of the heavy dependence of many less-developed countries upon this sector of their economies (Courtenay, 1965; Udo, 1965; Harrison Church, 1969). The importance of the spatial and economic overlap between the subsistence and cash sectors of the agricultural economy of many less-developed countries has frequently been commented upon and has proved an interesting area for geographical research. The fluctuating margin of plantation crop production by outgrowers who are basically subsistence farmers has been examined in Malaya (Ho, 1967) and Uganda (Smith, 1970); and several writers have investigated the spatial patterns of periodic markets and have shown how these provide a fascinating window on the socio-economic systems of the cash-subsistence interface (Hodder, 1965; Good, 1970; Jackson, 1971). Trade and development on a wider basis have received attention in Malaya (Courtenay, 1972), and elsewhere.

In most less-developed countries the agricultural economy is vulnerable both in terms of locally unreliable climatic conditions and in terms of changes in the supply/demand ratio which affects prices of agricultural produce both locally and internationally; these factors give rise to a degree of instability in the agricultural economy which may result in food shortages, seasonal hunger or changes in patterns of production (Hunter, 1967; Stouse, 1970; Udo, 1971). The promotion of spatial rather than purely temporal innovation is an important field for geographical research (Hagerstrand, 1967; Brown and Moore, 1969). In terms of practical agricultural development, within a broad context of rural development strategy, several studies from Africa have shown the importance of establishing a large number of small growth centres as foci of innovation, rather than a few large and relatively isolated improvement schemes (Republic of Kenya, 1969; Ajaegbu, 1970; Siddle, 1970; Adejuwon, 1971). Geographers working on problems of development have hitherto devoted their attention primarily to the increased understanding of the environmental and demographic framework of development and to the agricultural sector which is almost everywhere of basic importance. Rather less attention has, in general, been devoted to problems of industrial growth, transport expansion and the spatial aspects of political situations and changes. The contribution made by the industrial sector to the economies of most less-developed countries is so far limited by lack of resources, inadequate markets, problems of trade etc., yet one of the long-term aims of development is the gradual transformation of less-developed agriculture-based societies into more advanced industrially-orientated economies. If this transformation is to be achieved, it is essential that the spatial aspects of existing and proposed industrialisation programmes should be understood, so that planners may take account of these aspects in conjunction with economic, financial, technical and political considerations. An excellent basis for the further study of industrial patterns already exists in the broad studies of industrialisation and underdevelopment published by Mountjoy (1969) and Sutcliffe (1970). The available literature on the patterns and problems of industrialisation in the less-developed countries reflects the variety of the subject itself and includes material on extractive industry (Swindell, 1967; Mulchansingh, 1971), import-substitute industries (Hay, 1971), the emergence of industrial growth poles (Burley, 1962; Hoyle, 1963, Jansen, 1970; Siddle, 1971) and the development of industrial estates (Brown, 1969; Davey, 1970; Dickenson, 1970). Some attention has also been given to the growth of export-orientated manufacturing industries in developing countries; the experience of Hong Kong is often cited in this context (Shou-eng Koo, 1968), and discussion tends to focus particularly upon the development of marketing arrangements within groups of states, as in West Africa (Sakamoto, 1969; Barbour, 1972). The location of industry is one of the principal difficulties encountered in attempting to foster interestate economic integration between less-developed countries and the application of economic and geographical models and techniques may throw some light on the problems involved (Hance, 1961; Smith, 1969), although the relevance to LDC situations of industrial location theory as devised in Euroamerica may be questioned.

The study of transport problems is shared among several disciplines but is particularly relevant to economics and to geography. Economists are concerned primarily with the assessment of the demand for transport and with the fundamental problem of the cost of overcoming distance (Fromm, 1965; Prest, 1969). The geographer's interest in transport arises from its role as one of the chief factors influencing the location and distribution of economic and social activities. Much detailed research on the relationship between transport and economic development in the less-developed countries has been done by economists (Hazlewood, 1964; Owen, 1964 and 1968; Wilson, 1966; Brown, 1968); and in recent years the increasing use of models and quantitative techniques has brought a greater degree of precision to the spatial analysis of transport networks in the less-developed countries (Kansky, 1963; Taaffe, Morrill and Gould, 1963; Haggett and Chorley, 1969). Ideas concerning the relationship between transport and development have changed considerably (Gauthier, 1970; Hoyle, 1973); both economists and geographers have made important contributions to the theory and methodology of transport studies and, in recent years, a growing interdependence between the two disciplines in this context has begun to emerge. An attempt to relate transport network densities to development levels in statistical terms was made by Berry (1960) and the ideal-typical sequence of transport network development in a less-developed country postulated by Taaffe, Morrill and Gould (1963) has provided a basis for some later work (Rimmer, 1967; Ward, 1969; Hoyle, 1970; Stanley, 1970). Railways have provided a basic artery of communication in many less-developed countries and the relationship between railway network expansion and economic development has received some attention (O'Connor, 1965a and b). The critical role of the seaport as a permissive or restrictive factor in development has also been discussed (Hilling, 1966, 1969a and b; Peterec, 1967; Hoyle, 1967, 1968 and 1972; Hoyle and Hilling, 1970; Ogundana, 1971). Inadequate attention has hitherto been given by geographers to the problems of road transport, especially to 'the steady rise in the importance of road traffic, which first complements the railroad, then competes with it, and finally overwhelms it' (Taaffe, Morrill and Gould, 1963, p. 514). The capacity and connectivity of a road network is a very important factor, in regional and national development, and one of the chief objectives of transport policy in many less-developed countries today is therefore the improvement of rural road systems as part of a rural development strategy designed to reduce the degree of polarisation of development activity in the main urban-based growth centres.

Although general works on less-developed countries invariably recognise the importance of the political dimension in studies of development, the difficulty of obtaining reliable data is a particularly serious obstacle in this field. The importance of the chronological perspective in political development has been frequently demonstrated (Fordham, 1972; Stamp and Morgan, 1972) and numerous writers have analysed geo-political situations on a macro-scale within individual states and selected groups of less-developed countries (Dale, 1968; Fisher, 1968). Rather fewer attempts have been made, however, to analyse micro-scale political situations in spatial terms (Hodder, 1968; Langlands, 1972) or to apply statistical techniques in the field of electoral geography. The process of political modernisation has, however, been analysed in spatial terms at a national scale for several developing countries in tropical Africa (Soja, 1968; Riddell, 1970; Jones, 1971).

CONCLUSION

Any student of the development process must at some time ask the question: 'What is development for?' In some quarters there seems to be an unquestioned assumption that progress towards an industrial society, geared to mass consumption and materialism, is *ipso facto* a desirable objective. Elsewhere this assumption is questioned, as the disadvantages of rural and urban environmental pollution become increasingly apparent. It is perhaps important to recall the distinction made by Boudeville between the three concepts of growth, development and progress: 'Growth,' he stated, 'is merely a set of increases in quantities produced; development is growth plus a favourable change in production techniques and in consumer behaviour; progress is development plus a diminution of social tensions between groups within a society' (Boudeville, 1966, p. 168–9). If these definitions are accepted, it follows that growth without development, and development without progress, are possible situations which an area or a country may experience; implicit in Boudeville's definition is the related idea that whilst growth is an economic process, development is socio-economic in nature and progress is primarily a social phenomenon based upon economic growth. Within this continuum from growth to progress, geographers are primarily concerned with spatial aspects of development; but it is essential to preserve an awareness of the wider context within which these aspects are set.

A complementary view has been expressed by the President of Tanzania, Dr. Julius Nyerere: 'Development', he has written, 'means the development of people. Roads, buildings, the increase of crop output . . . are not development; they are only the tools of development . . . An increase in the number of school buildings is development only if these buildings can be, and are being, used to develop the minds and the understanding of people . . . Development brings freedom, providcd it is the development of people' (Nyerere, 1968, p. 123). The juxtaposition of these viewpoints on development—one basically an economist's approach, the other that of a socialist politician—demonstrates the importance of varied and complementary approaches to the study of development and to development objectives. Most people involved in studies of the less-developed countries now recognise the validity and importance of a multifaceted, interdisciplinary attack on development problems. It is generally recognised, too, that the laws of economics on the one hand and of practical politics on the other tend to influence very strongly the pattern and pace of development. Yet economic and political systems operate within an environmental and demographic framework which they do not themselves create; this framework is dynamic and powerful in the influence it wields. Within this framework there is ample scope for the detailed contributions of geologists, engineers, social scientists, geographers and many others with particular interests, skills and viewpoints. The important point, however, is that each specialist should not be exclusively preoccupied with his own field, but should be ready to contribute towards a common objective and if necessary to adapt his work accordingly.

In this context the geographer has an important role to perform, whatever his specialist interests (Steel, 1967; Farmer, 1971). At one level, it is primarily the geographer's task to explore, analyse and publicise the facts and implications of the *distribution* of resources and the changing *patterns* of development in spatial terms; decisions about development issues may be taken mainly by politicians and economists, but they should always be taken in the context of knowledge of the spatial disposition and interrelationships of the factors involved. At another level, specialist geographers have a positive contribution to make to the analysis of landforms, the appraisal of resources, the study of agricultural systems, industrial locations and to urban and regional planning. In these and many other fields the geographer's viewpoint is relevant and is increasingly welcomed. In theoretical terms there is ample scope for the application to the developing world, where data is adequately available, of techniques and methods of spatial analysis first evolved in advanced countries; and there is an important need for the evolution of new techniques, methods and models based on a considerable range of LDC experience and data rather than upon a limited number of isolated case studies. In this context geographers, with others, face the problems involved in the transfer of techniques from one environment to another, either within the less-developed world or between advanced and less-developed countries. Recent years have been marked by increasing knowledge of the complex patterns of continuity and change which characterise the less-developed world, and by the evolution of a wide variety of increasingly sophisticated techniques of spatial analysis which may help to unravel some of these complexities. It is important, however, to preserve within the 'new' geography some, at least, of the traditional values and viewpoints of the 'old',—particularly the emphasis upon spatial variety and spatial interrelationships. The best service that geography can offer to the less-developed countries today lies in the wide range of techniques, traditional and modern, that are available within the broad scope of the subject. The role of geography in development is thus positive and valid: as a provider of information necessary for development and in an essential dimension; as a field offering a range of useful techniques, some tried and tested, others experimental; and as a discipline open to collaboration with other disciplines, whilst preserving its own viewpoint. In these respects geography provides an important and worthwhile contribution towards that wider and deeper understanding of development problems which may in the long run contribute positively towards the overall improvement of living conditions in the countries of the less-developed world.

REFERENCES

Abler, R., J. S. Adams and P. Gould (1971) *Spatial organisation: the geographer's view* (Englewood Cliffs: Prentice Hall).

Adejuwon, O. (1971) 'Agricultural colonisation in twentieth-century Nigeria', *Journal of Tropical Geography* **33,** 1–8.

Ajaegbu, H. I. (1970) 'Food crop farming in the coastal area of southwestern Nigeria' *Journal of Tropical Geography* **31,** 1–9.

Antonini, G. A. (1971) 'Peasant agriculture in northwestern Dominican Republic', *Journal of Tropical Geography* **32,** 1–10.

Balandier, G. (1957) *Afrique ambigue* (Paris, Librarie Plon); trans. Helen Weaver (1966) *Ambiguous Africa* (London: Chatto & Windus).

Barbour, K. M. (1972) 'Industrialisation in West Africa: the need for subregional groupings within an integrated economic community', *Journal of Modern African Studies* **10,** 357–382.

Bauer, P. T. (1965) 'The vicious circle of poverty', *Weltwirtschaftliches Archiv* **95;** reprinted in I. Livingstone (ed.) (1971) *Economic policy for development* (Harmondsworth: Penguin Books).

Benneh, G. (1967) 'The role of agricultural geography in the economic development of Ghana', *Bulletin of the Ghana Geographical Association* **13,** 51–64.

Berry, B. J. L. (1960) 'An inductive approach to the regionalisation of economic development', *Essays on geography and economic development*, ed. N. S. Ginsburg (University of Chicago, Department of Geography, Research Paper 62).

Blaikie, P. M. (1971) 'Spatial organisation of agriculture in some north Indian villages', *Transactions of the Institute of British Geographers* **50,** 1–40.

Boudeville, J. R. (1966) *Problems of regional economic planning* (Edinburgh: University Press).

BRALUP (Bureau of Resource Assessment and Land-Use Planning) (1970) *Water development, Tanzania: a critical review of research* (University of Dar es Salaam, Research Paper 12).

Breese, G. (1966) *Urbanisation in newly developing countries* (Englewood Cliffs: Prentice Hall).

Breese, G. (ed.) (1968) *The city in newly developing countries* (Englewood Cliffs: Prentice Hall).

Bridges, R. C. (1963) 'The R.G.S. and the African Exploration Fund, 1876–80', *Geographical Journal* **129,** 25–35.

Bridges, R. C. (1973), 'Europeans and East Africans in the age of exploration', *Geographical Journal* **139,** 220–232.

Brown, J. A. (1969) 'Industrial estate development in India', *Pacific Viewpoint* **10,** 65–71.

Brown, L. A. and G. Moore (1969) 'Diffusion research in geography: a perspective', *Progress in Geography* **1,** 121–157.

Brown, R. T. (1968) *Transport and the economic integration of South America* (Washington: Brookings Institution).

Burley, T. M. (1962) 'Industrial expansion in the Federal District, Mexico', *Geography* **47,** 184–185.

Caldwell, J. C. and C. Okonjo (eds.) (1968) *The population of tropical Africa* (New York: Columbia University Press).

Chakravarti, A. K. (1970) 'Foodgrain sufficiency patterns in India', *Geographical Review* **60,** 208–228.

Chapman, M. (1969) 'Geography and the study of development', *Journal of Developing Areas* **3,** 319–338.

Clarke, J. I. (1969) 'Population policies and dynamics in Tunisia', *Journal of Developing Areas* **4,** 45–58.

Clarke, J. I. (1971) *Population geography and the developing countries* (Oxford: Pergamon Press).

Coppock, J. T. (1969) 'Agricultural geography in tropical Africa', *Environment and land use in Africa*, ed. M. F. Thomas and G. W. Whittington (London: Methuen), 9–20.

Courtenay, P. P. (1965) *Plantation agriculture* (London: Bell).

Courtenay, P. P. (1972) *A geography of trade and development in Malaya* (London: Bell).

Dale, E. H. (1968) 'Some geographical aspects of African land-locked states', *Annals of the Association of American Geographers* **58,** 485–505.

Dasmann, R. F., J. P. Milton and P. H. Freeman (1973) *Ecological principles for economic development* (London: Wiley).

Davey, J. A. (1970) 'Industrial development in Rajasthan and Madhya Pradesh, India', *Transactions of the Institute of British Geographers* **49,** 183–199.
Denevan, W. M. (1971) 'Campa subsistence in the Gran Pajonal, eastern Peru', *Geographical Review* **61,** 496–518.
Dickenson, J. P. (1970) 'Industrial estates in Brazil', *Geography* **55,** 326–329.
Dresch, J. (1967) 'Géographie et sous-développement', *Annales de Géographie* **76,** 641–643.
Dwyer, D. J. (1968) 'The city in the developing world and the example of southeast Asia', *Geography* **53,** 353–364.
Eighmy, T. H. (1968) 'Problems of census interpretation in developing countries: the Western Nigerian case', *Lund Studies in Geography*, Series B, 32.
Elkan, W. (1973) *An introduction to development economics* (Harmondsworth: Penguin Books).
Farmer, B. H. (1971) 'The environmental sciences and economic development', *Journal of Development Studies* **7,** 257–270.
Fisher, C. A. (ed.) (1968) *Essays in political geography* (London: Methuen).
Fordham, P. (1972) *The geography of African affairs* (3rd edition; Harmondsworth: Penguin Books).
Fromm, G. (ed.) (1965) *Transport investment and economic development* (Washington: Brookings Institution).
Gauthier, H. L. (1970) 'Geography, transportation and regional development', *Economic Geography* **46,** 612–619.
Gilbert, A. (1971) 'Some thoughts on the new geography and the study of development', *Area* (Institute of British Geographers) **3,** 123–128.
Ginsburg, N. S. (ed.) (1960) *Essays on geography and economic development* (University of Chicago, Department of Geography, Research Paper 62.).
Good, C. M. (1970) *Rural markets and trade in East Africa* (University of Chicago, Department of Geography, Research Paper 128).
Gottman, J. (1973), Review of M. Clawson (1971) *Surburban land conversion in the United States: an economic and governmental process* (Baltimore: Johns Hopkins Press), *Economic Geography* **49,** 86–88.
Grigg, D. (1970) *The harsh lands: a study in agricultural development* (London: Macmillan).
Grigg, D. (1973) 'Geographical studies of economic development with special reference to agriculture', *Studies in human geography*, ed. M. Chisholm and B. Rodgers (London: Heinemann, for the Social Science Research Council), 18–84.
Hagerstrand, T. (1952) *The propagation of innovation waves* (Lund Studies in Geography).
Hagerstrand, T. (1967) *Innovation diffusion as a spatial process* (Chicago: University of Chicago Press).
Haggett, P. (1965) *Locational analysis in human geography* (London: Edward Arnold).
Haggett, P. and R. J. Chorley (1969) *Network analysis in geography* (London: Edward Arnold).
Hall, S. A. and B. W. Langlands (eds.) (1968) *Uganda atlas of disease distribution* (Makerere University, Kampala, Department of Preventive Medicine and Department of Geography, Occasional Paper 12).
Hance, W. A. (1961) 'West African industry: an analysis of locational orientation', *Journal of International Affairs* **15,** 3–15.
Hance, W. A. (1970) *Population, migration and urbanisation in Africa* (New York: Columbia University Press).

Hance, W. A., V. Kotshcar and R. J. Peterec (1961) 'Source areas of export production in tropical Africa', *Geographical Review* **51,** 487–499.

Harrison Church, R. J. (1967) 'Urban problems and economic development in West Africa', *Journal of Modern African Studies* **5,** 511–520.

Harrison Church, R. J. (1969) 'The Firestone rubber plantations of Liberia', *Geography* **54,** 430–437.

Harrison Church, R. J. (1972) 'The case for industrial and general development of the smaller towns of West Africa', *La croissance urbaine en Afrique noire et à Madagascar* (Paris: Centre National de la Recherche Scientifique), 659–665.

Harvey, D. (1969) *Explanation in geography* (London: Edward Arnold).

Hay, A. M. (1971) 'Imports versus local production: a case study from the Nigerian cement industry', *Economic Geography* **47,** 384–388.

Hazlewood, A. (1964) *Rail and road in East Africa* (Oxford: Blackwell).

Hilling, D. (1966) 'Tema, the geography of a new port', *Geography* **51,** 111–125.

Hilling, D. (1969a) 'The evolution of the major ports of West Africa', *Geographical Journal* **135,** 365–378.

Hilling, D. (1969b) 'Container potential of West African ports', *Dock and Harbour Authority* **50,** 9–13.

Hirschman, A. O. (1958) *The strategy of economic development* (Yale University Press).

Ho, R. (1967) 'Rubber production by peasants of the Terachi valley, Malaya', *Transactions of the Institute of British Geographers* **41,** 187–201.

Hodder, B. W. (1965) 'Some comments on the origins of traditional markets in Africa', *Transactions of the Institute of British Geographers* **36,** 97–105.

Hodder, B. W. (1973) *Economic development in the tropics* (London: Methuen).

Hodder, B. W. (1968) 'The Ewe problem: a reassessment', *Essays in political geography*, ed. C. A. Fisher (London: Methuen), 271–283.

Horton, F. E., H. McConnell and R. Tirtha (1970) 'Spatial patterns of socio-economic structure in India', *Tijdschrift voor Economische en Sociale Geografie* **61,** 101–113.

Hoy, D. R. (1968) 'Geography's role in development planning in Guatemala', *The Professional Geographer* **20,** 336–360.

Hoyle, B. S. (1963) 'The economic expansion of Jinja, Uganda', *Geographical Review* **53,** 377–388.

Hoyle, B. S. (1965) 'Clement Gillman, 1882–1946: biographical notes on a pioneer East African geographer', *East African Geographical Review* **3,** 1–16.

Hoyle, B. S. (1967) 'Early port development in East Africa: an illustration of the concept of changing port hierarchies', *Tijdschrift voor Economische en Sociale Geografie* **58,** 94–102.

Hoyle, B. S. (1968) 'East African seaports: an application of the concept of *Anyport*', *Transactions of the Institute of British Geographers* **44,** 163–183.

Hoyle, B. S. (1970) 'Transport and economic growth in developing countries: the case of East Africa', *Geographical essays in honour of K. C. Edwards*, eds. R. H. Osborne, F. A. Barnes and J. C. Doornkamp (University of Nottingham), 187–196.

Hoyle, B. S. (1972) 'The port function in the urban development of tropical Africa', *La croissance urbaine en Afrique noire et à Madagascar* (Paris: Centre National de la Recherche Scientifique), 705–718.

Hoyle, B. S. (ed.) (1973) *Transport and development* (London: Macmillan).

Hoyle, B. S. and D. Hilling (eds.) (1970) *Seaports and development in tropical Africa* (London: Macmillan).

Hunter, J. M. (1967) 'Seasonal hunger in a part of the West African savanna', *Transactions of the Institute of British Geographers* **41,** 167–186.

Hutt, M. S. R. (1967) 'The geographical approach in medical research', *East African Geographical Review* **5,** 1–8.

Jackson, R. T. (1971) 'Periodic markets in southern Ethiopia', *Transactions of the Institute of British Geographers* **53,** 31–42.

Jansen, A. C. M. (1970) 'The value of the growth pole theory for economic geography', *Tijdschrift voor Economische en Sociale Geografie* **61,** 67–76.

Johnson, E. A. J. (1970) *The organisation of space in developing countries* (Cambridge, Mass.: Harvard University Press).

Jones, A. H. (1971) 'Spatial dimensions of modernisation in Uganda', *East African Geographical Review* **9,** 79–86.

Kansky, K. J. (1963) *Structure of transport networks* (University of Chicago, Department of Geography, Research Paper 84).

Keeble, D. E. (1967) 'Models of economic development', *Models in geography*, ed. R. J. Chorley and P. Haggett (London: Methuen), 243–302.

Kenworthy, J. M. (1964) 'Rainfall and the water resources of East Africa', *Geographers and the tropics: Liverpool essays*, ed. R. W. Steel and R. M. Prothero (London: Longmans), 111–137.

Kostrowicki, J. and W. Tyszkiewicz (1970) *Agricultural typology: selected methodological materials* (International Geographical Union, Commission on Agricultural Typology).

Lacoste, Y. (1965) *Géographie du sous-développement* (Paris: Presses Universitaires de France).

Langlands, B. W. (1971) *The population geography of Kigezi District* (Department of Geography, Makerere University, Occasional Paper 26). (This is the first of a series of seventeen studies on the population geography of each District of Uganda, Occasional Papers 26–42).

Langlands, B. W. (1972) *Inventory of geographical research at Makerere, 1947–72* (Department of Geography, Makerere University, Occasional Paper 50).

Lebon, J. H. G. (1968) 'The land and water use survey of north central Kordofan', *Geographical Journal* **134,** 546–550.

Lewis, W. A. (1955) *The theory of economic growth* (London: Allen & Unwin).

Mabogunje, A. L. (1965) 'Urbanisation in Nigeria: a constraint on economic development', *Economic Development and Cultural Change* **13,** 413–438.

Mabogunje, A. L. (1968) *Urbanisation in Nigeria* (London: University of London Press).

Mabogunje, A. L. (1971) *Regional mobility and resource development in West Africa* (Montreal: McGill-Queen's University Press).

Mabogunje, A. L. (1973) 'Manufacturing and the geography of development in tropical Africa', *Economic Geography* **49,** 1–20.

Maude, A. (1970) 'Shifting cultivation and population growth in Tonga', *Journal of Tropical Geography* **31,** 57–64.

McGee, T. G. (1971) *The urbanisation process in the third world: explorations in search of a theory* (London: Bell).

McMaster, D. N. (1962) *A subsistence crop geography of Uganda* (Bude: Geographical Publications Ltd., World Land Use Survey, Occasional Paper 2).

McMaster, D. N. (1968) 'The colonial district town in Uganda', *Urbanisation and its problems*, ed. R. P. Beckinsale and J. M. Houston (Oxford: Blackwell).

Morgan, W. T. W. and N. M. Shaffer (1966) *The population of Kenya: distribution and density* (Nairobi: Oxford University Press).

Mortimore, M. J. (1967) 'Land and population pressure in the Kano close-settled zone, northern Nigeria', *Advancement of Science* **23,** 677–686.

Moss, R. P. (ed.) (1968) *The soil resources of tropical Africa* (Cambridge University Press, for the African Studies Association of the United Kingdom).

Moss, R. P. (1969) 'The appraisal of land resources in tropical Africa', *Pacific Viewpoint* **10,** 18–27.

Mountjoy, A. B. (1969) *Industrialisation and underdeveloped countries* (London: Hutchinson).

Mountjoy, A. B. (ed.) (1971) *Developing the underdeveloped countries* (London: Macmillan).

Mulchansingh, V. C. (1971) 'The bauxite/alumina industry of Jamaica', *Journal of Tropical Geography* **33,** 20–30.

Myint, H. (1967) *The economics of the developing countries* (3rd edition; London: Hutchinson).

Myrdal, G. (1957) *Economic theory and underdeveloped regions* (London: Duckworth).

Myrdal, G. (1970) *The challenge of world poverty* (Harmondsworth: Penguin Books).

Nurske, R. (1953) *Problems of capital formation in underdeveloped countries* (Oxford: Blackwell).

Nurske, R. (1959) *Patterns of trade and development* (Stockholm).

Nyerere, J. (1968) *Freedom and socialism: a selection from writings and speeches, 1965–67* (Dar es Salaam: Oxford University Press).

O'Connor, A. M. (1965a) *Railways and development in Uganda* (Nairobi: Oxford University Press).

O'Connor, A. M. (1965b) 'New railway construction and the pattern of economic development in Uganda', *Transactions of the Institute of British Geographers* **36,** 21–30.

O'Connor, A. M. (1971a) *The geography of tropical African development* (Oxford: Pergamon Press).

O'Connor, A. M. (1971b) *An economic geography of East Africa* (2nd edition; London: Bell).

Odell, P. R. (1967) 'Economic integration and spatial patterns of economic development in Latin America', *Journal of Common Market Studies* **6,** 267–286.

Odell, P. R. and D. A. Preston, (1973) *Economies and societies in Latin America* (London: John Wiley).

Odingo, R. S. (1971) *The Kenya highlands: land use and agricultural development* (Nairobi: East African Publishing House).

Ogendo, R. B. (1972) *An industrial geography of Kenya* (Nairobi: East African Publishing House).

Ogundana, B. (1971) 'The location factor in changing seaport significance in Nigeria', *Nigerian Geographical Journal* **14,** 71–88.

Ollier, C. D. (1970) *Uganda land systems* (Christchurch: Military Engineering Experimental Establishment).

Oluwasanmi, H. A. and I. S. Dema, (1966) *Uboma: a socio-economic and nutritional survey of a rural community in eastern Nigeria* (Bude: Geographical Publications Ltd., World Land Use Survey, Occasional Paper 6).

Ominde, S. H. (ed.) (1971) *Studies in East African geography and development* (London: Heinemann).

Owen, W. (1964) *Strategy for mobility* (Washington: Brookings Institution).

Owen, W. (1968) *Distance and development: transport and communications in India* (Washington: Brookings Institution).

Pearson, L. B. (1969) *Partners in development: Report of the Commission on International Development* (New York: Praeger).

Perham, M. (1965) 'The President's Inaugural Address', *African Affairs*, Special issue to record the proceedings of the 1964 conference of the African Studies Association of the United Kingdom, 1–14; reprinted in M. Perham (1970) *Colonial sequence, 1949–69* (London: Methuen), 263–279.

Perroux, F. (1955) 'Note sur la notion de pôle de croissance', *Economie Appliquée* **8**, 307–320; reprinted in translation in I. Livingstone (ed.) (1971) *Economic policy for development* (Harmondsworth: Penguin Books), 278–289.

Peterec, R. J. (1967) *Dakar and West African economic development* (New York: Columbia University Press).

Prest, A. R. (1969) *Transport economics in developing countries* (London: Weidenfeld and Nicholson).

Prothero, R. M. (1964) 'Continuity and change in African population mobility', *Geographers and the tropics*, ed. R. W. Steel and R. M. Prothero (London: Longmans), 189–213.

Prothero, R. M. (1965) *Migrants and malaria* (London: Longmans).

Republic of Kenya (1969) *Development plan, 1970–74* (Nairobi: Government Printer).

Robinson, G. and K. B. Salih (1971) 'The spread of development around Kuala Lumpur: a methodology for an exploratory test of some assumptions of the growth pole model', *Regional Studies* **5**, 303–314.

Riddell, J. B. (1970) *The spatial dynamics of modernisation in Sierra Leone: structure, diffusion and response* (Evanston: Northwestern University Press).

Rimmer, P. J. (1967) 'The search for spatial regularities in the development of Australian seaports', *Geografiska Annaler* **49B**, 42–54.

Rogers, E. (1962) *The diffusion of innovation* (New York: Free Press of Glencoe).

Safier, M. (1969) 'Towards the definition of patterns in the distribution of economic development over East Africa', *East African Geographical Review* **7**, 1–14.

Sakamoto, J. (1969) 'Industrial development and integration of underdeveloped countries', *Journal of Common Market Studies* **7**, 283–304.

Seers, D. and L. Joy, (1971) *Development in a divided world* (Harmondsworth: Penguin Books).

Shou-eng Koo (1968) 'The role of export expansion in Hong Kong's economic growth', *Asian Survey* **8**, 499–515.

Siddle, D. (1970) 'Rural development in Zambia: a spatial analysis', *Journal of Modern African Studies* **8**, 271–284.

Siddle, D. J. (1971) 'A new national growth pole for Zambia', *Geography* **56**, 330–333.

Smith, J. (1970) 'The production of sugar cane by outgrowers in East Mengo District, Uganda', *East African Geographical Review*, **8**, 47–54.

Smith, R. V. (1969) 'Industrial development in Puerto Rico: locational factors and distributional patterns', *Southampton Research Series in Geography* **5**, 31–58.

Soja, E. W. (1968) *The geography of modernisation in Kenya: a spatial analysis of social, economic and political change* (Syracuse University Press).

Stamp, L. D. and W. T. W. Morgan (1972) *Africa: a study in tropical development* (3rd edition; London: Wiley).

Stanley, W. R. (1970) 'Transport expansion in Liberia', *Geographical Review* **60**, 529–547.

Steel, R. W. (1967) 'Geography and the developing world', *Transactions of the British Association for the Advancement of Science* **23**, 566–582.

Steel, R. W. (1970) 'Problems of population pressure in tropical Africa', *Transactions of the Institute of British Geographers* **49,** 1–14.

Steel, R. W. and R. M. Prothero (eds.) (1964) *Geographers and the tropics: Liverpool essays* (London: Longmans).

Stouse, P. A. D. (1970) 'Instability of tropical agriculture: the Atlantic lowlands of Costa Rica', *Economic Geography* **46,** 78–97.

Streeten, P. (ed.) (1972) *The frontiers of development studies* (London: Macmillan).

Sutcliffe, R. B. (1970) *Industrialisation and underdevelopment* (London: Addison-Wesley).

Swindell, K. (1967) 'Iron ore mining in West Africa: some recent developments in Guinea, Sierra Leone and Liberia', *Economic Geography* **43,** 333–346.

Taaffe, E. J., R. L. Morrill, and P. R. Gould (1963) 'Transport expansion in underdeveloped countries: a comparative analysis', *Geographical Review* **53,** 503–529.

Thomas, M. F. and G. W. Whittington (eds.) (1969) *Environment and land use in Africa* (London: Methuen).

Udo, R. K. (1965) 'Sixty years of plantation agriculture in southern Nigeria, 1902–62', *Economic Geography* **41,** 356–368.

Udo, R. K. (1971) 'Food deficit areas of Nigeria', *Geographical Review* **61,** 415–430.

United Nations (Department of Economic and Social Affairs) (1970) *Natural resources of developing countries: investigation, development and rational utilisation* (New York).

Vink, A. P. A. (1968) 'The role of physical geography in integrated surveys of developing countries', *Tijdschrift voor Economische en Sociale Geografie* **59,** 294–312.

Ward, M. (1969) 'Progress in transport geography', *Trends in geography*, ed. R. U. Cooke and J. H. Johnson (Oxford: Pergamon Press).

Wikkramatileke, R. (1972) 'The Jengka Triangle, West Malaysia: a regional development project', *Geographical Review* **62,** 479–500.

Williamson, J. G. (1965) 'Regional inequality and the process of national development: a description of the patterns', *Economic Development and Cultural Change* **13,** Part II, 84pp.

Wilson, G. W. (ed.) (1966) *The impact of highway investment on development* (Washington: Brookings Institution).

Young, A. (1968a) 'Mapping Africa's natural resources', *Geographical Journal* **134,** 236–241.

Young, A. (1968b) 'Natural resource surveys for land development in the tropics', *Geography* **53,** 229-248.

Young, A. (1969) 'Natural resource survey in Malawi: some considerations of the regional method in environmental description', *Environment and land use in Africa*, ed. M. F. Thomas and G. W. Whittington (London: Methuen), 355–384

Youngson, A. J. (ed.) (1972) *Economic development in the long run* (London: Allen & Unwin).

Chapter 2

The Appraisal of Land Resources

Anthony Young

Introduction

The central problem of resource appraisal in developing countries is the assessment of the productive potential of land. To realise this potential is the task of development planning. The productive potential is normally from agriculture *sensu lato*, including livestock products, together with forestry. Other types of land use, e.g. mining, urban development or wildlife reserves, are locally relevant. But in developing countries, the conflict between agriculture and other competing demands for land is not the major problem that it constitutes in the western world. Non-agricultural use covers a small proportion of the total area and, where they occur, the demand priorities are more clear-cut, the less intensive use giving place to that with higher productivity per unit area. The productive potential normally considered is, therefore, the capacity to grow crops, both food and non-food; to support livestock, by pasture or fodder crops; and to produce timber, from natural woodland or forest plantations. All depend ultimately on the capacity of the land to sustain plant growth. A distinction may be made between land and soil. *Land* comprises all the conditions of the physical environment, of which the soil is only one; the productivity of soils cannot be assessed in isolation.

There are three phases in planning a land development project: description, evaluation and development. The descriptive phase, in which the environment is surveyed and classified, is termed *natural resource survey*. In the evaluation, or land-classification phase, the resources offered by the environment are assessed in terms of their potential production. The development phase consists of planning the means whereby this resource potential can be translated into actual production. Each phase leads to the production of maps: from resource survey, maps of soils, landforms, etc;

Dr. Young is Reader in Environmental Sciences in the University of East Anglia.

from evaluation, land capability maps; and a map of land-use recommendations as part of the final development plan. These maps provide one means for the conveyance of information from one phase to another. The links between phases are not all in one direction; the economics of production inevitably bear upon evaluation, while the proposed land use will itself modify the basic resources.

NATURAL RESOURCE SURVEY

Surveys of environmental factors

Natural resource survey is concerned with eight factors of the physical environment: geology, landforms, climate, hydrology, soils, vegetation, fauna and disease. Information on these factors may be collected separately, and subsequently synthesised; or an integrated approach may be used, in which regional units of the total environment form the first stage of data collection. A survey of present land use is also necessary.

Environmental factors differ with respect to their distribution in space, variability in time, and stability. Distribution patterns of geology, landforms, groundwater, soils and vegetation can readily be represented on maps and show some degree of spatial co-variation. Climate, surface water, fauna and disease each exhibit spatial variations of a different nature. Geology, landforms and some properties of soils vary little with time; groundwater and vegetation show substantial but fairly predictable seasonal variations; climate and surface water exhibit larger and less regular variations in time.

Stability describes the resistance of an environmental factor to change induced by land use. Geology has a high stability in that its properties are altered little, if at all, by most forms of land use. Macro-climate is virtually unaffected, but vegetation clearance can substantially alter the micro-climate near the ground surface. Landforms are moderately stable, succumbing only to the severe impact of gullying caused by accelerated erosion. Groundwater reserves are lowered if drawn upon more rapidly than the natural recharge. River régimes are altered indirectly by vegetation clearance. Soil properties vary; some, such as texture, have a high stability whilst others, notably organic matter and properties dependent upon it, are substantially altered by land use. Vegetation and fauna are the most unstable, or fragile, factors, being relatively easily modified (as by burning of savannas) or totally destroyed. Thus most natural resources are not a static quantity to be drawn upon but the outcome of a dynamic interaction between natural processes and modification by man.

Resources may either be in a steady state or be decreasing in amount.

The use of mineral resources is necessarily by exploitation. Agricultural land use, however, may also draw upon resources without replenishment, as in soil deterioration caused by over-cropping or vegetation degeneration produced by over-grazing. The aim of development planning must be to establish a method and intensity of land use in which resources are maintained at an equilibrium level. Hence natural resource survey is not solely a matter of mapping static distributions. It is equally necessary to establish the nature and magnitude of variations in time, to determine the functional relations between present land use and resources, and to estimate the effects on the resource base of planned future changes in use. Some aspects covered, and information collected in relation to purpose, in resource survey are as follows (Young, 1968):

GEOLOGY. *Hydrogeology*: occurrence and disposition of water-bearing strata; amount, quality and flow rates of groundwater. *Engineering geology*: foundation properties of rock and regolith, for road construction; availability of road-making materials (crushed rock, sand and gravel, laterite); potential dam sites. Systematic geological surveys are normally conducted separately from other resource surveys, and are not an essential prerequisite for agricultural planning.

LANDFORMS. Slope angles, in relation to soil conservation works and general land use planning. Location of stream channels and areas liable to flooding, for road alignment purposes.

CLIMATE. Means and confidence limits for monthly and annual rainfall; drought severity/frequency curves; monthly balance of rainfall and potential evapotranspiration. Radiation and day length, in relation to crop varieties.

HYDROLOGY. *Groundwater*, c.f. hydrogeology above. *Surface water*: river flow régimes; flood severity/frequency curves. *Irrigation potential* of groundwater and rivers. *Soil moisture*: seasonal variation, drought hazard.

SOILS. *Soil survey*: soil types present, extent and distribution. For each type, physical properties (e.g. limiting horizons, structure), chemical properties, organic matter, moisture régime, erosion hazard. Present state of soils as compared with natural conditions.

VEGETATION. *Pasture resources*: species composition, fodder value, carrying capacity and present ecological status of pastures. *Forest inventory*: species and volume of extractable timber; regrowth rate. Ecological surveys of a more general nature may also be conducted, as an indication of agricultural potential as well as with respect to grassland and timber resources .

FAUNA AND DISEASE. Insect vectors and faunal carriers of human and cattle diseases. In special circumstances, wildlife surveys.

Integrated surveys

The method of integrated surveys, or land systems method, is based on mapping units of the total physical environment. It is an outcome of the use of air-photograph interpretation, and was arrived at independently by different organisations employing this method (Brink *et al.*, 1966; Christian and Stewart, 1968). The primary mapping unit is the *land system*, originally defined as 'an area with a recurring pattern of topography, soils and vegetation' (Christian and Stewart, 1953), and subsequently as 'an area with a recurring pattern of genetically linked land facets' (Brink *et al.*, 1966). Both are *post hoc* definitions, the working practice being that they are those regional divisions which become apparent in the course of photo-interpretation. *Land facets* are areas within which, for practical purposes, environmental conditions are uniform; in practice these are the smallest areas that can be distinguished on air photographs. Some surveys map only land systems; in others, facets are described, with an indication of their relative extent, but are mapped only for sample areas. The land system is not a collection of areas that happen to be contiguous; the facets within a system are causally linked, for example by geomorphological origin or present-day groundwater movement. Each land system is known by a local name and is unique.

Survey is carried out by a combination of photo-interpretation and field observation. The optimum sequence of operations is: (1) preliminary inspection of air photographs; collection of background information; (2) field reconnaissance, comparing photograph images with ground features; (3) main photo-interpretation, drawing boundaries of provisional mapping units; (4) planning of field survey, selecting traverses and sampling points on the basis of provisional mapping units; (5) main field survey; (6) revision of photo-interpretation, determining final mapping units, combining or subdividing provisional units where indicated by field survey results; (7) collection of supplementary data (e.g. climate) and compilation of key. Boundaries are based mainly on landforms and to a lesser extent, particularly in swamplands, on vegetation.

The nature of the information given about a land system, and the manner of its presentation, are illustrated in Figure 2.1. The relation between the land facets, in this case a catenary one, is indicated by a cross-section, and their spatial pattern is illustrated by a map. A common alternative is a block diagram. Tabular information on individual environmental factors is given for the system as a whole and for each facet.

The main advantage of the method is the speed and cheapness with which large areas can be covered at the reconnaissance scale of survey. This arises from the economies of replacing much field activity by photo-interpretation as a basis for stratified sampling. Boundary location, which takes a large part of the time in field soil survey, is accomplished from the photographs. There is a further economy in that some areas, e.g. mountains, can be eliminated from field survey entirely. The amount of information and accuracy of boundaries is certainly greater than could be achieved at the same cost by field traverses only. It is commonly supposed that the method becomes less useful for detailed surveys, although this has not been formally demonstrated. A second advantage lies in the manner of data presentation using a regional basis with two scales, that of the system and that of the facet. By this means the essentials of any land system can be assimilated more readily than from separate maps and accounts of each factor.

Limitations of the method arise from the use of geomorphology as the main, but not exclusive, basis. In particular, its utility depends on the extent to which landforms can be used to predict soils. It is more useful in regions of varied relief than on plains. For example, in both savanna and rainforest environments it is common to find gently undulating relief underlain by rocks varying from acid to basic composition, with indistinguishable landforms but major differences in soil type. In such environments, which include much of the richer agricultural land, field soil survey is necessary.

It is not self-evident that the natural landscape is ordered into units primarily at the two scale levels represented by the land system and facet. The size of land systems varies widely between different surveys, from orders of 10 km² to 1000 km² (Young, 1969); this is in part a true reflection of the nature of landscapes, but to what extent is difficult to test, owing to the lack of specification of the degree of internal variation permitted within a system. Some surveys have employed 'land units' as a mapping unit at an intermediate level (e.g. Bawden, 1965). With respect to landforms alone there is an hierarchy of units at different scales (Thomas, 1969; Young, 1969). The land facet is not a natural landscape unit, in the sense of a valley or slope. There is a tendency for the system/facet arrangement to be treated as a Procrustean bed, into which natural landscape units of differing size and nature are forced. There is a lack of logical rigour in that each factor is itself part of the definition of units of the total environment. Since no single factor will exhibit complete co-variation with the mapping units, the latter are a loosely-defined base from which to proceed. It is possible to find the extent of correlation of any one factor, e.g. soils, with land systems or facets, but this is the only type of statistical question that can be asked. Such objections could be met if boundaries were based solely on landforms. This is common working practice at the photo-reconnaissance stage, other factors being subsequently added before constructing a combined key. It would provide a firmer basis if this practice were explicitly recognised; the definitions of land systems and facets would then be given in terms of landforms only, other factors being listed as associated features.

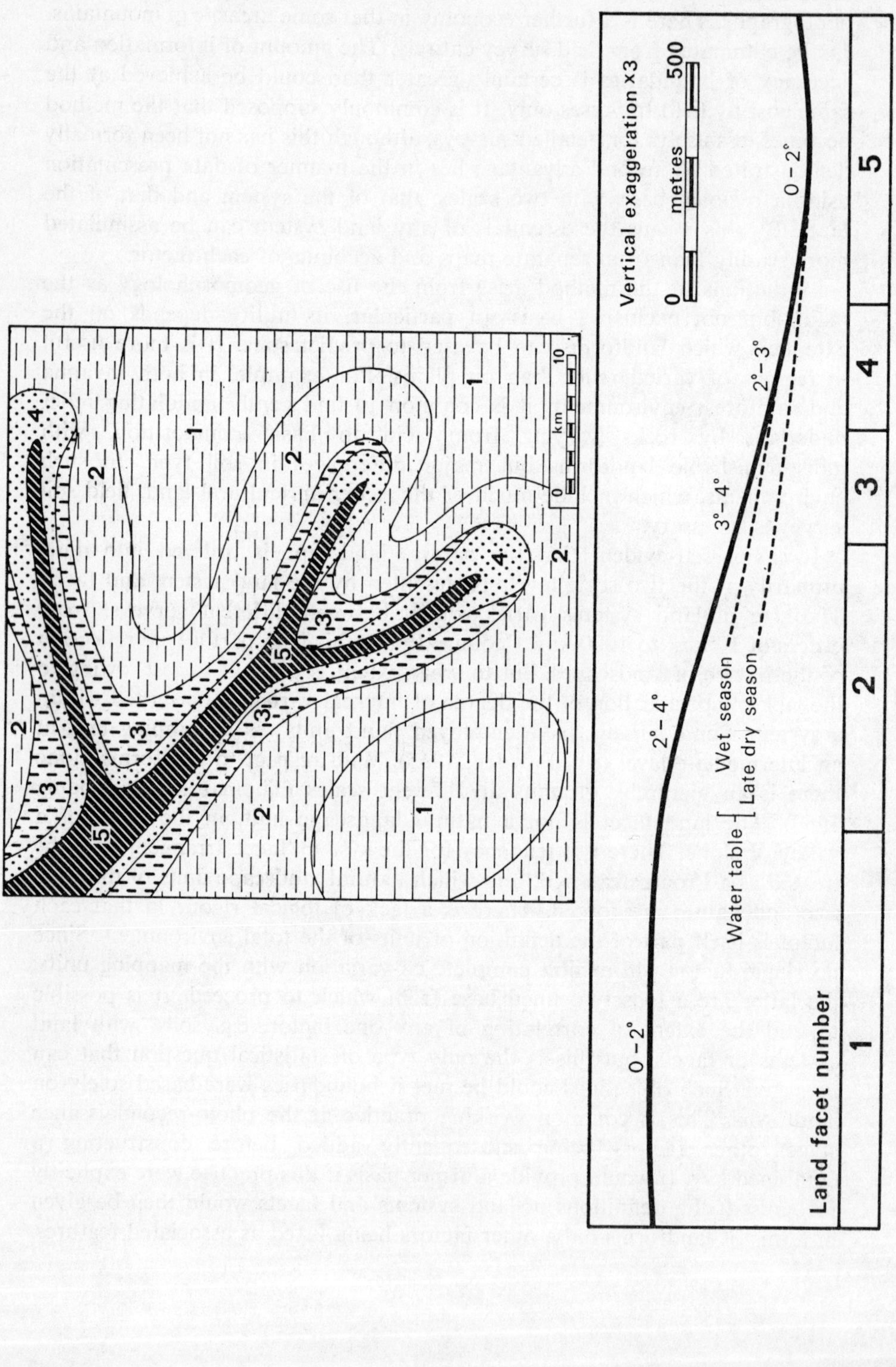
0 km 10
Vertical exaggeration ×3
0 metres 500
0–2°
2°–4°
3°–4°
2°–3°
0–2°
Water table
Wet season
Late dry season
Land facet number
1
2
3
4
5

Figure 2.1 The Lilongwe land system, Malawi (Jackson and Young, 1960; Brown and Young, 1964; Young 1965)
Location and extent: covers much of Lilongwe District and parts of Dowa and Dedza Districts, Central Province, Malawi; approximately 3600 km²
Geology: Basement Complex rocks, of intermediate to basic composition, including charnockitic gneiss.
Landforms: gently undulating plain at 1000–1300 m; local relief 50–100 m; low drainage density, maximum valley-side slopes 2½°–4°, valley floors broadly concave.
Climate: savanna type with single wet season (Aw in Köppen classification); mean annual rainfall 750–900 mm, of which 85–90 per cent falls in November-April; mean temperatures, July 16 °C, December 24 °C.
Hydrology: minor valley floors frequently have no stream channel; flow is by subsurface and surface seepage with intermittent floods; streams cease flow shortly after start of dry season and rivers of moderate size have very low or no discharge by August.
Agriculture: one of the most productive agricultural regions in Malawi, almost entirely under cultivation except in valley floors; annual cropping of maize, groundnuts, tobacco, pulses, millet, etc., frequently as mixed stands.

LAND FACETS

Facet no.	*Landforms*	*Soils and hydrology*	*Vegetation and land use*
1.	Level to gently sloping convex crests, 0°–2°.	Dark red ferruginous sandy clay; freely drained.	Almost entirely cultivated. Residual trees of *Combretum-Acacia-Piliostigma* woodland.
2.	Gently sloping convexities and upper valley sides, 2°–4°.	Yellowish red ferruginous sandy clay; freely drained.	As facet 1.
3.	Gently sloping lower valley sides, 3°–4°.	Dark brown sandy clay, mottled in depth, with iron concretions; seasonal drainage impedance.	Mainly cultivated. Vegetation similar to facet 1, with *Acacia* spp. common.
4.	Concave lower valley sides, with lobate expansions at valley heads, 2°–3°.	Mottled to dark grey loamy sand or sandy loam; subsurface seepage in wet season.	Grazing; vegetable patches. Hydromorphic grassland.
5.	Concave valley floors, with or without stream channel, 0°–2°.	Very dark grey clay. Wet season, water table at surface, intermittent flooding; dry season, water table remains within rooting depth.	Grazed normally throughout the year. Hydromorphic grassland, *Hyparrhenia* spp.

Notes on the extent to which each factor was believed to correlate with the landform units could be added. This would make clear the differing basis of the distributions shown for each factor, which at present appear with equal status.

The most substantial limitation of the method is its failure to collect the information needed for land-use planning. The fact that much surplus, i.e. non-useful, data is collected is not in itself a fault; such information may be necessary to define the mapping units themselves. But arising from its emphasis on the relatively stable factor of landforms, the approach is a static one, based on present conditions and existing land use. To meet the needs of development planning it needs to be supplemented—Moss (1969) has argued that it should be replaced—by studies of the dynamic relations between land use and the less stable factors of hydrology, soils and vegetation.

Survey procedures and planning needs

In many types of resource survey the information needed to make a scientific description of a particular factor, and to classify and map its distribution, differs in part from that needed to plan land development; and different survey operations are needed to collect these two types of information. There is a tendency for resource surveys to concentrate on descriptive and mapping operations to the neglect of activities directed towards obtaining the information needed for land-use planning. This may be illustrated by examples.

In soil survey, most of the classification and all of the mapping must be based on morphological properties observable in the field. One major diagnostic property, colour, has no effect on plant growth, whilst others, e.g. texture, structure and limiting horizons, are both observable and of agricultural significance. It is then necessary to establish correlations between the mapped soil classes and their chemically-determined properties. Most soil-survey reports give only a brief, general appraisal of the agricultural value of the soil.

If a soil survey is to be of full use in development planning it should provide, first, estimates of crop yields under specified management practices and, secondly, evidence of the probable response of each soil type to changes in land use. Such information cannot be obtained from the field operations conventionally carried out in soil survey. Crop yield estimates require comparison between soils of the surveyed area and soils outside this area for which agronomic data are available. To obtain information on responses to land-use practices is more difficult; one method is a study of soil conditions (e.g. organic matter levels) in conjunction with existing farming systems. This is not to deny that a soil map of the conventional type forms a necessary basis for land-use planning. It needs, however, to be supplemented by further information, and to obtain this requires a broadening of the field operations carried out in soil survey (Young, 1973).

A similar position occurs in ecological surveys of pasture resources. A necessary first stage is to map the distribution of plant communities; this is done by air-photograph interpretation and field survey, the latter comprising physiognomic descriptions and species-frequency counts. The main requirement needed for planning livestock management, however, is the carrying capacity of each community, i.e. the number of cattle-equivalents that can be supported without degeneration of the pasture (natural or improved) together with such information as growth rates and fodder content of grasses. It would be justifiable to devote at least as much survey time to obtaining such information as to the initial mapping operation. The main technique at present available is again comparison with stock-carrying land outside the survey area (Condon, 1968).

In defence of current practices it is commonly argued that a soil, vegetation or land-systems map is an interim document, to be followed by work of an applied nature; such work is normally carried out by different individuals—crop or pasture agronomists—to those responsible for field survey. This custom may be criticised on two counts. First, it causes the surveyor to concentrate on map production to the neglect of field operations directed towards appraisal. Secondly, it assumes that relations between the resource base and land use are of cause and effect in one direction only. The field surveyor, whether of soils, vegetation or other environmental factors, e.g. hydrology, should be directly concerned both with appraisal and with assessing the impact on the environment of proposed changes in land use.

Non-cartographic presentation

Resource-survey information need not be presented in map form. For example, forest inventory is normally carried out by enumerating areas randomly sampled from the whole survey region; the data presented is an estimate of the total timber resources. It is possible to observe soils and landforms on a similar basis, producing estimates of the proportions of the different types of land, but not a map of their distribution. In pre-investment and feasibility studies, which precede more detailed surveys for development planning, such estimates may be sufficient and can be obtained at a considerable saving in survey time and cost (Robertson and Stoner, 1970).

Resource survey data lends itself well to computer-based data handling. Each observation of any factor is stored with its locational coordinates.

Maps can then be produced on demand, specifying the combinations of properties on which they are to be based and the classes to be mapped. For example a map showing slope angle, topsoil texture and nitrogen status could be called for. This method supplements rather than replaces maps of the conventional type. It has attractive possibilities in permitting the processing of data prior to printing. For example, from agronomic data, equations relating maize yield to climatic and soil parameters and to management practices could be obtained. These could be applied to the survey area to produce maps of yield estimates under various specified management practices; means and upper and lower confidence limits could be given, taking account both of climatic variability and of the error of estimate in the original equations. If, subsequently, some additional property were shown to be limiting yields, revised maps incorporating it could be produced.

Costs and returns

It is sometimes suggested that the additional costs of detailed resource surveys are not justified by the returns. This can be examined by a consideration of the orders of magnitude involved. Suppose that a soil survey is to be carried out in an area of annual cropping and that the main anticipated cash crop is groundnuts. A detailed soil survey, on a scale of 1:25 000, can be completed for a cost of the order of $1·00/ha (Bie and Beckett, 1970). One hectare of groundnuts yields c. 1000 kg, valued at $120. It is reasonable to assume that gains of all kinds from soil survey, e.g. selection of planting areas and variation of fertiliser treatment with soil type, will raise net production by at least 10 per cent. This gives a return of $12, considerably above the cost of survey, in the first year alone. Perennial crops, which have a higher value of output per unit area, will give still greater returns. Thus even at high discount rates, the benefit/cost ratio of soil surveys is likely to be high. Klingebiel (1966) gives examples with benefit/cost ratios ranging from 45:1 upwards.

LAND CAPABILITY CLASSIFICATION

The evaluation phase in development planning involves taking the resources present and assessing their productive potential. To do this, the results of resource survey are compared with the technical requirements of various types of land use. A common practice is to produce a *land-capability classification* map, which shows land in terms of its 'suitability' for one or more types of use. Following upon this phase, economic and other factors are taken into consideration, leading to decisions on land-use recommendations.

In the simplest form of classification, land is allotted to one of a series of *capability classes* (I, II, III . . .) of decreasing productive potential. In the scheme of the US Department of Agriculture (USDA) Soil Conservation Service (Klingebiel and Montgomery, 1961), Classes I–IV contain cultivable land and V–VIII are non-cultivable. This system is based on the assessment of *limitations* to land use, namely erosion hazard, excess water, soil root-zone limitations and climatic limitations. With a few exceptions, the original scheme does not give quantitative limits (e.g. slope angle, soil depth, rainfall) for the severity of each limitation necessary to relegate land to a given capability class. This does not preclude the adoption of such limits in local adaptations of the system. For example it might be decided that Class I land must have a slope of less than 2° and a soil depth exceeding 2 m, Class II land 5° and 1·5 m respectively, and so forth; the presence of any one limitation exceeding the specified value is sufficient to place land in the lower class. Such adaptations have been widely used in surveys of developing countries.

In the example given in Figure 2.1, facet 1 would fall into USDA Class I, land with few limitations. Facets 2 and, probably, 3 would fall into Class II, on the basis of slope angle and the consequent need for soil conservation works. Facet 4 has limitations caused by slope, the sandy texture of the soil and drainage impedance, and would fall into either Class III (severe limitations) or IV (very severe limitations). Wet valley-floor land, as in facet 5, is placed in Class V, a class conventionally reserved for wetlands unsuited to arable use but suited to pasture.

Capability classifications can be applied with respect to various forms of land use. The two most common are suitability for arable use in general (i.e. non-specific with regard to crops) and for irrigation. The US Bureau of Reclamation (1953) scheme is designed for the latter purpose and incorporates quantitative limits for each class. Alternatively, capability may be separately assessed with respect to various types of land use, e.g. perennial crops, padi, other annual crops, pasture and forestry (Haantjens, 1965). Land-use recommendations can then be arrived at by comparing the economics of each type of use; there is a choice between giving priority either to the most profitable type or to the most intensive, i.e. that which gives highest gross production per unit area. A further stage is to classify separately with respect to individual crops.

General-purpose capability classification systems have inherent limitations (a fuller discussion is given in Young (1972)). Because of varying technological requisites, the order of relative suitability is not always the same for different types of use. To take an extreme example, swampy valley floor land may be highly suitable for *padi*, conditionally suitable (given drainage works) for other crops and unsuitable (owing to extraction vehicles coming up against soft ground) for forestry, whereas on adjacent freely-drained slopes the order is reversed. Most systems do not provide the basic information needed for economic analysis, namely crop yield estimates, although there have been attempts at this (Steele, 1967).

The fundamental limitation of land capability classifications is their temporary or static nature. They only hold good as long as particular technological and economic conditions prevail and they are intrinsically unsuited to take account of future changes in these conditions. To regard capability classification maps as having the permanence of soil maps leads to an inflexibility in planning. Changes in conditions need to be referred back to the primary resource data, leading to re-evaluation.

THE TECHNOLOGY OF RESOURCE USE

To give substance to the framework provided by land classification systems, it is necessary to consider the technology of productive land use. Each of the main types of use has certain needs in common. However, within each type, specific requirements vary widely, e.g. as between different crops or different forms of irrigation. These requirements arise from two main sources: the physiological needs of plants (crops, grasses or trees) and the techniques of production.

Annual crops

A basic requirement for the cultivation of annual crops is to match the length of the growing season (in the tropics, the period in which the soil moisture content within the root zone remains above wilting point) to the physiological requirements of the crop. This produces, for example, a zonation among the major cereals, from maize and finger millet in the moist savannas (mean annual rainfall over *c.* 800 mm) through sorghum to bulrush and panicum millets at the savanna/steppe margin. Among root crops, cassava is drought-resistant and tolerates a wide range of climates, whereas yams are a crop of the rainforest/savanna transition zone. Cotton and groundnut cultivation can be extended into the drier savannas, and cultivars (cultivated varieties) of cotton with physiological growing periods appropriate to different climatic zones have been developed. As the drier climatic limits for a given crop are approached, soil moisture retention becomes of critical importance; hence in such climatically marginal zones, heavy-textured but well-structured soils acquire an added significance.

A feature common to all forms of annual cropping is that bare soil is necessarily exposed for all or part of the year. There is consequently an erosion hazard, the severity of which varies with slope angle. Grassed bunds or other types of conservation works must be placed at closer intervals as the slope becomes steeper; the erosive power of surface wash increases more than linearly with angle of slope and hence the cost of conservation works varies similarly. It is for this reason that, in areas of non-irrigated annual cropping in the tropics, slope angle has a far greater effect on the assessed land class than that of soil type. A third major problem in annual cropping, the maintenance of soil organic matter, is discussed below.

Each crop has specific requirements. For example, maize grows best on medium to heavy-textured soils. It has a high nitrogen demand, a crop of 2000 kg/ha removing 40 kg/ha N. Maintenance of a good soil organic matter status is necessary, in addition to nitrogenous fertilisers. It is a soil-exhausting crop and should not be grown two years in succession, a practice all too common. Groundnuts are best suited to sandy loams; compact soils inhibit peg penetration and clays adhere to the nuts causing difficulties in harvesting. Being a legume it returns nitrogen to the soil, and is therefore an excellent component in rotations. For tobacco, free drainage throughout the profile is essential. Fire-cured tobacco prefers clay loams or sandy clays; flue-cured tobacco is unusual in requiring a low nitrogen supply in the latter part of the growing season and is suited to coarse-textured soils, e.g. sandy loams. Many tobacco soils have duplex textural profiles, a sandy topsoil overlying a heavier subsoil, making them highly prone to erosion. These specific agronomic features affect the yield of each crop on a given soil type.

Land-use planning of annual cropping involves decisions on crop rotations, including both lengths of fallows and crop sequences. It should be possible to determine such rotations separately for each soil type, to specify cultivation practices, e.g. fertilisers, and to estimate yields.

Perennial crops

In the cultivation of perennial crops, a critical requirement is that the soil moisture content shall not reach wilting-point down to the maximum depth reached by roots. This can be calculated by comparing drought severity/frequency curves, monthly potential evapotranspiration and the available moisture in the soil. Most perennials also suffer depression in yields from moisture stress at low moisture levels above wilting-point (Williams and Joseph, 1970). The greatest problems, both of crop loss and yield lowering, again arise with sandy soils.

Whereas in annual cropping, organic matter conservation is the major soil problem, with perennial crops it is the maintenance of nutrient supplies. Crop rotation is not available as a means of balancing nutrient demands, hence heavy and continuous fertilisation is needed. For reasons of moisture supply, perennials are grown where rainfall is high and leaching intense; there is consequently a particular need for potassium, which occurs as an exchangeable cation and is thus readily removed by leaching. Oil palm is the most demanding crop in this respect. It is the second most productive of all crops, in terms of calories per unit area per year (being exceeded by sugar cane); it gives highest yields where there is no dry season and, hence, strong leaching; and it has heavy demands for both nitrogen and potassium. In West Africa, yields vary closely with exchangeable potassium status. A mature plantation yielding 6000 kg/ha (fruit) may respond to dressings of as high as 600 kg/ha potassium (Hartley, 1968). The need for fertilisation is increased by the replacement of indigenous palms by high-yielding varieties. Oil palm also has the special requirement that a system of harvesting roads (or, on level ground, rail tracks) must be constructed to convey the fresh fruit bunches to the processing plant on the day of picking. Subsoil drainage impedance damages the palm, hence on level ground deep drainage ditches are necessary.

Rubber grows in similar climates to oil palm but has considerably lower nutrient demands; hence although economic conditions are less favourable, it may be preferred to oil palm on shallow soils where there is a limiting horizon of laterite. Cocoa does not grow well in strongly leached soils, preferring an exchangeable cation saturation of over 35 per cent; consequently strongly ferallitic soils and permanently humid regions are avoided, and it is grown in West Africa in the 1000–1800 mm rainfall zone. Heavy fertilisation is again essential for sustained high yields. Tea requires a deep but freely-drained soil; being a crop of which the leaf is harvested, it has high nitrogen requirements. It is frequently grown on steep slopes characteristic of dissected rainforest relief, and severe erosion can ensue unless terracing or other conservation works are maintained.

Other types of land use

Road construction provides an example of the requirements of a different form of land use. In the humid tropics soils on basic rocks, whilst rich agriculturally, make poor road foundations, owing to the deep regolith, high clay content and swelling properties of montmorillonitic clay materials. Poorer agricultural soils, for example over sandstone or quartzite, are preferable for road foundations. The closely-spaced dendritic pattern of stream dissection characteristic of rainforest environments, coupled with the occurrence of flash floods, causes problems of alignment: either roads must be devious or there must be frequent bridges, in either event giving higher construction and maintenance costs.

Other examples could be given from irrigation, forestry and other forms of production. Each type of land use, and each crop, has differing requirements, which affect the relative evaluation of environmental features for different purposes.

THE IMPACT OF LAND USE ON RESOURCES

The possible effects of changes in land use upon resources are well recognised in the case of soil erosion. Other examples are the prevention of vegetation degeneration under pastoral use and the question of recharge where groundwater resources are drawn upon for irrigation. The principle in all such cases is the same: either a land-use system is devised that maintains the resource at a constant level or the resource is drawn upon. The latter course, inevitable with mineral reserves, can only be a short-term solution with respect to land resources.

This general question is illustrated by the example of the maintenance of soil resources under annual cropping. Under natural vegetation there is a continuous cycling of organic matter and plant nutrients between vegetation and soil; clearance of the vegetation for annual cropping breaks this cycle, cutting off the inflow to the soil store but not arresting, and possibly accelerating, the outflow from it; consequently, cultivation causes an inevitable depletion of soil organic matter and nutrients. This is usually discussed in the context of shifting cultivation systems, in which restitution of the soil is accomplished by a period of bush fallow, allowing the natural mechanisms of plant succession to build up the soil plant cycle. The fallow: cultivation ratio necessary to maintain equilibrium under such conditions has been estimated at between 3:1 and 6:1 in the rainforest zone and 10:1 or more in the savannas (Nye and Greenland, 1960). Systems of permanent cultivation of annual crops are subject to the same constraint. Either the losses sustained during cropping must be made good, or the farming will be based on 'soil mining', drawing on the stores built up in the soil during the past.

Losses from the soil during cultivation may be considered under two heads: plant nutrients and organic matter. Apart from small amounts brought down by rainfall, the ultimate source of all nutrients except nitrogen is from the weathering or rock materials. Thus orthoclase feldspar releases potassium and apatite releases phosphorus. Hence soils containing weatherable minerals in the profile, within rooting depth, receive a steady renewal of nutrient supplies from rock weathering. This accounts in part for the unique ability of soils derived from basic rocks, in particular the ferrisol (basisol) class, to sustain continuous cropping without fallows. In ferallitic soils this source of nutrient renewal is absent.

The main immediate source of nutrients under natural conditions is the decomposition of organic matter. For each nutrient there is a cycle, soil–plant–litter–soil, together with gains (rainfall, rock weathering, atmospheric nitrogen fixation) and losses (leaching, oxidation, erosion). The cyclic flows are substantially larger than the gains and losses. Hence the losses, when crops replace natural vegetation and nutrients are removed in the harvest, are relatively rapid. The absolute amounts removed, however, are not large. For example a 2000 kg/ha crop of maize removes of the order of 40 kg nitrogen, 5 kg phosphate and 10 kg potash. To restore such losses wholly by artificial fertilisers would require some 200 kg/ha of sulphate of ammonia plus smaller amounts of mixed fertiliser. Dressings of this order of magnitude are technically practicable, i.e. the weights involved are not excessive, and give an economic return.

The position is different with respect to the soil organic matter. The annual loss during cultivation, under both rainforest and moist savanna environments, is of the order of 2000 kg/ha. There are two problems in replacing such quantities: the labour involved and the source of material. Contributions can be made by ploughing in crop residues and farmyard manure, bringing in cut grass, or growing and ploughing in a green manure crop (e.g. pigeon pea). Nevertheless in the lowland tropics it is usually impracticable to maintain soil organic matter levels except by recourse to fallows.

To plan a farming system that will maintain the soil at an equilibrium level, organic matter losses during cultivation must be balanced against net gains during fallow plus additions (manure, etc.). The loss (mainly by bacterial oxidation) depends in part on the amount initially present. In the model proposed by Nye and Greenland (1960) it is assumed that a percentage, the decomposition constant, of the initial organic matter is lost in one year's cultivation; they give values for the decomposition constant of 1·8–4·7 per cent for lowland rainforest and 2·5–6·6 per cent for savanna environments. This is obtained by an indirect process of calculation, and may under-represent initial cultivation losses. A period of over-cropping, with fallows too short to replace cultivation losses, will cause a decline in the soil organic matter level; assuming the decomposition constant model to be valid, this will decrease the absolute amount lost. Such soil deterioration is at present widespread. Either soil erosion will supervene or a position will ultimately be reached at which losses equal gains. This will occur when soil organic matter is considerably below its natural level; crop yields also stabilise, but are low. This is the condition of low-level equilibrium; Ruthenburg (1971) reports a case from the Indian Deccan where wheat yields have stabilised at the extremely low level of 350 kg/ha.

Any appraisal of land potential needs to take into account the long-term effects of projected land use on the soil. A first step is to ensure the maintenance of organic matter at a reasonable equilibrium level, say 75 per cent of that under natural conditions. This will in itself assist in nutrient problems and can then be supplemented by fertilisers. At the present we are far from being able to draw up the balance sheet of soil gains and losses necessary to plan rotations in this way. Monitoring soil changes under existing farming systems would be one method of approach.

Many of the relations between environment and land use are more complex than the simple maintenance of resources at an equilibrium level. Thus in the forest-savanna transition zone, fallows may be either of woody species or, less favourably for soil regeneration, of grass. Which will occur depends on the frequency of cultivation and burning, and the soil moisture status, the latter being related to texture. Grassland is less conserving of moisture than forest and hence if a grass fallow becomes established it will not be invaded by woody species. Such ecological changes can lead to changes in land-use systems. Moss (1968) had described some complex interactions in this zone in West Africa, involving relations between soil, soil moisture, vegetation, insect pests, economic conditions and land-use systems. Similar interactions could be described for pastoralism, forestry and irrigated land use. These examples indicate that a one-way land-evaluation procedure, from resources via capability to development, is insufficient. Changes in economic conditions alter land-use systems, which in turn affect resources, and such interactions need to be taken into account in land appraisal.

THE PRODUCTIVE POTENTIAL OF LAND

The problem of estimating the productive potential of land, or the population which it can support, can only be answered in terms of levels of technology and living standards. One approach is to base assessment on existing farming practices. This was employed by Allan (1965) in the context of shifting cultivation systems in Zambia. Let

C = cultivated area per capita, in a given year
L = length of cultivation period, in the cultivation–fallow cycle
F = length of fallow, in the cultivation–fallow cycle
P = percentage of total area that is cultivable

If there is no unutilised land, $L/L + F$ is equivalent in spatial terms to the proportion of cultivable land that is cultivated in any one year. The total area needed per capita, A, is given by

$$A = C \cdot \frac{L+F}{L} \cdot \frac{100}{P}$$

For example if $C = 0{\cdot}25$ ha, $L = 3$ years, $F = 15$ years and $P = 50$ per cent, then $A = 0{\cdot}25 \times 18/3 \times 100/50 = 3$ ha per capita. The constants are determined separately for each of several mapped vegetation-soil

Table 2.1 Estimated potential productivity of Lilongwe land system (Figure 2.1) under improved technology. Based in part on agronomic data in Brown and Young (1964)

Crop	Area	Yield	Production	Calorific value	Subsistence food production	Price	Income
	ha	kg/ha	kg	kcal/kg	Mcal	$/kg	$
Maize	2	2000 (grain)	4000	3500	14 000	—	—
Groundnuts	1	1000 (nuts in shell)	1000	—	—	0·12	120
Tobacco	1	600 (cured leaf)	600	—	—	0·70	420
Fallow	4	—	—	—	—	—	—
Pasture	2	(milk, meat)	—	—	(milk, meat)	—	—
Total	10	—	—	—	14 000	—	540

Costs:			
400 kg fertiliser	$50	Gross cash income:	$540
Hybrid seed, implements, etc.	$40	Costs:	$ 90
Total:	$90	Net cash income:	$450

Note. To simplify calculations it is assumed that all maize is consumed for subsistence and all other crops sold; meat and milk production for pasture, which will supplement food supplies, is left out of consideration.

classes within a region, *A*, and calculated for each, and the 'critical population density' of the region is found.

The advantage of this approach is its realism. It is based on existing conditions and assumes there will be no technological improvements. It gives no indication of land potential under improved technology and is only applicable where the system is in a state of equilibrium. Many annual cropping systems, however, have passed this stage, and are currently in a condition of semi-permanent cultivation in which, with shortened fallows, soil reserves are being drawn upon (Ruthenburg, 1971).

In the moist savanna zone, overall rural population densities are commonly of the order of two persons per hectare of cultivable land. In densely settled parts of the rainforest zone and on ferrisol (basic rock) areas of the savannas, densities of five persons per hectare are usual, whilst in the drier savanna to steppe zone, one person per hectare is typical. Such figures are necessarily highly generalised, but give some standard for comparison when considering rural peasant communities at present-day levels of technology.

An alternative approach is an assessment based on establishing the productivity of known resources under specified management practices. There is insufficient knowledge of the controls over crop yields to be able to achieve this with precision, but the possibilities may be illustrated by an example. Consider the land system shown in Figure 2.1. Cultivable land (facets 1–3) occupies 80 per cent of the total area, and pasture (facets 4–5) 20 per cent. Assume it is empirically found that, with dressings of farmyard manure, a cycle of four years cultivation and four years fallow maintains organic matter levels. A suitable cropping rotation for this relatively fertile soil is maize–groundnuts–maize tobacco, which includes a legume and gives both food and cash crops. Suppose the productivity is to be estimated for improved technology, i.e. methods at present in use by more progressive farmers, involving the use of hybrid maize seed and fertilisers (100 kg/ha sulphate of ammonia for maize and tobacco, plus 100 kg/ha single superphosphate for tobacco). Given these inputs, the areas cultivated, estimated yields and consequent food production and cash income for an area of 10 ha are given in Table 2.1.

These assumptions give a production of 1400 Mcal[1] plus $45 per hectare. One adult requires 1000 Mcal per year; therefore 10 hectares provides subsistence for 14 adults or 3 farm households each of 3 adults and 3 children. If this were in fact the population, there would be a cash income of $25 per capita, or $150 per household, plus subsistence food, on farms of 3·3 ha. Farm population density would be 1·8 persons per hectare, or 180 per square kilometre of productive land.

Existing farming is at lower levels of technology. Details of two sample areas from a 1968/69 sample survey of agriculture which lie mainly within this land system are given in Table 2.2. Area 13 is part of a special development region, has better farming methods than most other sample areas and has the highest maize yields of any part of Malawi. A comparison of its situation with the preceding assessment suggests that currently it has about the maximum population density consistent with maintaining a stable farming system. Area 17 has a higher population density and low yields; income is heavily supplemented by non-farm sources. Even if improved technology led to a doubling of yields income would still be low. The evidence suggests that this area is over-populated, that is, the farm population density is too high, in relation to the land potential, to be supported at current Malawi standards of living.

Table 2.2 Actual production (1968/69) from areas situated mainly on Lilongwe land system

Source: Malawi Government (1970)

Sample survey area	Proportion of holdings applying fertiliser	Average maize yield	Farm population /cultivated area	Average farm size	Net farm income	Cash from non-farm sources
no.	per cent	kg/ha	persons/ha	ha	$/ha	$/capita
13	26	1780	1·85	2·73	15·00	4·0
17	3	750	3·10	1·60	5·00	5·9

This example is of a soil type with above-average fertility. Under a similar climate on the highly-weathered ferallitic soils characteristic of much of the plateau surfaces of Africa, there would need to be a cultivation/fallow ratio of less than 1 : 1 and yields for similar inputs would be 33–50 per cent lower. Hence a density of one person per hectare of productive land is the greatest that could be supported at a similar standard of living.

CONCLUSION

The problem of land appraisal presents itself at various scales: those of the farm, district, nation and the developing world as a whole. At the level of the individual farm it is faced annually by the farmer, with motivation varying from the commercial aim of maximising profit, to, at the opposite extreme, the 'survival algorism' of subsistence farming, obtaining minimal food requirements with the greatest possible assurance. The district or local level is the scale at which much development planning takes place, e.g. as land settlement and reorganisation schemes; land appraisal is here the concern of government or quasi-governmental organisations. This is the scale at which most practical surveys for land assessment are carried out. At the national level, land productivity is the concern of governments in two main respects: to provide a food supply and as a source of export earnings. At the world scale there are the problems, to be faced by the end of the twentieth century, of whether agricultural production can match growth in population and at what level decreasing marginal returns from land resources will enforce a halt to the growth. We are far from having the knowledge to make such estimates, although the completion of the *Soil Map of the World* (FAO/UNESCO, 1971) will mean that for the first time some basic standardised data is available.

In the processes of land appraisal there needs to be a better integration of the phases of survey, evaluation and development planning. Resource survey procedures require modification to include operations directed specifically towards appraisal. In planning resource development, account should be taken of the impact of projected changes in land use upon the resources. The static model of a fixed set of resources, which can be successively described, assessed and developed, is inadequate; the relation between land use and environment is one of mutual interaction, and this fact needs to be taken into account in survey and assessment procedures. Fundamentally, land appraisal depends on a knowledge of the nature and functioning of the physical environment and of the technology of methods of production.

NOTE

1. One megacalorie (Mcal) = 1000 kilocalories (kcal; also written Calories) = 1 000 000 calories.

REFERENCES

Allan, W. (1965) *The African husbandman* (London: Oliver and Boyd).

Bawden, M. (1965) 'A reconnaissance of the land resources of eastern Bechuanaland', *Journal of Applied Ecology* **2**, 357–365.

Bie, S. W. and P. H. T. Beckett (1970), 'The costs of soil survey', *Soils and Fertilisers* **33**, 203–217.

Brink, A. B. A. *et al*, (1966) *Report of the working group on land classification and data storage* (MEXE no. 940; Christchurch, Hants: Military Engineering Experimental Establishment).

Brown, P. and A. Young (1964) *The physical environment of central Malawi with special reference to soils and agriculture* (Zomba, Malawi: Government Printer).

Christian, C. S. and G. A. Stewart (1953) *General report on survey of Katherine–Darwin region*, 1946 (Land Research Series No. 1; Melbourne: Commonwealth Scientific and Industrial Research Organisation).

Christian, C. S. and G. A. Stewart (1968) 'Methodology of integrated surveys', *Proceedings of a conference on aerial surveys and integrated studies, Toulouse*, 1964 (Paris: UNESCO), 233–280.

Condon, R. W. (1968) 'Estimation of grazing capacity on arid grazing lands', *Land evaluation*, ed. G. A. Stewart (Melbourne: Macmillan), 112–124.

FAO/UNESCO (1971) *FAO/UNESCO soil map of the world, 1:5 000 000. Vol. IV. South America* (Paris: UNESCO).

Haantjens, H. A. (1965) *Agricultural land classification for New Guinea resources surveys* (Technical Memorandum 65/8; Canberra: Commonwealth Scientific and Industrial Research Organisation).

Hartley, C. W. S. (1968) 'The soil relations and fertiliser requirements of some permanent crops in west and central Africa', *The soil resources of tropical Africa*, ed. R. P. Moss, (Cambridge: Cambridge University Press), 155–183.

Jackson, G. and A. Young, (1960) 'Soil surveys: Maonde, Dedza District', *Annual report of the Department of Agriculture for the year 1958/59, Part II* (Zomba, Malawi: Government Printer), 143–147.

Klingebiel, A. A. (1966) 'Costs and returns of soil surveys', *Soil Conservation* **32,** 3–6.

Klingebiel, A. A. and P. H. Montgomery (1961) *Land-capability classification* (Agriculture Handbook 210; Washington DC: US Department of Agriculture, Soil Conservation Service).

Malawi Government (1970) *National sample survey of agriculture 1968/69* (Zomba, Malawi: Government Printer).

Moss, R. P. (1968) 'Land use, vegetation and soil factors in south-west Nigeria: a new approach', *Pacific Viewpoint* **9,** 107–127.

Moss, R. P. (1969) 'The appraisal of land resources in tropical Africa. A critique of some concepts', *Pacific Viewpoint* **10,** 18–27.

Nye, P. H. and D. J. Greenland (1960) *The soil under shifting cultivation* (Farnham Royal, Bucks: Commonwealth Agricultural Bureau).

Robertson, V. C. and R. F. Stoner (1970) 'Land use surveying: a case for reducing the costs', *New possibilities and techniques for land use and related surveys*, ed. I. H. Cox, (Berkhamsted: Geographical Publications).

Ruthenburg, H. (1971) *Farming systems in the tropics* (Oxford: Clarendon Press).

Steele, J. G. (1967) *Soil survey interpretation and its use* (Soils Bulletin no. 8; Rome: FAO).

Thomas, M. F. (1969) 'Geomorphology and land classification in tropical Africa', *Environment and land use in Africa*, ed. M. F. Thomas and G. W. Whittington (London: Methuen), 103–146.

U.S. Bureau of Reclamation (1953) *Bureau of Reclamation manual. Volume V, Irrigated land use, Part 2, Land classification* (Washington DC: US Department of Interior).

Williams, C. N. and K. T. Joseph (1970) *Climate, soil and crop production in the humid tropics* (Kuala Lumpur: Oxford University Press), chapter 3.

Young, A. (1965) Map: *Malawi, natural regions and areas. Sheet 2. Central Malawi* (London: Directorate of Overseas Surveys).

Young, A. (1968) 'Natural resource surveys for land development in the tropics', *Geography* **53,** 229–248.

Young, A. (1969) 'Natural resource survey in Malawi: some considerations of the regional method in environmental description', *Environment and land use in Africa*, ed. M. F. Thomas and G. W. Whittington (London: Methuen), 355–384.

Young, A. (1972) 'Rural land evaluation', *Essays in applied geography*, ed. J. C. Doornkamp and J. C. Dawson (London: Arnold), 5–33.

Young, A. (1973) *Geographical Journal* **139,** 53–64. 'Soil survey procedures in land development planning', 5–33.

Chapter 3

Ecological Modelling of a Tropical Watershed: A Guide to Regional Planning

Gustavo A. Antonini, Katherine Carter Ewel and John J. Ewel

Introduction

Increased pressure on the limited resources of the earth make the study of resource depletion one of the most urgent needs of our society today. The study of man–environment relationships is imperative because the stress of human occupancy upon the biosphere has reached a critical stage. The need is particularly great in tropical regions, for it is there that population growth is most rapid and resource management is, in general, least technically developed. This chapter is concerned with the need to develop new geographical and predictive techniques to evaluate resource depletion in the tropics. As such, the chapter describes the application of time-sequential models as new tools for measuring the impact of past land use changes on the physical landscape and predicting the resulting environmental consequences. The objective is to discuss these new methods which can assess quantitatively the productive potential and relative environmental stress of traditional exploitative tropical life-support systems, using the central Dominican Republic as a case study.

Study region

The Dominican Republic is a microcosm of emerging nations, attempting to achieve accelerated economic development under the twin pressures of rapid population growth and rising per capita income. Though agriculture is the dominant economic sector, with 60 per cent of the total labour force,

---

Dr. Gustavo Antonini is Associate Professor of Geography and Latin American Studies at the University of Florida, Gainesville, and is also Coordinator of Research for Latin American Studies at that institution. Dr. Katherine Ewel is associated with the Florida State Museum at the University of Florida. Dr. John Ewel is Assistant Professor of Botany at the University of Florida.

farm output accounts for only 24 per cent of the gross domestic product (Inter-American Development Bank, 1971, p. 192). To accelerate expansion of the farm sector, to achieve a more equitable income distribution and to stem the tide of rural migrants to the cities, the Dominican Republic has embarked upon a number of coordinated regional development programmes scheduled for completion under the government's National Development Plan for the 1971–74 period (ONAPLAN, 1970).

O greatest significance to the country's economic development is the construction of a series of hydroelectric power and water storage installations at Tavera on the Yaque del Norte River and near Sabana Iglesia on the Jagua River (SOGREAH, 1968; Lahmeyer International GMBH, 1967). Valued at US $60 million and scheduled for completion about 1975, this project will comprise a total installed power capacity of 135 000 kilowatts and in addition will sustain the irrigation requirements of a 400 km² tract downstream in the semi-arid but potentially fertile north-western region of the country (Antonini, 1972, p.253).

The ecological study described in this chapter covers the 407 km² area located upstream from the Jagua dam site. Since 1950, population has doubled in this mountainous region, yet agricultural productivity during the same period has declined. Furthermore, much of this upland, formerly in pine forest, has been subjected to slash-and-burn farming and extensive grazing (United Nations, 1971). These factors, juxtaposed with the cessation of lumbering operations in 1967, have produced an out-migration of some 4000 labourers, principally to New York City.

In this populated upland region, the links between man and nature are so direct that clearly the continued practice of marginal farming and illegal logging will cause increased erosion of the soil and depletion of the region's resource base. Increased demand for downstream irrigation water requires that sedimentation be halted and erosion controlled. The ecological incompatibilities arise because the more intense the land use the greater the hazard from erosion and siltation.

Modelling procedures

To study and evaluate the dynamic interplay between man and the land, the geographer must be able to integrate the virtual infinitude of variables and attributes of the natural landscape with the cultural features of evolving land use. Fortunately, these are not haphazardly arranged, but are organised into man-land ecosystems in which agents, materials and processes are intimately related in well-defined geographical patterns. Each land-use system is a man-managed activity that is fitted with varying degrees of conformity to the physical resource base. Departures of land-use systems from the natural order are due to cultural investments. In this study, a technique was required to examine those systematically related natural variables and cultural landscape features that are functionally involved in land use and which affect ecological balance in the region. Energy flow modelling (Odum, 1969 and 1971) was used as the integrating mechanism for evaluating cause-effect relationships in the watershed area between evolving land use and ecological change.

All systems carry on energy transfer processes. Similarly, energy flow is a common denominator which can be used to unite interactions of all kinds. All viable systems maintain energy flow through themselves and culturally determined systems are no exception to this rule. In the Jagua–Bao region, man's relationship to the land is a fairly direct one in which the population is almost entirely dependent upon local energy sources—sun, water, available nutrients—rather than energy derived from fossil fuels. Consequently, the feedback processes between man and the land are direct and the total population that can be sustained in a given community depends almost entirely on their direct relationship to the land. The construction of the Jagua dam is an additional constraint within the ecological system since the viability of the hydropower facilities is dependent on the management practices in the upper watershed which in turn affect the hydrological characteristics of the basin itself. Modelling was used to relate present population growth patterns and evolving land-use practices to their corresponding effects on the dam over the 1945–2015 year period.

Two models were developed. First, a land-use model was established to examine the relationship between population growth and conversions of forest to various classes of agricultural use. Second, a water balance model was used to examine rates of runoff produced by the temperature and rainfall régimes in the area under different soil characteristics. The final analysis integrated the two models by imposing different run-off coefficients on each of the land-use practices to quantitfy the effects of evolving land-use patterns on future siltation in the downstream reservoir.

The modelling process entailed the following sequence of steps: (1) construction of an energy flow diagram; (2) determination of flows and transfer coefficients; (3) conversion of the energy flow diagram to a set of differential equations; (4) scaling and analog computer programming of the differential equations, according to standard procedures (see, for example, Peterson, 1967); and (5) simulation on the analog computer. Throughout the process, the analog computer was used because a variety of simulations can be explored almost immediately and communication between the computer and the modeller is nearly instantaneous. The models described could, however, have been simulated equally well using digital computers.

Data base

Owing to the size and scope of the study, various data collection techniques were employed. The general lack of published documentary material on contemporary land use required that heavy reliance be placed on extensive personal interviews with local inhabitants as a primary research method in reconstructing the cultural fabric of daily and yearly activities.

Table 3.1 Categories of information obtained by field interviews

I	Classification of administrative (mapping) units	
II	Population	
	A.	Census
	B.	Labour force
	C.	Internal/external migration
III	Land-use activities	
	A.	Farming
	B.	Grazing
	C.	Logging
	D.	Other activities
	E.	Historical land-use information
IV	Land tenure	
	A.	Nature and scope of rights in land
	B.	Owned land
	C.	Sharecropped land
	D.	Common (squatted) land
V	Transportation	
	A.	Trafficability
	B.	Road maintenance
	C.	Mode of transportation
VI	Diet	
	A.	Food consumption
	B.	Imported foods
VII	Vegetation changes	
	A.	Types
	B.	Reasons
VIII	Local energy sources	
	A.	Functions
	B.	Types and consumption of fossil fuels
IX	Mass Media and education	
	A.	Newspapers and periodicals
	B.	Education

Interviewing was performed during the summer of 1971 and the sampling procedure was structured to obtain cultural data from all twenty-one settlements at the smallest definable territorial level. Categories of information obtained by field interviewing are presented in Table 3.1. The interview schedule classified differences in the spatial arrangement of human activities based on cultural and physical elements of the settlement's resource base.

Another primary source of information on changing land use was aerial photography available for the years 1948, 1958 and 1966. The three sets of photographs were examined under stereoscopy and land-use maps were compiled at a 1:50 000 scale. For each map four basic categories were delineated: forest, pasture, short-cycle mixed farming and coffee. Where photo-scale precluded identification of individual units, combinations of the basic divisions were plotted using a ratio of major to minor land use (e.g. forest greater than pasture, 70 per cent forest and 30 per cent pasture). Next, composite land-use change maps were compiled for the years 1948–58 and 1958–66. These maps indicated the change in land use for any given area of the study region. The purpose of this map compilation phase was to establish a basis for planimetering land-use areas in each time dimension and obtaining transfer coefficients of land-use change between the various mapped units for later model simulation. Table 3.2 is a summary of map measurements for total area under forest, pasture, mixed farming and coffee for the 1948, 1958 and 1966 periods.

Table 3.2 Summary of map measurements for total area under forest, pasture, mixed farms and coffee for the 1948, 1958 and 1966 periods

Land use	Area 1948 km^2	 1958 km^2	 1966 km^2
Forest	227	172	147
Pasture	98	119	163
Mixed farming	60	82	69
Coffee	22	34	28
Total area	407	407	407

LAND-USE MODEL

The energy flow diagram of the land-use model is shown in Figure 3.1. Population is represented by the hexagon. This symbol, denoting a self-maintaining consumer, has inputs from outside subsidies and agricultural production, and outputs expressed as agricultural work (farming, animal husbandry, land clearance), as well as losses which include death and

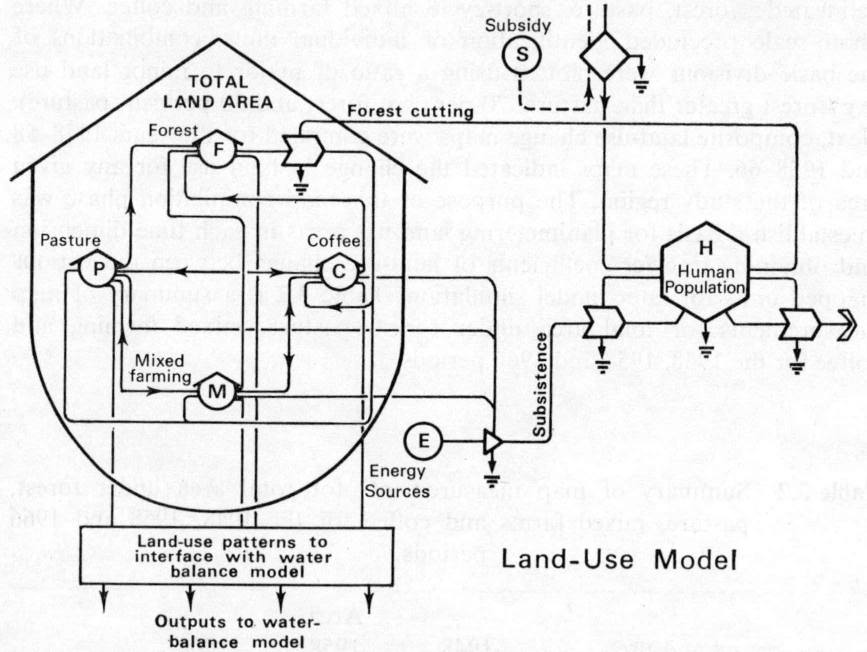

Figure 3.1 Energy-flow model of land-use pattern and population growth in the Jagua–Bao watershed (symbols after Odum, 1969 and 1971)

emigration. The lines indicating losses from the population and energy expended in agricultural tasks, as well as forest clearing, flow into arrow-shaped rectangles called work gates. These work gates imply that one energy flow interacts with a second, in such a way that the first permits the second to occur and magnifies (in a manner often assumed to be multiplicative) its effect. The work gate is similar to the triangular symbol, called a 'constant gain amplifier'. In this case, however, the rate of flow from one component increases the rate of flow from another by some constant factor, while not constituting a significant energy drain on the first component. In Figure 3.1, the interaction of various amounts of agricultural land with environmental energy sources (e.g. sun, water, nutrients) is an example of the use of the constant gain amplifier. The amount of crops that can be produced is proportional to the amount of land cultivated, provided that the necessary factors for growth of the crops are present.

Two circles are shown on the diagram, one labelled 'energy sources' and the other labelled 'subsidy'. The energy circle is generally used to represent energy from outside the system of immediate interest and implies the presence of a continually available forcing function. Solar energy, mineral nutrients and rainfall are examples of possible energy sources. The subsidy circle represents money (or its energy equivalent) sent into the area by emigrants for their families. The flow of money (shown by a dashed line) from this outside subsidy is shown to pass through a diamond-shaped symbol, called an economic transactor, which indicates the exchange of money for an equivalent energy flow.

The left-hand side of Figure 3.1 shows a large tank-shaped symbol (passive energy storage) containing four smaller storage symbols labelled 'forest', 'coffee', 'pasture' and 'mixed farming'. Each storage represents the area devoted to each land use at any given time. Their enclosure within the larger storage symbol, labelled 'total land area', implies that the sum of the four land-use storages is constant. This is believed to be the first application of such an areal constraint within the context of the energy flow language, a factor which enhances the technique's adaptation to problems of regional analysis.

In the model land is removed from forest through man's overt action. It can become farmland, a coffee plantation or pasture. Likewise, as land is abandoned, flows are shown going from each agricultural land use back into forest. In addition, any of the three agricultural land units is inter-convertible with any other. This should be drawn as two lines—an input and an output—connecting each one of the three to the others, but, because of the confusion which results from so many lines, each of these transfers is shown as a single line with arrows pointing in both directions. This should not, however, be interpreted as implying that the rates of flow are the same in both directions.

At the lower left-hand side of Figure 3.1 is a box labelled 'land-use patterns to interface with water-balance model'. What is indicated here is that the area in each land unit was entered as a variable into the water-balance model used ultimately to predict sedimentation rates.

The differential equations which express the rates of change of each component shown in Figure 3.1 are:

$$\mathrm{d}H/\mathrm{d}t = k_1\,(k_2M + k_3P + k_4C)\,H + k_5SH - k_6H - k_7H^2$$
$$\mathrm{d}F/\mathrm{d}t = k_8K + k_9P + k_{10}C - k_{11}FH$$
$$\mathrm{d}C/\mathrm{d}t = k_{12}FH + k_{13}M + k_{14}P - k_{15}C$$
$$\mathrm{d}P/\mathrm{d}t = k_{16}FH + k_{17}M + k_{18}C - k_{19}P$$

where k represents various transfer coefficients and flow constants, discussed below, and

H = human population
P = land in pasture
F = land in forest
C = land in coffee
M = land in mixed farming = total land area − ($P \quad F \quad C$)
S = subsidy sent by emigrants to families residing in the region

Calculation of rates

The rates of change between the various land-use categories are shown in Table 3.3. The total loss from each compartment is shown, as well as the partitioning of the loss to each of the other land uses. These are the transfer coefficients derived from the planimetered areal measurements off the composite land-use change maps, described earlier. The flows into and out of the population component of the model are flows of energy, those leading from the three producing land-use categories—mixed farming, coffee and pasture—through the constant gain amplifier representing the energy yield from each of the land units. Energy flow from each land-use category was evaluated according to the following standard procedure.

To calculate the energy flow from mixed farming, the total yield per unit

Table 3.3 Rates of land use change per year, averaged over 18 years. Values represent per cent change per year and each one corresponds to a coefficient in the land-use model

		To				
		Forest	Pasture	Mixed farming	Coffee	Total loss
From	Forest	—	k_{16}:2·26	0·91	k_{12}:0·36	k_{11}:3·53
	Pasture	k_9:2·58	—	2·78	k_{14}:0·68	k_{19}:6·04
	Mixed farming	k_8:1·83	k_{17}:5·17	—	k_{13}:1·57	8·57
	Coffee	k_{10}:2·22	k_{18}:3·45	3·10	—	k_{15}:8·77

area of each crop was determined from interview schedules conducted in the region. Yield was converted to an equivalent caloric value of 2·7 kilocalories (kcal) per gram of edible material; this value was derived by averaging the caloric values, according to Wu Leung and Flores (1961), of each crop grown in the study area. Poultry production was included in the mixed farm production estimate; 4·5 kcal g^{-1} (Golley, 1961) was used as the conversion factor for the meat. A net rate of 8·45 x 10^6 kcal ha^{-1} $year^{-1}$ was calculated for this flow.

The net yield of meat, primarily beef, was calculated similarly, using the conversion factor of 4·5 kcal g^{-1}. This value was about an order of magnitude less than that of crops: 6·53 x 10^5 kcal ha^{-1} $year^{-1}$. Coffee has little caloric value itself, yet the return of dollars for its sale represents a fairly sizeable income. Therefore, the amount of money which was exchanged for coffee was converted to an equivalent number of kilocalories. To do this, it was assumed first that money earned from the sale of cash crops produced is normally exchanged for an equivalent caloric input and, second, that in a non-fossil-fuel economy money can be used as barter in the exchange of energetic inputs and outputs. A day's work should therefore bring a wage allowing a day's subsistence. Similarly, money earned from the sale of meat should accordingly be exchanged for an equivalent number of calories of another necessary food item, or for items such as clothing or farm implements which would facilitate the production of an equivalent amount of food. By dividing the average caloric value of crops by the average price received for them at market, the value of one dollar was calculated to be worth approximately 13 000 kcal. This value is similar to the conversion rate of 10 000 kcal to the dollar determined independently by Odum (1967) for a fossil-fuel-based economy. The net yield of coffee, on this basis, was 3·17 x 10^5 kcal ha^{-1} $year^{-1}$.

Information gleaned from the interviews indicated that about one-third of the total population, primarily men, were engaged in some type of agricultural activity. A daily expenditure of 2000 kcal per person was estimated for these workers. Additional seasonal inputs of women and children in the harvesting of coffee were also included. The expenditure of labour for each type of agricultural activity was subtracted from the yield to give a net flow of energy into the population. The labour involved in agricultural activity included cutting down the forest, even though it is shown on the diagram as a separate flow.

The return of money to the Jagua–Bao region from emigrants was a very significant input in a number of the communities. However, it was discovered during earlier simulations that the energy returning in the form of dollars essentially compensated for the loss of energy by the emigration process itself; these rates were therefore included in the remaining net inputs and outputs of the population component. Because not all vital statistics were known, the coefficients used to describe changes in the population, k_1, k_5, k_6 and k_7, differed slightly from those calculated. The model demonstrated very early, for example, that the doubling of the population between 1950 and 1970 was a rate which could not be sustained for very long. The physical resources of the Jagua–Bao region, according to the land-use model, would not even remotely be capable of maintaining that kind of continued population increase without massively increased subsidies from outside the basin. Therefore, the population curve was programmed in such a way that the growth rate decreases in time, although the population itself is still increasing in the year 2015. In this sense, the model is optimistic with regard to future population trends within the region.

WATER BALANCE MODEL

Land management practices within the watershed are intimately related to the usefulness of the downstream reservoir; the two systems, land and water use, are so tightly coupled as to be inseparable. However, unlike the evolving patterns of land use, data on the water inputs and outputs to the watershed, as well as the effects of land use on these flows, were not readily available.

Given accurate data on soil characteristics, total rainfall inputs, storm intensities, storm frequencies, slope, ground cover and stream channel morphology, hydrologists can predict with great accuracy the runoff and sediment loads likely to occur from a watershed. For example, the series of studies edited by Pereira (1962) are excellent examples of detailed hydrological studies on tropical-based watersheds. Very often, however, as was the case in the Jagua–Bao region, the data required for a detailed accurate analysis of the relationship of land use to runoff and sedimentation are lacking. In such cases, regional planners can choose (1) to abandon the problem as being insoluble; (2) to suggest, optimistically, that the development project should stop while they gather their ten-year baseline data on soils, runoff and rainfall inputs; or (3) to proceed to make the best estimates possible based on available information and general relationships among climate, land use and hydrology. Aware of its limitations, but dismayed by the alternatives, the latter approach was chosen.

The purpose of this phase in the analysis was to develop a water-balance model, compatible with the land-use model, to permit coupled simulation and utilisation of the limited available watershed data to the fullest extent possible. To achieve this end, the energy flow language was adapted to Tosi's (1968, 1971) method of calculating water balances, which in turn is closely linked to the Holdridge life zone classification (1947, 1967;

Holdridge *et al.*, 1971). A life-zone map of the Dominican Republic (Tasaico, 1967) made this a particularly useful starting point for moisture budgeting within the Jagua–Bao region. Tosi's water balance calculations are keyed to Holdridge's (1959) assumption regarding the linear relationship between potential evapotranspiration and biotemperature, biotemperature being the average temperature calculated by the substitution of 0 for values below 0 °C and, tentatively, above 30 °C. The calculation of water balances usually involves the construction of an input–output budget table, with an entry for each variable each month. Rather than divide the year into a series of discrete months, however, it was necessary to express the water régime as a continuous flow process. Thus, Tosi's method was translated first into the energy flow language and then into the water balance model shown in Figure 3.2.

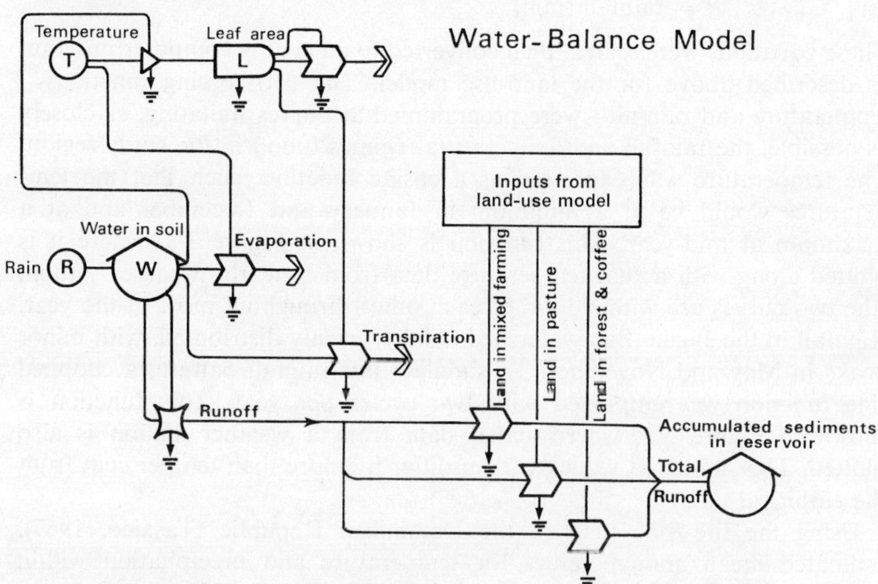

Figure 3.2 Energy-flow model of water budget and sediment accumulation in the Jagua–Bao watershed (symbols after Odum, 1969 and 1971)

This model has two variable forcing functions: temperature and rainfall. Rainfall is the input to soil moisture, from which water can be lost either as evaporation (controlled by temperature), transpiration (controlled by leaf area) or runoff. The switching mechanism controlling runoff prevents this loss until the amount of water in the soil exceeds field capacity. Leaf area, the area of leaf surface per area of soil surface, is affected by the interaction of soil moisture and temperature and has a loss term which increases as the square of the leaf area present. Two flows shown in Figure 3.2 are assumed to be critical to certain processes, but are insignificantly small when considered as drains on the storages from which they originate. These two flows—the metabolic water involved in plant growth and the cost of transpiration to the leaf area component—do not appear as loss terms in the differential equations. The equations which express the relationship shown in Figure 3.2 are:

$$\mathrm{d}L/\mathrm{d}t = k_{20}TLW - k_{21}L^2$$

$$\mathrm{d}W/\mathrm{d}t = R - W(k_{22}T + k_{23}L) - \text{runoff}$$

$$\text{where runoff} = W - (W(k_{22}T + k_{23}L) + \text{field capacity})$$

L = leaf area (area of leaf surface per area of soil surface)
T = temperature (°C)
W = water in soil (mm)
R = rainfall (mm)

These equations were scaled, then converted to an analog computer program as described above for the land-use model. The two forcing functions—temperature and rainfall—were programmed as curves imitating, as closely as possible, the rainfall and temperature régimes found in the study region. The temperature was expressed as a cosine function, such that the temperatures would be at a minimum in January and December and at a maximum at mid-year. This function is shown in Figure 3.3, where it is plotted along with actual temperature data from a nearby weather station. The two curves are within 1 °C of each other throughout most of the year. Rainfall in the Jagua–Bao watershed is fairly evenly distributed, with minor peaks in May and November. To simulate this rainfall pattern, a chopped sine function was employed with two cycles per year. This function is shown in Figure 3.3, where actual data from a weather station is also plotted. The simulated values seldom differ by more than ten per cent from the empirical data.

Using the life-zone map of the Dominican Republic (Tasaico, 1967), estimated mean annual values for temperature and precipitation within the watershed were calculated to be about 22 °C and 1550 mm per year, respectively. The magnitude of each of the forcing functions was set to produce these values for temperature and rainfall. Because of the moderately high elevation of the basin, it was not necessary to correct temperature data for differences between air temperature and biotemperature. Above elevations of about 250 m at the latitude of the Dominican Republic, air temperature can be taken as approximately equivalent to biotemperature (J. Ewel and Whitmore, unpublished).

Another parameter which exerts considerable influence on water balance is the water-holding capacity of the soil. Based on the known parent material, topography and geomorphology of the study region, an attempt was made to characterise a single 'most typical' soil for the entire watershed. The Las Piedras Puerto Rican Soil Series was selected as being a soil for which physical characteristics were available and which was apparently very similar to the principal soils found within the Jagua–Bao region. Using the data of Lugo-López (1953), field capacity of this soil was calculated to be 147 mm, assuming an effective depth of one metre. The model was run for a simulated one-year period and the results of that simulation are shown in Figure 3.3. The temperature and rainfall simulations were discussed above: evapotranspiration (i.e. evaporation +

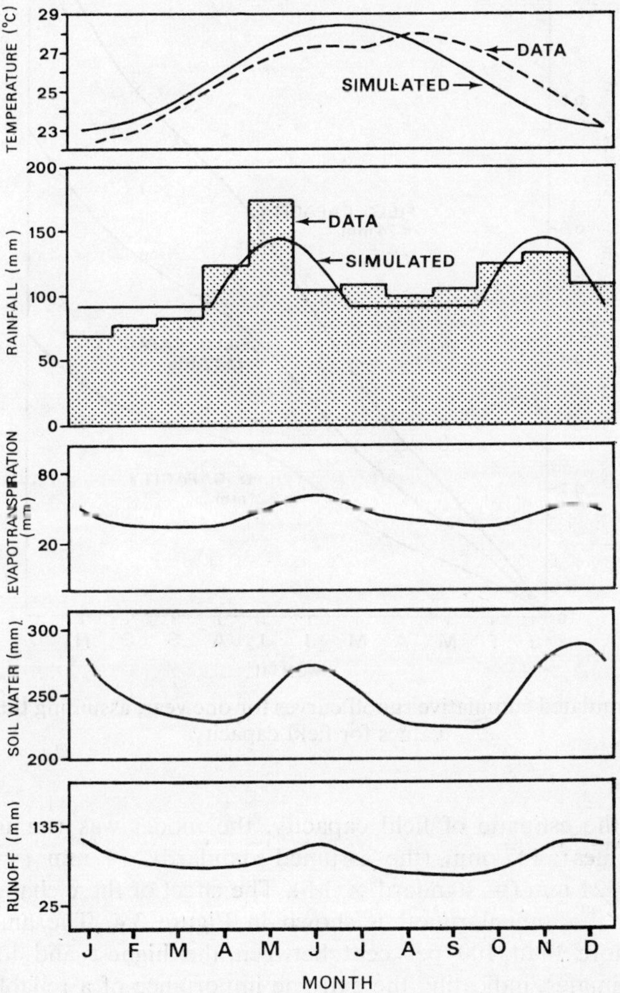

Figure 3.3 Simulation outputs of water balance model for one year

transpiration), water in the soil and runoff all show two peaks during the year, as would be expected with a bimodal rainfall distribution. Soil-moisture recharge lags behind rainfall, indicating that runoff only occurs after evapotranspiration needs have been met and the soil is recharged to field capacity. Cumulative runoff during the simulated one-year period was obtained by integrating the continuous runoff data. To evaluate the effect

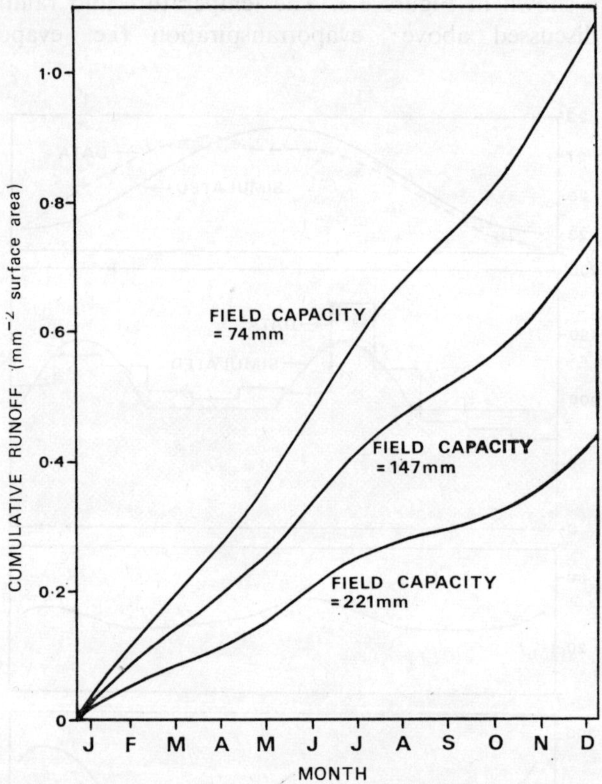

Figure 3.4 Simulated cumulative runoff curves for one year, assuming three different values for field capacity

of varying the estimate of field capacity, the model was run using three different values: 147 mm (the assumed standard), 74 mm (= standard × 0·5) and 221 mm (= standard × 1·5). The effect of these changes in field capacity on the annual runoff is shown in Figure 3.4. The annual totals differ by more than 100 per cent between the highest and lowest field capacity estimates, indicating the extreme importance of a reliable measure of soil water holding capacity for an accurate prediction of runoff. In subsequent simulations, the intermediate, and assumed standard, value of 147 mm was used.

INTEGRATED LAND-USE – WATER-BALANCE MODEL SIMULATIONS

The empirical land-use model and the water-balance model based on extrapolated data were integrated to examine the interaction of changing land use on runoff and sediment accumulation in the downstream reservoir. To achieve this goal, model simulation was run from a 1945 baseline to take advantage of the antecedent land use and population data from the area. At 1975, the approximate date of completion of the Jagua dam, runoff data were coupled with the land-use model and the simulation was then continued up to the year 2015. The rate of sediment accumulation behind the dam was obtained by interfacing each simulated land-use category—forest, coffee, pasture and mixed farming—with the runoff output from the water-balance model. Each land use was assigned a coefficient to express the relative amount of erosion which might occur under that land unit. Values were conservatively assigned with 1·0 for forest and coffee, 2·0 for pasture and 3·0 for mixed farming. The values probably underestimate the differences in soil erodibility among the various land-use categories. Penman (1963), for example, reviewed considerable erosion data and reported relative erosions under a variety of land uses which differed by as much as three orders of magnitude. Pereira (1962) found that the sediment load from a farmed watershed in tropical Africa was four times greater than from a comparable forested watershed. In a Puerto Rican situation very similar to the Jagua–Bao region, Noll (1953) reported that erosion from mixed farming areas was ten times greater than from forest; coffee plantations, which in Puerto Rico were clean-cultivated, produced twice as much sediment as forest. The simulated erosion values for the four land-use categories were then summed and integrated over the period from 1975 to 2015 to obtain a curve of sediment load over time for each run of the model.

Eight simulations of the integrated land-use–water-balance model, using a variety of coefficients, rates and assumptions involving future trends and public policies, are shown in Figure 3.5. Population growth and land-use changes are shown as they evolve from 1945 to the year 2015; changes in land use are expressed as percentages of the 407 km² total area. The relative cumulative sediment loads which would be deposited in a reservoir at the base of the Jagua–Bao watershed are plotted in Figure 3.6 for each of the land-use patterns predicted by the simulations. The topmost curve (line 'A') is the accumulation of sediments that would result if all land

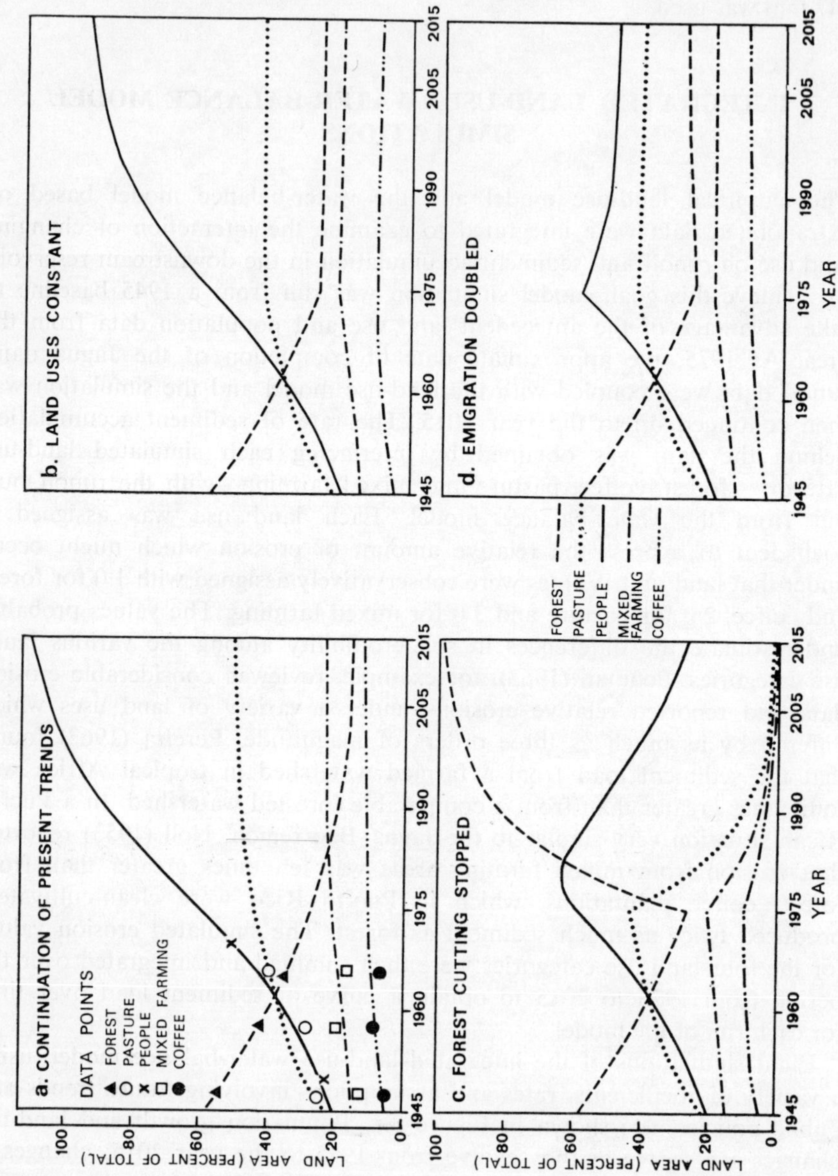
a. CONTINUATION OF PRESENT TRENDS
DATA POINTS
FOREST
PASTURE
PEOPLE
MIXED FARMING
COFFEE
b. LAND USES CONSTANT
c. FOREST CUTTING STOPPED
d. EMIGRATION DOUBLED
FOREST
PASTURE
PEOPLE
MIXED FARMING
COFFEE
LAND AREA (PERCENT OF TOTAL)
POPULATION (THOUSANDS)
YEAR
1945
1960
1975
1990
2005
2015

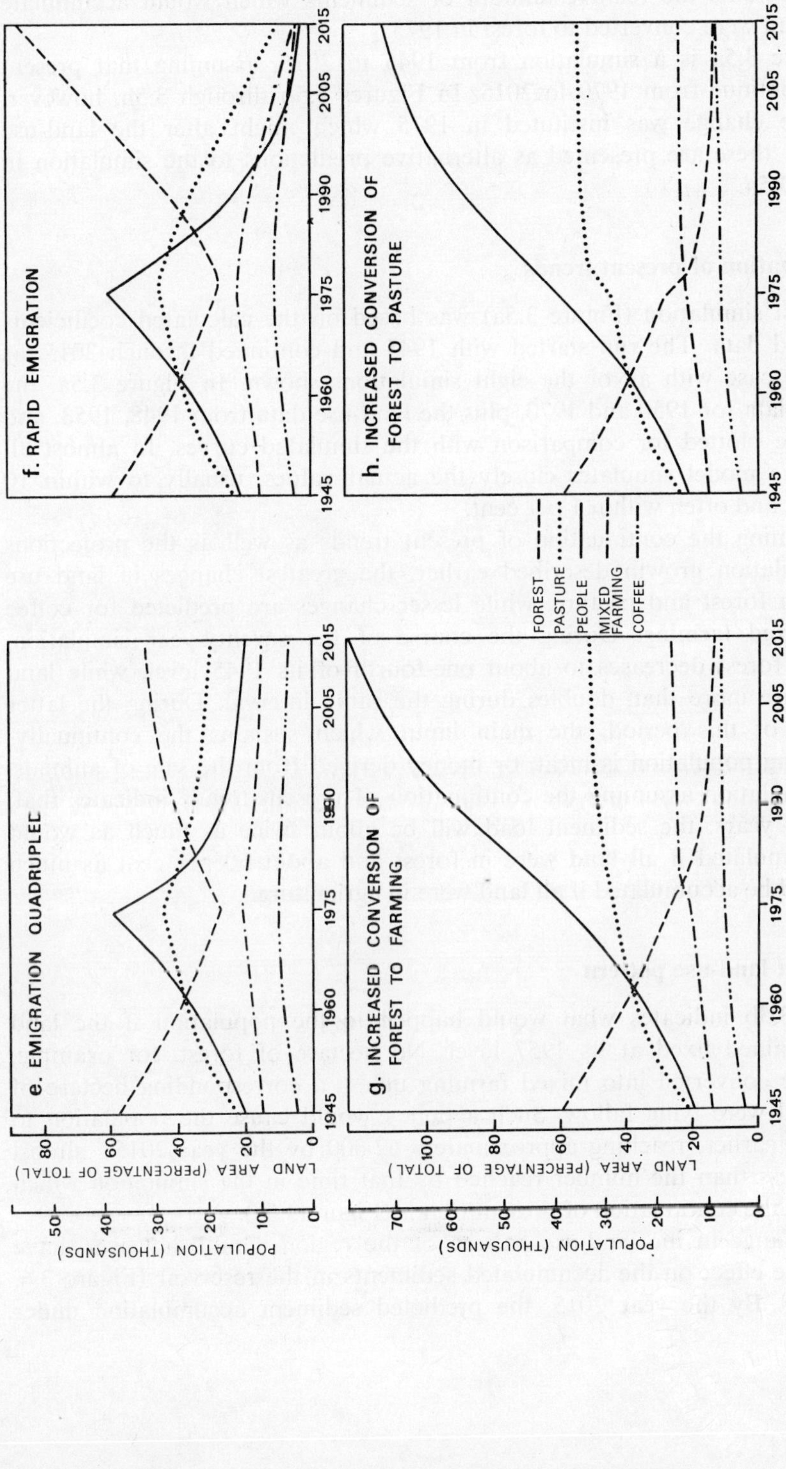

Figure 3.5 Simulation results of the land-use model for the period 1945–2015

were immediately put into mixed farming in 1975; the lowest curve (line 'I') represents the relative amount of sediments which would accumulate if all land were converted to forest in 1975.

Figure 3.5a is a simulation from 1945 to 2015, assuming that present rates continue from 1970 to 2015. In Figures 3.5b through 3.5h, however, a single change was instituted in 1975 which might alter the land-use pattern; these are presented as alternative predictions to the simulation in Figure 3.5a.

Continuation of present trends

The first simulation (Figure 3.5a) was based on the calculated coefficients and field data. The run started with 1945 and continued through 2015, as was the case with all of the eight simulations shown. In Figure 3.5a, the census data for 1950 and 1970, plus the land-use data from 1948, 1958, and 1966 are plotted for comparison with the simulated curves. In almost all cases the model simulates closely the actual values, usually to within 10 per cent and often within 1 per cent.

Assuming the continuation of present trends as well as the projections of population growth described earlier, the greatest changes in land use occur in forest and pasture, while lesser changes are predicted for coffee and mixed farming. During the course of the seventy-year simulation, land in forest decreases to about one-fourth of its 1945 level, while land in pasture more than doubles during the same interval. During the latter portion of this period, the main input which sustains the continually-increasing population is meat, or money derived from the sale of animals. The simulation assuming the continuation of present trends indicates that, after 40 years, the sediment load will be about twice as much as would be accumulated if all land were in forest and about 60 per cent as much as would be accumulated if all land were in agriculture.

Constant land-use pattern

Figure 3.5b indicates what would happen to the population if the land use remained fixed at its 1957 level. No hectare of forest, for example, could be converted into mixed farming unless a corresponding hectare of farmland were lying fallow. Such a policy would cause the population to level off earlier, reaching approximately 62 000 by the year 2015: almost 10 000 less than the number reached by that time in the simulation which assumed the continuation of present trends (Figure 3.5a).

No change in the land-use pattern of the region would, however, have very little effect on the accumulated sediments in the reservoir (Figure 3.6, line 'D'). By the year 2015, the predicted sediment accumulation under

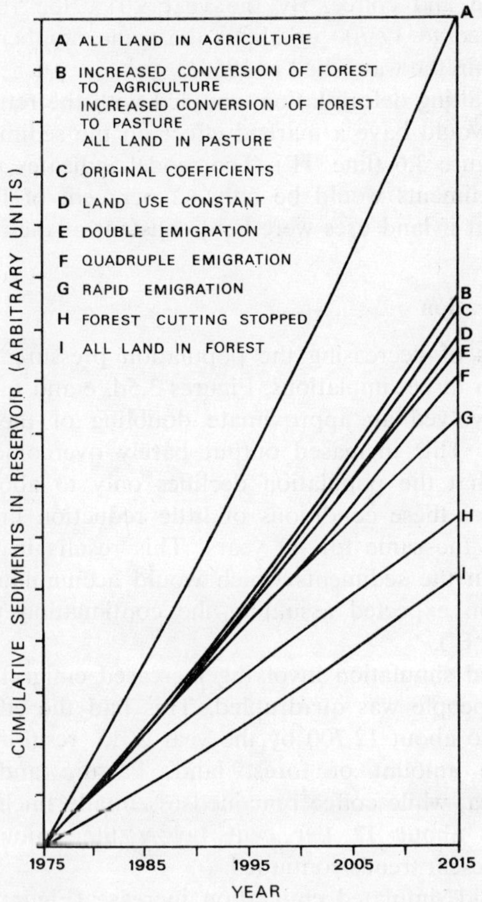

Figure 3.6 Simulated accumulation of sediments from the Jagua–Bao watershed for the period 1975–2015 for eleven different land-use patterns

an instituted policy of fixed land use would be only 4 per cent less than that which would occur if present trends were to continue.

Forest cutting terminated

The next simulation, Figure 3.5c, assumed that all forest cutting would be stopped in 1975 and that the pre-1975 rates of reversion from agricultural uses to forest would continue in the future. Under these conditions, almost all of the land area would revert to forest by the year 2015. The population would continue to increase until about 1982, after which it would begin to decline due to the reduced amount of land available for farming, pasture and coffee. By the year 2015, the region's population would be reduced to 17 700 or about the same number of people which were settled within the watershed in 1955.

A policy of halting deforestation, coupled with the return of agricultural land to forest, would have a marked effect on the sediment accumulation, as shown in Figure 3.6 (line 'H'). The model indicates that, by 2015, the accumulated sediments would be only 62 per cent of the amount which would be present if land uses were held constant at their 1975 levels.

Increased emigration

The possibilities of decreasing the population pressure on the watershed were explored in three simulations, Figures 3.5d, e and f. The first of these, Figure 3.5d, involved an approximate doubling of the present regional emigration rate. This increased output barely overcomes the population expansion, so that the population declines only to about 31 000 by the year 2015. Under these conditions of little reduction in population, land use stays about the same for 40 years. This results in only a seven per cent reduction in the sediments which would accumulate, compared with the accumulation expected assuming the continuation of present trends (Figure 3.6, line 'E').

For the second simulation involving increased emigration (Figure 3.5e), the outflow of people was quadrupled. This had the effect of decreasing the population to about 12 700 by the year 2015, resulting in a significant increase in the amount of forest land. Pasture and mixed farming decreased in area, while coffee remained essentially unchanged. Sediments would decrease about 12 per cent below the amount which would accumulate if present trends continued.

The most rapid simulated emigration increase (Figure 3.5f) resulted in less than 1000 people remaining within the region by the year 2015. In this case, all agricultural land use decreases as a result of the population decline and most of the land reverts to forest. The resulting sediment accumulation after forty years (Figure 3.6, line 'G') would be only 81 per cent of that which would result from the continuation of present trends. This is still greater, however, than the sediment accumulation which would result if forest cutting were curtailed in 1975 and present rates of reversion from coffee, pasture and farmland to forest were to continue (Figure 3.6, line 'H').

Modified land-use trends

Two final simulations involved increases in the rate of conversion from forest to mixed farming, and from forest to pasture, shown in Figures

3.5g and 3.5h respectively. Both these changes result in the same amount of accumulated sediments after 40 years; this is also the amount of sediment which would be expected if all land were converted to pasture in 1975. Line 'B' in Figure 3.6 shows that this sediment accumulation rate would, after forty years, result in an increase of only about 3 per cent more sediments than would accumulate if present trends were to continue. However, the populations which could be supported by each of these changes are different. An increase in the conversion from forest to agriculture would result in a 2015 population of 77 000, whereas the increased conversion from forest to pasture would produce a 2015 population of 75 000: a difference of about 3 per cent.

CONCLUSION

The purpose of this chapter has been to discuss an interdisciplinary systems approach to land resource analysis that can evaluate the feasibility of manipulating critical land-use elements governing the ecosystem for programming planned change. With regard to the case study, the impact of evolving land use on a depleting resource base and the longevity of a reservoir in the central Dominican Republic have been considered. Specific results of the systems analysis and computer simulations are summarised in Figure 3.7 by projecting the predictable outcome from eleven public policies over the 1975–2015 year period.

Each bar in Figure 3.7 is a distinct policy and the length of the bar represents the time required to attain the expected sediment load in the year 2015, maintaining as constant the 1975 land-use pattern (see the mid bar in Figure 3.7). Bars shorter than the mid-bar indicate more rapid rates of sediment accumulation, while longer bars indicate slower rates of sedimentation. Over a wide range of land management policies, the time required for each to reach the same sediment load differs by less than 10 years. The conversion of all land to mixed farming results in a decrease in time required to reach the standard sediment load by about 15 years, while curtailing deforestation and conversion of all land to forest in the year 1975 increases the time to reach that sediment load by 29 and 44 years respectively. Because of the conservative relative erosion rates assigned to each of the land-use categories in the integrated model, Figure 3.7 should represent *minimal* differences to be expected as a result of instituting one of a variety of land management decisions within the Jagua–Bao basin.

Energy flow models have been used to make an assessment of changes in the physical environment as a result of increasing population pressure over time, to evaluate the effects of these trends on evolutionary changes

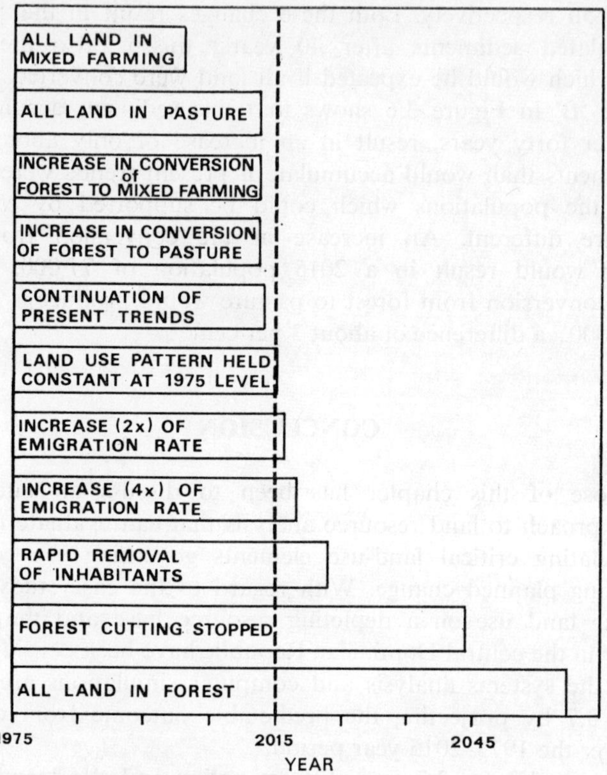

Figure 3.7 Time required for various land-use patterns to produce the same cumulative sediment load that would be produced in 40 years under conditions of no future change in the 1975 land-use pattern

of forest lands and to obtain insights into the central problem of the relation between tropical environment and resource depletion. In this latter case, modelling and computer simulation have generated predictable outcomes for alternative public policies which can help avert erosion and control sedimentation.

Ecological similarities with other tropical regions make possible a wide application of the modelling techniques developed. The solution to population–resource problems usually requires that regional planners have at their disposal requisite information on predictable policy outcomes, in order to arrive at developmental decisions. This study of the central Dominican Republic may be considered a test case where problem-solving techniques have been developed and applied in order to achieve a greater understanding of the problems involved in watershed management in the tropics.

REFERENCES

Antonini, Gustavo A. (1972) 'Dominican Republic', *Britannica Book of the Year*, 252–253.

Golley, F. B. (1961) 'Energy values of ecological materials', *Ecology* **43,** 581–584.

Holdridge, L. R. (1947) 'Determination of world plant formations from simple climatic data', *Science* **105,** 367–368.

Holdridge, L. R. (1959) 'Simple method for determining potential evapotranspiration from temperature data', *Science* **130,** 572.

Holdridge, L. R. (1967) *Life zone ecology* (rev. ed.; San José, Costa Rica: Tropical Science Center).

Holdridge, L. R., W. C. Grenke, W. H. Hatheway, T. Liang and J. A. Tosi, Jr. (1971) *Forest environments in tropical life zones, a pilot study* (Oxford: Pergamon Press).

Inter-American Development Bank (1971) *Socio-economic progress in Latin America* (Washington: Inter-American Development Bank).

Lahmeyer International GMBH (1967) *Proyecto multiple Rio Jagua/Rio Bao: estudiò de pre-inversion* (Frankfurt, Germany: Lahmeyer International GMBH).

Lugo-López, M. A. (1953) *Moisture relationships of Puerto Rican soils* (Technical Paper 9; Rio Piedras, P.R.: University of Puerto Rico Agricultural Experiment Station).

Noll, J. J. (1953) *The silting of Caonillas Reservoir, Puerto Rico* (Technical Paper 119; Washington, DC: Soil Conservation Service USDA).

Odum, H. T. (1967) 'Energetics of world food production', *The world food problem* (A report of the President's Science Committee; Washington, DC), Vol. 3.

Odum, H. T. (1969) *An energy circuit language for ecological and social systems: its physical basis* (Appendix I of Progress Report to USAWC; Contract AT—(40-1)—3666).

Odum, H. T. (1971) *Environment, power and society* (New York: Wiley–Interscience).

ONAPLAN (1970) *Primer plan nacional de desarrollo* (Santo Domingo: Oficina Nacional de Planificación).

Penman, H. L. (1963) *Vegetation and hydrology* (Technical Communication No. 53; Harpenden: Commonwealth Bureau of Soils).

Pereira, H. C. (ed.) (1962) 'Hydrological effects of changes in some East African catchment areas', *East African Agricultural and Forestry Journal* **27,** 1–131.

Peterson, G. R. (1967) *Basic analog computation* (New York: MacMillan Co.).

SOGREAH (1968) *Surveys for the multipurpose development of the Yaque del Norte and Yaque de Sur* (8 vols.; Grenoble, France: Société Grenobloise d'Etudes et d'Applications Hydrauliques).

Tasaico, H. (1967) *Reconocimiento y evaluación de los recursos naturales de la República Dominicana: estudio para su desarrollo y planificación* (con Mapa Ecológico de la República Dominicana, 1:250 000) (Washington, D.C.: Pan American Union).

Tosi, J. A., Jr. (1968) 'Cálculo del balance hídrico', *Zonas de Vida de Venezuela*, by J. Ewel, A. Madriz and J. A. Tosi, Jr. (Caracas: Ministerio de Agricultura y Cria), 42–48.

Tosi, J. A., Jr. (1971), 'The monthly water balance', *Forest environments in tropical life zones, a pilot study*, by L. R. Holdridge, W. C. Grenke, W. H. Hatheway, T. Liang and J. A. Tosi, Jr. (Oxford: Pergamon Press), 49–58.

United Nations (1971) *Inventariación y fomento de los recursos forestales: República Dominicana* (FO: SF: DOM.8; working document No. 1; Santo Domingo: Programa de las Naciones Unidas para el Desarrollo).

Wu Leung, Woot-Tsuen and Marina Flores (1961) *Food composition table for use in Latin America* (Guatemala City, C.A. and Bethesda, Maryland: Institute of Nutrition of Central America and Panama, and the Interdepartmental Committee on Nutrition for National Defense, National Institute of Health).

ACKNOWLEDGEMENTS

Financial support for field work was obtained from the Center for Latin American Studies, University of Florida, Gainesville. The authors wish to express their appreciation to Professor Howard T. Odum, for valuable suggestions made during the study. In the field, agronomist Frank García, Industrias Portela de Navarrete, measurably assisted in interviewing and data collection. Florida graduate students in Geography, Glenn M. Fish, Howard Tupper and Michael Stearman, provided significant assistance in compilation of cartographic materials, data analysis and water balance–runoff calculations respectively. Analog computer facilities were provided by the Department of Nuclear Engineering Sciences, University of Florida, Gainesville.

Chapter 4

Bioclimatology and Land Evaluation in Uganda

L. W. Hanna

Introduction

Land evaluation for agricultural development must be based on all the components of the physical environment but difficulties of data evaluation and classification mean that investigation and discussion must focus upon a few relevant criteria. In Uganda the high rate of evapotranspiration and the uncertainty of the rainfall combine to make water the most important single control over agriculture and explain the considerable interest in rainfall and in methods of conserving soil water (Russell, 1968). Most work has traditionally been concerned with the probability of receiving a critical rainfall total. However, more recent work on potential evapotranspiration has shown that a single rainfall value is inadequate to satisfy the wide range of atmospheric demands between the wetter lake shores of southern Uganda and the drier north. Moreover each crop makes individual demands of water both in total and seasonally. The scale of the agricultural enterprise is a relevant consideration as this determines the frequency with which the minimum water requirement must be received to avoid economic failure. Bioclimatology is concerned with the complex interactions between the climatic environment and plant life and can go a long way to remove the shortcomings of evaluations based upon rainfall alone. Water-balance studies related to agricultural production is an aspect of bioclimatology which has considerable importance to land evaluation. However, the characteristics of the water needs of a single crop are unique so that any map based upon water balance statistics cannot have direct application to agricultural land use or productivity. As yet insufficient relationships have

---

Dr. Hanna is Lecturer in Geography in the University of Newcastle upon Tyne and was formerly Lecturer in Geography at Makerere University, Uganda.

been demonstrated to allow the development of models useful in land-use strategies so that decisions must be based upon partial evaluations of the environment. Classifications of all physical resources share the problem of the reliability of the data, the predictive ability of the criteria, and their relevance to the proposed use (Goodall, 1966). In this chapter these aspects are discussed in relation to two crops, sugar cane and tea, both of considerable importance in the Ugandan economy. The choice of two commercial crops is not entirely arbitrary as expansion requires processing facilities and thus high capital investment. In the case of tea, expansion of existing processing facilities together with the addition of four new factories (bringing the total to thirty-six) is projected by 1976. Sugar cane is also to be increased in southern Uganda, where it is presently concentrated, but ambitious schemes for Bunyoro open the possibility of expansion of cane growing into new and untried environmental conditions. Sugar cane illustrates the problems associated with the water balance of an annual or biannual crop which has a distinct pattern of water needs between planting and harvesting. Tea, on the other hand, grows continually unless it is pruned, and its water needs like most tree crops are related only to atmospheric variables.

Inevitably the identification of parameters important to productivity must be made within the experience of agriculture. Since the well-established areas for sugar cane and tea are in the moister regions of southern Uganda and Toro the dominant environmental controls may well not be as important to the same crops in northern Uganda. There is a considerable difference between an annual total of 1500 mm of rainfall evenly spread throughout the year and a similar total confined to eight months of the year. Thus relationships derived from bioclimatic studies in southern Uganda must be applied to other areas with some caution. There is also a practical problem of applying the relationships derived from a few years' intensive climatic study at a single site to the vast scale of the whole country for which a long run of records is needed to predict the occurrence of the relevant climatic events. Daily water balances give the necessary precision for the analysis of the crop performances but could not be applied spatially because of the highly localised nature of tropical rainfall. A compromise must be achieved between the daily data and longer period totals of, say, one month which is more amenable to statistical analysis over a large number of years.

WATER-BALANCE AND CROP SYSTEMS

Manning's work on rainfall probability assumed an optimum rainfall for the cotton crop and attempted regional estimates of the potential yield of cotton in Uganda (Manning, 1956). Manning recognised the approximations used in this assumption and suggested that all water requirements should be based on evapotranspiration for the crop and the exploitable soil water. Since then great advances have been made in agroclimatology, and water balances can be estimated to provide a quantitative assessment of the land and its potential. However, the broad water-balance studies based upon mean annual rainfall and actual evapotranspiration (Piexoto and Obesi, 1965) are difficult to apply because of sparse data. It has been claimed that an index using the difference between mean annual rainfall and potential evapotranspiration could be useful to agriculture (Davies and Robinson, 1969) but it is difficult to see its value except in delimiting natural vegetation zones which probably adjust to the mean annual moisture situation. However, the potential water deficit has the merit that it is purely climatological and does not depend on the plant cover, the type of plant or its stage of growth. Each plant bears an individual relationship to the potential evapotranspiration and thus the seasonal demands for water depend on land use. Any water balance study which ignores the land-use factor is a vague generalisation. Water deficits are not always harmful and can be beneficial (Owen and Watson, 1956; Winter *et al.*, 1970); so there is no need to assure a water supply which can satisfy the potential evapotranspiration.

The problem of soil water storage is difficult to solve on a national scale because of the great variability in tropical soils. Thornthwaite's attempts to consider storage by weighting the surplus in his humidity index (Thornthwaite, 1948) can only be useful in a general way. Some tropical crops root deeply, especially where dry seasons encourage root extensions. Grass roots can reach depths of 6 m and, contrary to evidence from India, tea in East Africa has been shown to extract water from depths of 4·6 m in dry periods (Nyasaland 1956/57) and will even penetrate weathered rock at depths of 6 m (Kerfoot, 1962). Absolute limits do exist and are usually imposed by the water table, solid rock or laterite horizons, which in areas of ridges and valleys can change over very short distances. Variable annual rainfall can alter the maximum storage from year to year so that in a dry year the soil profile can be dried throughout the root range even in wetter lake-shore areas. The subsequent wet season may not completely replenish the soil water with the result that the drought hazard is increased. For these reasons water-balance studies should take into consideration the effects of different levels of soil water storage. The need for a statistic which provides information on the distribution of moisture deficits in the soil has been emphasised elsewhere (Russell, 1960) and this requires analysis of seasonal régimes.

Generalised models of crop water need have been used in Australia (Fitzpatrick and Arnold, 1964; Slatyer, 1960) and in Rhodesia (Whyte, 1966) and have been successfully fitted to the pattern of expected rainfall.

Annual crops, because of the lower water demands early and late in their life cycle, can grow part of the time with precipitation less than potential evapotranspiration. Planting dates can be adjusted to suit high water demands later in the cycle, which has been recommended for cotton in Uganda (Manning, 1960), and by fully utilising scarce water resources this method improves yields. By planting imported varieties of maize which mature in 120 days, instead of the 170 day season for local varieties, a more reliable yield has been made possible in areas of short or highly variable wet seasons (Dagg, 1965; Dowker, 1963). Tropical crops exist in such a variety of forms that each has a wide climatic spectrum (Huxley, 1965), and within areas which are climatically suitable for a crop there is a variety of performance. When dealing with cash crops, survival is not enough and minimum levels of productivity must be achieved. The relationship between climatic factors and crop yields is an extremely complex one, and cannot be described adequately by approximate measurements of the climatic environment unless a single factor is dominant (Watson, 1963). In Uganda, water is the major limiting factor and for this reason quantitative assessments of water availability are most useful in studies of crop potential.

In East Africa the agro-climatic network is more satisfactory than in many other tropical areas. Uganda in particular has developed a close network in the Lake Victoria zone (Rijks, 1968). Sansom's classification of the climate of East Africa (1954) using the temperature-based method of Thornthwaite (1948) was proved to be in error by later evaporation studies (Pereira *et al.*, 1962). The Penman method (1948), with its better theoretical basis incorporating the drying power of the air with the energy for evaporating, has proved more valuable. The weighting given to net radiation makes its accurate measurement vital to the successful application of the formula. The network of fully instrumented stations is now sufficiently wide to map Penman E_o for all three East African countries (Woodhead, 1968 a and b; Rijks *et al.*, 1970). E_o, or the evaporation from an open water surface, is similar to Thornthwaite's potential evapotranspiration. A climatic classification using the more accurate Penman estimate has been drawn up for the rangelands of Kenya (Dagg, 1965). Since it was designed for rangeland all moist regions were grouped in one class while the dry areas were subdivided according to Thornthwaite's index to facilitate rangeland management but the omission of the humid subdivision limits its value for crop studies and it is doubtful if the index is sufficiently fine to apply to crop yields.

It would appear that climatic classification has a role in dealing with the management of natural vegetation communities, but no such simple answer can solve the problem of assessing crop performance. Any quantitative assessment of the environment for agriculture should include an output term in the form of the expected economic yield of the specific crop. Models of crop ecosystems are possible though the complexity of the relationships makes it difficult to take into account all régimes and interactions (Nix, 1968). Yields of crops are not only related to the physical environment but are dependent on the cultivation practices and the physiology of the plant.

Attempts have been made to relate crop yields to the amount of deficit experienced during the plant's life or for a critical period of growth. For example, correlation between the accumulated monthly deficits of the vegetative period of sugar cane and final yields have been proved for Mauritius (Halais, 1966). The vegetative period of Mauritius is limited to eight months of the year when rainfall is adequate. Planting and harvesting are arranged to make use of this rainy season. In the equatorial conditions of southern Uganda the vegetative period is continuous and a more variable water pattern results. In East Africa two rainy seasons occur annually but the incidence and amount of rain are highly variable. Yields of sugar cane in Uganda can only be accounted for by careful daily water balance calculations. Alternating wet and dry periods within monthly periods can cause errors in the water balance. Monthly totals of precipitation and evapotranspiration used to estimate the net gain or loss of soil moisture during the month can only give an approximation of the total soil moisture available at the end of each month. Errors, caused by irregular amounts of rainfall unevenly distributed, accumulate in the monthly totals of soil water storage. Thus, over several months, monthly soil water estimated under sugar cane were ± 30 or 40 mm in error. Using daily meteorological data it is possible to estimate fairly accurately the movements of water in the soil profile. This reduces greatly the possibility of large errors but also gives details of changes within the monthly period.

Studies of sugar cane suggest that the optimum for growth is soil-moisture at low tensions (Clements, 1964) and even slight differences in soil-moisture stress reduce total yields (Hardy, 1966). Thus both the duration and intensity of stress must be accounted for in a moisture statistic. Because growth rates are reduced as water falls below field capacity and the effect increases up to wilting point the most useful index must use a weighting factor for periods when the whole profile is reduced to zero. Such an index derived from the sum of stress days when available soil moisture in the profile is reduced below the 50 per cent level and twice the sum of days when the profile is completely dry has been successfully used to explain sugar cane yields at Kakira, Uganda (Hanna, 1971*a*). The index, called accumulated stress, assumes that those days when rainfall exceeds evapotranspiration are non-stress days, when growth is not limited by lack of water. Such an index is more difficult to calculate than simple moisture deficits derived from total rainfall and potential evapotranspiration. The interpretation of correlations between yield and climate derived from either experimental plots or from analysis of estate data is made difficult by the unique conditions of the site and the cultivation practices. To be useful, the environmental parameters must be integrated into a simple index to represent a complete cycle which in the case of sugar cane can be 720 days long. This was the reason for using the accumulated stress index to explain sugar cane yields. However, there is no reason to suppose that because of the high correlation with yields at Kakira, other factors (in particular temperatures) would not be more significant in other areas of Uganda. In the absence of trials or indeed reliable estate yield data, except in the narrow lake-shore region, it is impossible to get over this difficulty; the associations developed at Kakira could give, at least, some indication of yield potential over the country as a whole. Since in southern Uganda the vegetative season is continuous and since there is no fixed planting and harvesting period, the moisture régimes of cycles of cane at Kakira are found to be different depending on the incidence of the main rain periods within the cycle. Accumulated stress could not account for the distribution of stress within the cycle which no doubt has an important influence on yield and in some areas, where dry seasons are longer and occur with more certainty, planting and harvesting would have to be more rigid. The estimation of the water balance at Kakira was based upon daily meteorological readings from the estate over a period of fifteen years and it would be unrealistic to attempt such a detailed study on a national basis. Moreover the soil moisture status would have to be summed over a period of days to enable any reliable prediction to be made from long term records. Ten days has been suggested as the minimum period for valid comparison between years in East Africa (Manning, 1960) and although ten-day meteorological data has been published since 1963 for agricultural research stations, long period records from East African Meteorological Department stations are only published as daily or monthly figures.

The calculation of a water balance statistic useful in predicting crop yields requires a more detailed treatment than the difference between monthly rainfall (P) and the monthly water needs of the crop (E_t). To estimate accumulated stress for sugar cane which gives an indication of moisture conditions within a monthly period requires some simulation of the wetting and drying process in the soil. In the case of sugar cane it is important to know for each month the time that available soil moisture (A.W.) is below 50 per cent of available water capacity and the time it is reduced to zero. Assuming deep, well-drained soils, available water capacity within the root zone would be about 250 mm. Thus the level 50 per cent A.W. (W_k) approximates 125 mm when roots have reached maximum extension but is only 50 mm and 100 mm respectively in the first two months of growth. The state of soil moisture during any month depends on available soil moisture at the beginning, W_1, available soil moisture at the end, W_2, and on the relationship between total P and total E_t. The simplest case occurs when soil moisture is adequate both at the beginning and end of the month ($W_1 > W_k$ and $W_2 > W_k$). Whether or not $P > E_t$, stress will not be experienced, for even in the absence of rain soil-water reserves must be adequate. This assumes that the soil water reserves are not depleted below W_k at any time during the month. If precipitation was evenly distributed or all coming early in the month then this will be so; however, an exception to this rule was the month of February 1964. No stress was apparent from monthly figures, but from daily figures 60 per cent of the month's rainfall was concentrated in the final five days of the month restoring soil moisture to 48 mm above W_k. Earlier in the month, depletion of soil moisture had only been temporarily halted by light falls of rain and available moisture fell to 85 mm on the twenty-fourth of the month causing eighteen days to have soil moisture below W_k. In East Africa where much of the precipitation comes not from random convection but from disturbances associated with convergence or incursions of westerly air, large totals of precipitation can be concentrated in a few consecutive rain days. Rainfall exceeding the daily requirement for sugar cane, i.e. about 5 mm, may fall on less than 10 per cent of the month during dry periods and exceed 30 per cent of the month only in wet seasons. Estimation of the amount of stress requires a knowledge of daily rainfall distribution. To assume no stress when W_1 and W_2 exceed W_k can often be an error, unless the soil profile is at field capacity or if available water is at least 200 mm at the beginning of the month. If the available soil water is below W_k at the beginning or end of the month then stress does occur. To estimate its duration it is necessary to consider two cases, one when soil water is increasing during the month and another when soil water is being reduced. In both cases it must be assumed that this increase or decrease is at an even rate. Thus the rate of decrease must be E_t/d mm/day (d = number of days in the month) while it must be increased at the rate of $(P - E_t)/d$ mm/day. The former case must have a smaller error since daily variations in E_t are not so large as variations in daily rainfall.

The case of decreasing available soil water

If $W_1 > W_k > W_2$ it can be assumed, provided there is no precipitation, that available moisture will be reduced at the rate of E_t/d mm/day and will fall to W_k in $\dfrac{(W_1 - W_k)}{E_t/d}$ days. For the rest of the month there will be water stress. The same argument can be used when precipitation occurs during the period, except that precipitation will extend the time of adequate moisture by $\dfrac{P}{E_t/d}$ days. Thus the duration of water stress in a period of d days becomes—

$$d - \frac{d}{E_t}(P + W_1 - W_k) \qquad (1)$$

The case of increasing available soil water

If $W_1 < W_k < W_2$ soil moisture will be raised to W_k in

$$\frac{d(W_k - W_1)}{P - E_t} \text{ days} \tag{2}$$

and this will be the duration of stress.

Both these cases consider that soil water is adequate either at the beginning or end of the month. When it is insufficient both at the beginning and at the end then the only relief from stress can be precipitation. If rainfall is sufficient to raise storage above W_k then equation (1) will apply. But if storage and rainfall together are less than W_k then the following case must be considered.

The case of inadequate available soil water

This occurs if total water from rainfall and storage is less than W_k and then is reduced to zero at the end of the month.

If $W_1 = W_2 = 0$ and $P < W_k$, or $P + W_1 < W_k$ and $W_2 = 0$, some soil moisture will be in storage on $\frac{P + W_1}{E_t/d}$ days and the rest of the period will be completely dry and suffer severe stress, i.e.

$$\frac{d - d(P + W_1)}{E_t} \text{ days} \tag{3}$$

It was not possible in the first two cases to apportion the rain days between days when soil storage was below W_k and days when it exceeded W_k. However when all days in the month are below W_k the effect of added precipitation cannot be ignored. Thus the period when severe stress occurs, owing to the lack of available soil moisture within the root range and the absence of any precipitation, can be estimated.

It has already been shown that in the case of adequate soil moisture at the beginning and end of the month, stress can occur if precipitation is irregular in amount and incidence and thus fails at times to make good the more consistent losses by evapotranspiration. To assume no stress during these wetter months is to underestimate the actual amount of stress experienced by each cycle. On the other hand, when soil moisture is low, the formulae tend to overestimate stress owing to the error in estimating the rate of change of soil-water storage, particularly when this is increasing (second case above).

While these are examples of errors in estimating the amount of stress from totals of precipitation and evapotranspiration, monthly estimates of W_1 and W_2 may be wrong, since in calculating them we assume all precipitation will become available for evapotranspiration. If precipitation is concentrated at a period in the month when soil moisture levels are high, the profile can be filled to field capacity and the surplus will percolate beyond the root zone. This was the case in August 1957 (Table 4.1). At the beginning of the month the profile was at field capacity (250 mm A.W.) so that precipitation on August 1 and 3 caused a surplus of almost 30 mm. Subsequently soil moisture was reduced until at the end of the month it was 201 mm A.W. Monthly totals of P and E_t suggest a net loss of 20 mm and an estimate of 230 mm A.W. at the end of the month. Thus the accumulated monthly soil water estimates remain higher than accumulated daily totals from August to December 1957; for example, the monthly figures for December 1957 suggest thirteen days of stress while the daily moisture balance estimates twenty-four days. Later in the cycle another error in the monthly figures appears. At the end of March 1958 the profile was dry throughout the root range and though a total of 246 mm rain fell in April, the month was dry until April 6 and more than one half of the total fell after April 21. Because of this actual evapotranspiration was well below the potential rate of 145 mm. Daily calculations allowing for the delay in arrival of the rainfall estimated that 120 mm A.W. had accumulated at the end of April. Monthly estimates, assuming uninterrupted evapotranspiration, suggested 101 mm A.W. after the same period. This difference between the monthly and daily calculations was maintained until the end of August when the profile was again reduced to zero.

Errors in calculating the monthly loss or gain of soil water will accumulate in the monthly totals of soil water storage. Thus between May 1957 and July 1958 the cumulative water storage was affected for a total of nine months, or almost 50 per cent of the crop cycle. For the period from August to December 1957 monthly calculations were 30 mm higher, while from April to July 1958 monthly calculations are 20 mm lower than totals computed from day to day water balances. Thus monthly calculations can be higher or lower than the more precise daily calculations, and the amount of the deviation cannot be consistent as it will depend upon the pattern of wet and dry periods within the cycle and the distribution of rain within each month. This has a very great influence on the estimation of stress. However, the effect of cancelling out positive and negative differences reduces the disparity in the total stress for each cycle calculated by the two methods. For all forty-three cycles of virgin cane at Kakira there was a close relationship between the two sets of estimates ($r = +0.97$, $n = 43$).

Table 4.1 Moisture balance for a cycle of sugar cane at Kakira, Uganda (Cane planted 30/4/57, harvested 21/5/59)

		1957								1958						
		May	Jun	Jul	Aug	Sep	Oct	Nov	Dec	Jan	Feb	Mar	Apr	May	Jun	Jul
Monthly E_t (mm)		44	68	89	121	147	140	137	150	161	145	159	145	145	135	127
Monthly P (mm)		177	147	135	101	47	52	204	107	16	108	66	246	118	137	72
P–E_t (mm)		133	79	44	−20	−100	−88	67	−43	−145	−37	−93	101	−27	2	−55
(1) Cumulative water storage in soil (mm)		100	247	250	230	130	42	109	66	—	—	—	101	74	76	21
(2) Cumulative water storage in soil (mm) (daily figures)		100	247	250	201	103	14	81	39	—	—	—	120	99	99	44
Days of stress from (1)	(a)*	3	—	—	—	—	29	4	13	16	21	13	5	18	11	25
	(b)									15	7	18				
Days of stress from daily balance	(a)	2	—	—	—	6	25	21	24	8	14	3	16	13	23	24
	(b)									22	9	22	5			

(a) days with soil moisture below *Wk* but more than zero; (b) days with no soil moisture in the root range.
Note. Only the first fifteen months are included since in August 1958 soil moisture was reduced to zero and the profile remained dry until harvest. There was no appreciable difference between monthly and daily calculations for this period.

PREDICTION OF SUGAR CANE YIELDS FOR UGANDA

These models were applied to monthly data from eighteen stations in Uganda which had thirty years of monthly rainfall data and reliable estimates of potential evapotranspiration. Estimates of potential evapotranspiration were available for a few stations for the period 1960 to 1968, but some had shorter periods of records (Rijks, Owen and Hanna, 1970). However, monthly evapotranspiration in Uganda varies little from year to year and the seasonal pattern shown by the mean monthly figures can be taken as a reasonable estimate of potential evapotranspiration for years prior to 1960. Rainfall is more variable and its influence on the water balance much greater. It is difficult to use the confidence limits of monthly rainfall developed for East Africa (Manning, 1956) to provide an estimate of soil water conditions throughout a growing season with any degree of certainty. This is because the water balance at any one time within the year depends on the storage term which is dependent on an indefinite period beforehand which in dry areas could extend into the preceding year. For this reason a continuous water balance was calculated from 1931 to 1969 from which probability estimates could be made. The available water stored in the soil was calculated month by month and the stress estimated using the models already described. The effect of soil depth and type was examined by comparing stress when available water capacity was 100 mm, 200 mm and 250 mm. No allowances were made for lower rates of evapotranspiration early in the cycle since in the wetter areas of Uganda at least, planting can be in almost any month of the year. On the basis of the annual accumulated stress, yields were estimated for twenty-four-month crops using the regression calculated from sugar cane cycles at Kakira, Uganda (Hanna, 1970):

$$y = 94{\cdot}58 - 0{\cdot}058x$$

(where y = yield in tons/acre and x = accumulated stress).

The yields in this regression refer to cycles of 720 days and shorter cycles must be expected to produce substantially lower yields. Also the practice of ratooning, where a second and third crop is produced from the old roots after harvesting, produces lower yields than the first or plant crop but they have shorter cycles, eighteen rather than twenty-four months, which partly offsets this effect on the long term yield pattern. The practice of resting the ground, or planting a green crop for four or five months after the uprooting of the second and final ratoon can benefit the soil water storage for the newly planted cane. Thus the yields estimated over the thirty-nine-year period must be considered high as they refer to plant crops from twenty-four-month cycles. However the continuous growth simulated in this programme makes greater demands on soil water than if there were a resting period before each plant crop which normally occurs once in five years. Also, the greater frequency of cycles produced by the shorter ratoon crops reduces evapotranspiration in the early establishment phase for each crop; and the total length of full canopy when the crop can transpire at the potential rate would be shorter. Thus the accumulated stress is probably also overestimated which would have the effect of depressing the yields. While it would be wrong to suggest that precision could be achieved by cancellation of errors, these opposing effects must reduce the error and they are probably less important than the errors inherent in the water-balance calculation.

Figure 4.1 shows the range of yields within 90 per cent confidence limits. Sugar cane is grown with success in southern Uganda at Lugazi and the 40–75 per cent tons per acre predicted for this area justifies the planned expansion of the capacity of Lugazi and Kakira. Similar high yields would appear possible in an extensive area in northern Uganda and in the west, where sugar cane is not yet established on a commercial basis. Although the upper 90 per cent limit of 75 tons per acre in Gulu is similar to the Lugazi area, the lower limit is higher, 52 tons per acre, and would suggest a good commercial prospect. The national sugar works at Kinyala,

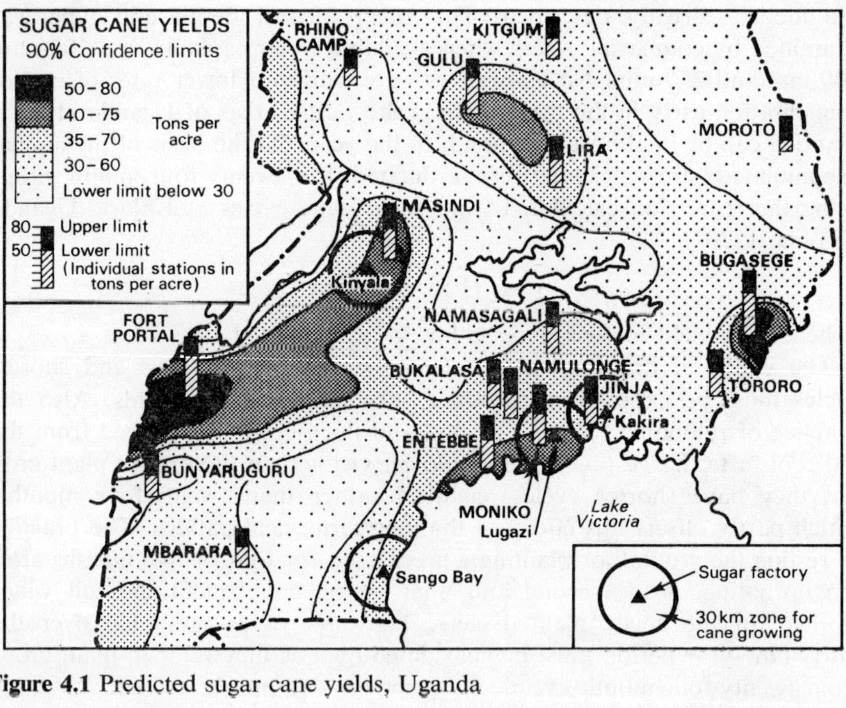

Figure 4.1 Predicted sugar cane yields, Uganda

Bunyoro, should stimulate both estate and outgrower production in this area during the mid-1970s; from 1973 production is scheduled to represent up to 30 per cent of the 1970 output from Lugazi and Kakira (Uganda Government, 1972). Factories, no matter how large their capacity, can economically process only cane grown within a distance of about 50 km. Thus there are promising areas in the west (around Fort Portal), in the east (in the Elgon foothills) and in the Gulu–Lira area of the north which cannot be developed without the construction of more sugar refineries. In the northern area the climatic régime is more akin to other sugar areas, for example Mauritius and Natal, where there is a distinct wet season and a lengthy drought. While it ought to be possible to complete a cycle of sugar cane in one season and use the dry period for harvesting, the short growing period would limit yields and it would be more economical to extend growth to include a second wet season. This is common in Natal following a low rainfall year. High temperatures during the cycle are likely to be more serious here than at Kakira where even for very dry cycles mean maximum temperatures infrequently exceeded 28·4°C. Maximum temperatures in excess of 31°C, which has been considered harmful in other sugar areas (Clements *et al.*, 1952), would probably reduce these yield levels which have been estimated from water availability alone.

The effect of deep soils in increasing the available water capacity is generally to reduce stress by prolonging the period when water is available in the profile. The wetter areas of Uganda clearly show this effect (Table 4.2). Southern Uganda as represented by Entebbe and Moniko, western Uganda as at Fort Portal, the Bugasege area of Elgon and the Gulu area show more favourable water balances when available water capacities are 250 mm rather than 100 mm. However, in the drier areas, more stress, and therefore lower yields, occurred with the higher available water capacities.

Table 4.2 90 per cent confidence limits of sugar cane yields according to available water capacity

	A.W. 100 mm		A.W. 200 mm		A.W. 250 mm	
	Lower	Upper	Lower	Upper	Lower	Upper
Entebbe	44·55	71·46	44·73	75·74	45·18	79·19
Moniko	43·83	77·61	42·81	83·23	41·91	85·94
Fort Portal	61·22	86·71	66·51	93·35	70·76	96·36
Bugasege	57·99	79·82	60·20	84·07	61·15	87·01
Gulu	51·67	74·41	52·70	76·95	52·49	77·98
Jinja	29·53	63·60	28·87	62·74	28·51	63·93
Kitgum	30·13	58·66	27·87	59·54	27·59	59·71
Moroto	19·17	45·23	18·31	44·89	18·53	44·44

This is due to low rainfall failing to achieve full profile except in the shallower soils. Thus the 100 mm A.W. regime can have adequate soil water for much longer periods than the 250 mm régime where the profile has lost more than half its available water for very long periods in the cycle. It is for this reason that the yields have been mapped for a 100 mm available water capacity regime.

It has been suggested that planting dates are not fixed in the wetter lake shore areas but they may have significance to the water balance in northern Uganda where a more consistent dry season occurs. To test this a simplified model of water needs for a twenty-four-month cycle of sugar cane was applied to the rainfall and potential evapotranspiration data for Gulu from 1931 to 1969. E_t was considered equal to E_o during the vegetative period or full canopy and during the first four months of growth E_t/E_o was taken as 0·4, 0·5, 0·7 and 0·9 respectively. Water balances were calculated for twenty-four-month periods commencing in January 1931 and run continuously to 1969. Another sample of water balances was calculated commencing in January 1932. Thus the yields were estimated for cycles planted in January of each year in the thirty-eight-year period. This procedure was followed for planting in each of the twelve months of the year and the results compared. As expected, planting outside the wet season reduced the stress experienced largely because of the lower water demands in the establishment phase of the crop. Planting in the middle of the wet season produced yields little different from those calculated using a constant E_t/E_o of unity. However, though February and November planting produced higher yields the distributions for all months satisfied the 90 per cent probability limits shown on the map. The yields for crops planted in February were highest and for those planted in June were lowest; but the difference between these two was only significant at just over the 1 per cent level. It is important to recognise that this assumes that stress will be equally effective on yields. The limiting effect of stress early in the cycle would probably reduce the difference between planting dates even further. This suggests that calculations using water needs equal to E_o throughout the year and the assumption that planting can be in any month provides a valid estimate of yields even in areas with a pronounced dry season.

THE PREDICTION OF TEA YIELDS

Tea is productive continuously as long as climatic factors are favourable and for this reason it is more difficult to apply a single index which can adequately integrate the environmental controls on yield. In existing tea areas, that is, the lake shore area of southern Uganda and in Toro, tea is constantly in production except immediately after pruning which is in a three-year cycle. This revitalises the bushes and stimulates renewed flushing so that it exerts a considerable effect on yields. The recovery from pruning, when the plant is less affected by climate, is rapid over a six-month period when yields increase at about 26 lb of made tea per acre per month. The altitude difference between Toro and the Lake Victoria area has little or no effect on this recovery. Yields indicated in Figure 4.2 refer to fully

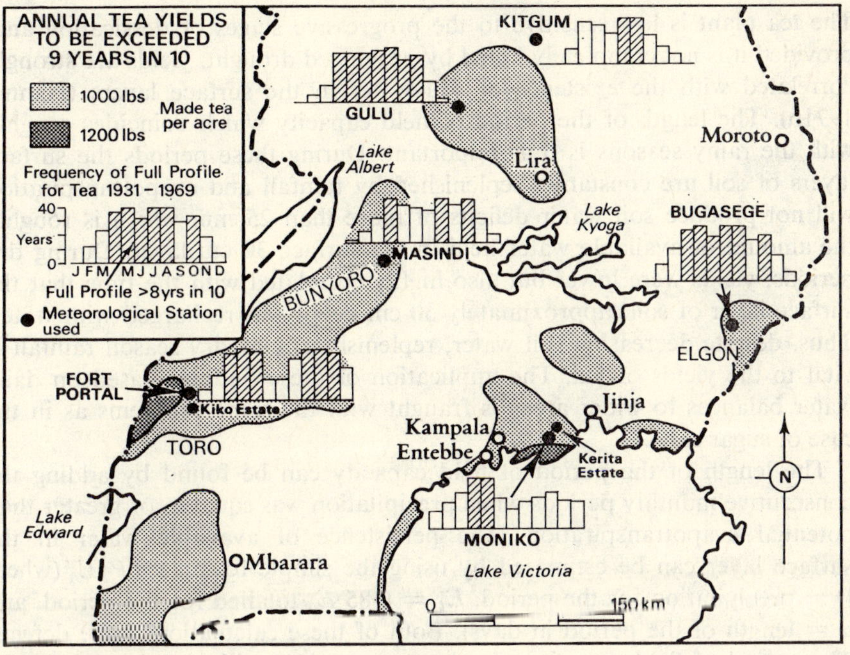

Figure 4.2 Predicted annual tea yields, Uganda

productive fields and in practice an estate would have about one-third of its tea pruned at any one time and have this area below full production for six months. Age is another complicating factor for not all estates, nor indeed all bushes on a single estate, are comparable in age.

Nutrition is more effective as the bush matures and the deeper more extensive roots reach larger reserves of soil moisture and thus become less susceptible to drought. At one estate in Toro, differences between three- to four-year-old tea and ten- to fifteen-year-old tea were significant at just above 1 per cent. Yields improved rapidly up to eight years, then more slowly to fifteen years. Tea bushes, planted over thirty years ago on an estate in southern Uganda, showed no trend over a five-year period and can be considered fully mature. It is obviously impossible to compare performance between young and old tea bushes at any time and, for tea less than fifteen years old, comparisons between years must take age into account. The yield increment for each day of water availability at the Toro estate was 3·56 lb of made tea per acre compared to 4·42 lb of made tea per acre for the fully mature tea in southern Uganda.

Yield prediction for tea

The tea plant is less sensitive to the progressive stages of dessication and, provided it is not completely killed by prolonged drought, yields are strongly correlated with the existence of soil water in the surface layers (Hanna, 1971b). The length of the period of field capacity which coincides roughly with the rainy seasons is thus important. During these periods the surface layers of soil are constantly replenished by rainfall and evapotranspiration will not produce soil water deficits of more than 25 mm which is roughly the amount of available water held in the surface 30 cm layer. During dry periods, yields were lower but also highly correlated with the time that the surface layer of soil (approximately 30 cm deep) was recharged with water. Thus, despite decreasing soil water, replenishment by dry season rainfall is vital to the yields of tea. The application of these findings based on daily water balances to wider areas is fraught with the same problems as in the case of sugar cane.

The length of the period of field capacity can be found by adding the consecutive monthly periods when precipitation was equal to or greater than potential evapotranspiration. The persistence of available water in the surface layer can be estimated by using the simple formula $d.P/E_t$ (where P = precipitation for the period, $E_t = 0{\cdot}85\ E_o$ totalled for the period, and d = length of the period in days). Both of these calculations have defects. The period of field capacity may have some days with deficits despite the fact that each month had total rainfall equal to, or greater than, evapotranspiration. Also dry season water availability would be in error if daily rainfall exceeded 25 mm when the excess would percolate downwards thus allowing subsequent drying to remove available water from the critical surface layer. In this case stress would be underestimated. More positively the formula will give the number of days that moisture is available in a surface layer, the depth of which depends on rainfall intensity. However in dry season conditions in Uganda daily rainfall totals are not likely to be excessive and will usually only replenish the surface layers of soil which are particularly important to tea production.

Using these monthly calculations for estimating water availability at Kiko estate in Toro, correlation between yield and water availability explained 98 per cent of variation in yields during wet periods and 63 per cent variation during dry periods. This compares favourably with the correlations

Table 4.3 90 per cent confidence limits of tea yields for selected Uganda stations

	Lower limit	Upper limit	Sequences of periods at full profile in months	Total time at full profile in months*
Fort Portal	1172	2087	3, 3	6
Moniko	1150	2145	2	2
Jinja	894	1959	2	2
Bugesege	1001	2008	2, 2	4
Masindi	1065	2051	1, 1, 1	3
Gulu	1000	1880	7	7
Kitgum	893	1734	2, 2	4

* These months had full profile in more than 80 per cent of the years 1931–69.

between yield and water availability as estimated from daily water balances at Kerita in southern Uganda. This was applied to the monthly rainfall and evapotranspiration data for the stations used in the sugar cane water balance study and the yields predicted from regressions based on daily yields and water-balance studies at Kerita (Hanna, 1971b):

(1) For period of field capacity

$Y = 62{\cdot}43 + 4{\cdot}42X$

where Y = yield in pounds of made tea per acre

X = no. of days at field capacity

(2) For dry seasons

$Y = 4{\cdot}09X + 69{\cdot}52$

where Y = yield in pounds of made tea per acre

$X = d.P/E_t$ (where d = no. of days in the month)

Yields from both periods of field capacity and for dry seasons were totalled for each calendar year and 90 per cent confidence limits calculated (Figure 4.2). The incidence of periods of full profile are equally important as they will determine the pattern of production throughout the year. The seasonal régime is suggested in the graphs on Figure 4.2 which show the monthly frequency of full profile under a tea crop for the period 1931 to 1969. The Gulu–Lira area in northern Uganda, while showing reasonable prospects of successful tea production, has a less favourable régime since the high-yielding period of full profile is eight years in ten confined to the period April to October, and in fifteen of the years from 1931 to 1969 the period of full profile was interrupted by one, two or three months with deficits exceeding 25 mm. More serious however is the five-month period of dessication during which moderate yields depend on the periodic replenishment of surface soil water. However, the period between dry season falls could be important, particularly in the establishment phase when young shallow rooting bushes could be killed. Thus without supplementary irrigation the yields predicted in these areas could hardly be realised. Most of the other areas, with annual prospects of at least 1000 lb made tea per acre nine years in ten, have already been successfully planted with tea, with the exception of the Elgon area which would appear to justify development.

CONCLUSION

Much of this chapter has been concerned with the problem of estimation of soil water statistics from meteorological data and their spatial interpolation. The problem in bioclimatology lies in finding significant relationships between climate and crop production. The application of these criteria to evaluate the prospects for crop production throughout the country must be concerned with their predictions, which in the variable conditions of East Africa require long periods of records. For statistical reasons it is necessary to use a time period of a month rather than a day. It is necessary, therefore, to attempt assessments of conditions during monthly periods, which can in some measure account for the very important fluctuations from day to day. The water balances calculated from rainfall data over periods of almost 40 years have been used to estimate yields of sugar cane and tea within 90 per cent confidence limits. From these predictions, the prospects for good crops of sugar cane in western Uganda, Bunyoro, Mount Elgon and the Gulu–Lira area of the north are encouraging for future expansion into these areas. The most important contrast in northern Uganda is the long dry period, but this ought not to prevent success, even if a two-year crop is planned. Calculations suggest that any one date of planting will not give significantly better soil water conditions in the plants' two-year cycle. However, planting in a long dry spell would not allow germination and for this reason alone planting may be more seasonal than is the case at present in southern Uganda.

The water balance for tea crops also suggests the suitability of tea to the same areas of northern Uganda and Elgon which have not so far been developed for this crop. The long period of dessication may be a serious drawback. In the very dry year 1967, many tea estates in Toro suffered serious losses through die-back of bushes. If this occurred more frequently, as is possible in northern Uganda, establishment without supplementary irrigation would be difficult. The drought period would also produce a more seasonal pattern of cropping than is known in the existing tea areas.

These evaluations can only hope to be approximations, for the very nature of the terrain imposes great contrasts within small areas, not only in the water balance, but in many related aspects. Radiation and temperature contrasts seriously influence these yield assessments in areas away from the climatic regions in which they were evolved. They can however provide a framework for planning the controlled trials which are vital before any extensive new areas are planted with these crops.

REFERENCES

Clements, H. F. (1964) 'Interaction of factors affecting yield', *Annual Review of Plant Physiology* **15,** 412.

Clements, H. F., G. Shiguera and E. K. Akamine (1952) 'The factors affecting the growth of sugar cane', *Technical Bulletin* (University of Hawaii Agricultural Experimental Station) **18,** 59.

Dagg, M. (1965) 'A rational approach to the selection of crops for areas of marginal rainfall in East Africa', *East African Agricultural and Forestry Journal* **30,** 296–300.

Davies, J. A. and P. J. Robinson (1969) 'A simple energy balance approach to the moisture balance climatology of Africa', *Environment and Land Use in Africa*, ed. M. F. Thomas and G. W. Whittington (London: Methuen), 23–56.

Dowker, B. D. (1963) 'Rainfall reliability and maize yields in the Machakos district', *East African Agricultural and Forestry Journal* **28,** 34–38.

Fitzpatrick, E. A. and J. A. Arnold (1964) *General report on lands of the West Kimberley area, West Australia* (Land Research Series No 9; Commonwealth Scientific and Industrial Research Organisation), 88–93.

Goodall, D. (1966) 'Classification, probability and utility', *Nature* **211,** 53–54.

Halais, P. (1969) 'A classification of vegetative seasons for cane production', *Mauritius Sugar Industries Research Institute Annual Report*, 110–112.

Hanna, L. W. (1971a) 'Climatic influences on yields of sugar cane in Uganda', *Transactions of the Institute of British Geographers* **52,** 41–59.

Hanna, L. W. (1971b) 'The effects of water availability on tea yields in Uganda', *Journal of Applied Ecology* **8,** 791–813.

Hardy, M. (1966) 'Water consumption of the cane plant', *Mauritius Sugar Industries Research Institute Report*, 95.

Huxley, P. A. (1965) 'Climate and agricultural production in Uganda', *Experimental Agriculture* **1,** 81–97.

Kerfoot, O. (1962) 'Root systems of forest trees, shade trees and tea bushes', *East African Agricultural and Forestry Journal* **27** (special issue), 24.

Manning, H. L. (1956) 'The statistical assessment of rainfall probability and its application to Uganda agriculture', *Proceedings of the Royal Society* **B 144,** 460–480.

Manning, H. L. (1960) 'Cotton crop potential and rainfall expectation in Uganda', *Tropical Meteorology in Africa*, ed. D. J. Bargman (Munitalp), 402–414.

Nyasaland (1956/57) *Annual Report of the Tea Research Station* (Nyasaland: Department of Agriculture).

Nix, H. A. (1968) 'The assessment of biological productivity', *Land Evaluation*, ed. G. A. Stewart (Melbourne: Macmillan) 77–87.

Owen, P. C. and D. J. Watson (1956) 'The effect on crop growth of rain after prolonged drought', *Nature* **177,** 847.

Penman, H. L. (1948) 'Natural evaporation from open water, bare soil and grass', *Proceedings of the Royal Society*, **A 193,** 120–145.

Pereira, H. C., J. S. G. McCulloch, M. Dagg, P. H. Hosegood and M. A. C. Pratt (1962) 'Assessment of the main components of the hydrological cycle', *East African Agriculture and Forestry Journal* **27** (Special issue), 8–15.

Piexoto, J. P. and G. O. P. Obasi (1965) 'Humidity conditions over Africa during the I.G.Y.', *Scientific Report No. 4*, General Circular Project; Massachusetts: Massachusetts Institute of Technology.

Rijks, D. A. (1968) 'Agrometeorology in Uganda—a review of methods', *Experimental Agriculture* **4**, 263–274.

Rijks, D. A., W. Owen and L. W. Hanna (1970) *Potential Evaporation in Uganda.* (Uganda: Water Development Dept., Uganda Government).

Russell, E. W. (1960) 'Agricultural uses of meteorological data', *Tropical Meteorology in Africa*, ed. D. J. Bargman (Munitalp), 390–401.

Russell, E. W. (1968) 'Some agricultural problems of semi-arid areas' *The soil resources of tropical Africa*, ed. R. P. Moss (Cambridge University Press), 121–133.

Sansom, H. W. (1954) 'The climate of East Africa according to the Thornthwaite classification', *East African Meteorological Department Memoir*, **3,** 2.

Slatyer, R. O. (1960) 'Agricultural climatology of the Katherin area, Northern Territory', Regional Survey Technical paper No. 13; Australia: Commonweath Scientific and Industrial Research Organisation.

Thornthwaite, C. W. (1948) 'An approach towards a rational classification of climate', *Geographical Review* **35,** 55–94.

Uganda Government (1972) *Third Five-Year Development Plan, 1971/72 to 1975/76* (Entebbe).

Watson, D. J. (1963) 'Climate, weather and plant yield', *Environmental control of plant growth*, ed. L. T. Evans (London: Academic Press), 337–349.

Whyte, R. O. (1966) 'The use of arid and semi arid lands', *Arid lands: a geographical appraisal*, ed. E. S. Hills (London and Paris: Methuen and UNESCO), 301–358.

Winter, E. J., P. J. Salter and E. G. Cox (1970) 'Limited irrigation in crop production', *The role of water in agriculture*, ed. J. A. Taylor (University College of Wales, Aberystwyth, Memorandum No. 12 (1969); Oxford: Pergamon Press), 147–160.

Woodhead, T. (1968a) *Studies of Potential Evaporation in Tanzania* (East African Common Services Organisation and Water Development and Irrigation Division, Government of Tanzania).

Woodhead, T. (1968b) *Studies of Potential Evaporation in Kenya* (Water Development Dept. (Ministry of Natural Resources), Republic of Kenya).

Chapter 5

Problems of Water Resource Management in a Semi-Arid Environment: The Case of Irrigation Agriculture in the Central Jordan Rift Valley

Ian R. Manners

Introduction

The Jordan Rift Valley, which extends for a distance of nearly 400 kilometres from the slopes of Mount Hermon in the north along the length of the Jordan Valley and the Wadi Araba to the Gulf of Aqaba in the south, presents one of the most dramatic landscapes in the Middle East. Steep, highly-faulted escarpments mark the margins of the depression throughout most of its length; during the long summer drought, only the lush greenery of the irrigated areas disrupts the overall aridity. The central section of the depression, extending from Lake Tiberias to the Dead Sea, is referred to locally as *El Ghor* (Figure 5.1). Within the Ghor, the River Jordan follows a meandering course along the central axis of the depression, the present floodplain of the river (*El Zor*) being incised to depths of up to 50 metres below the main Ghor terrace. A number of tributary wadis, carrying a small perennial streamflow, flow into the depression from the bordering massifs and these, together with the River Jordan, represent the major sources of water for irrigation agriculture. This region exemplifies many of the problems confronting the development of irrigation agriculture in a semi-arid environment, and, as a result, the Ghor has been selected to illustrate not only the nature and dimensions of these problems but also the contribution that geographical analysis can make to their solution. In particular, this chapter will attempt (1) to identify the particular problems

---

Ian R. Manners is Assistant Professor of Geography at the University of Texas at Austin.

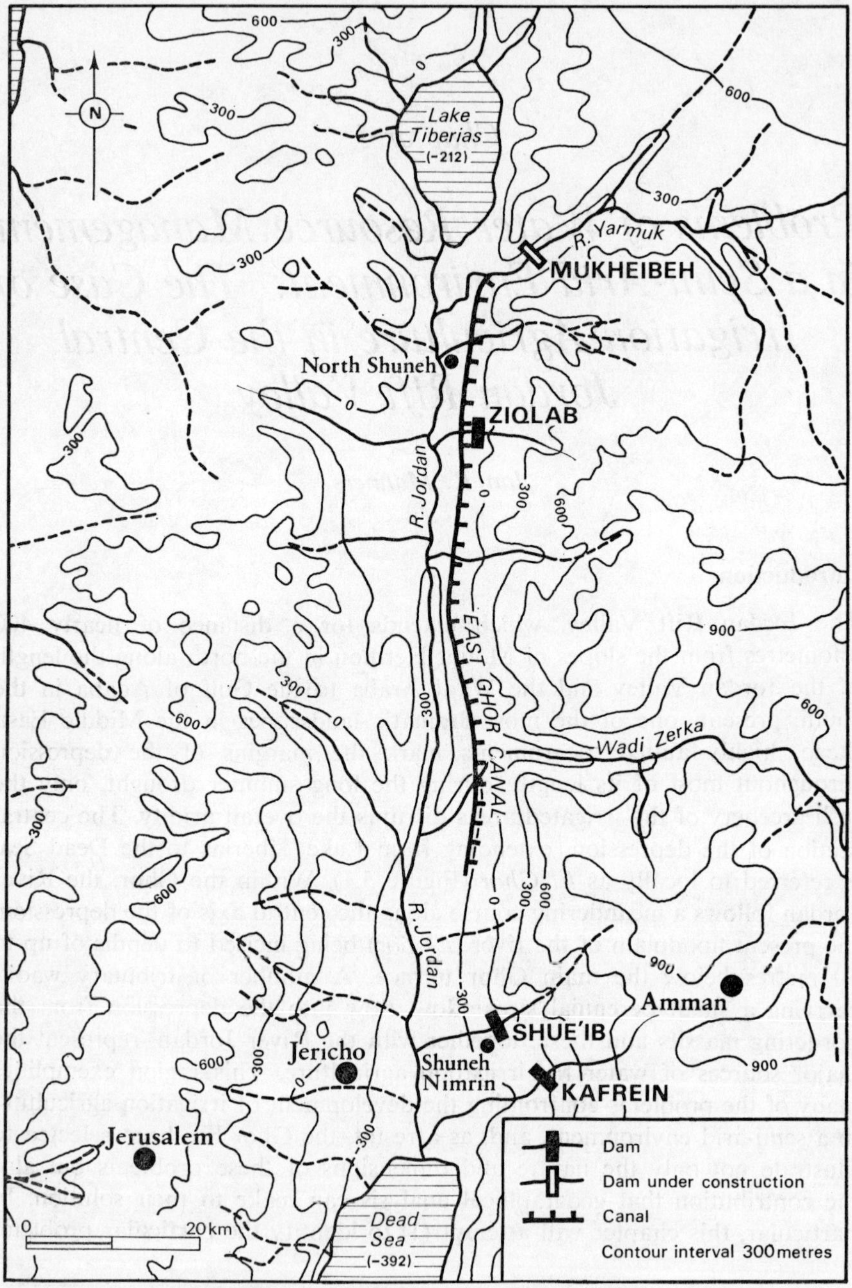

Figure 5.1 The central Jordan rift valley (the Ghor)

that confront development planners and irrigation farmers in a semi-arid environment, and (2) to illustrate the contributions that geographical analysis can make towards achieving the efficient and productive use of available resources, focusing specifically on the evaluation and estimation of available water resources and on the analysis of those processes and perceptions that influence the level of technology that can be applied to the development of those resources.

Within the Central Jordan Rift Valley, as throughout the arid zone and its semi-arid borderlands, the general threat of moisture deficiency is immediately apparent. The variety of irrigation techniques, ranging from qanats to pump wells, that can be found even within the comparatively small area of the Ghor reveal the common concern with water, its sources and its availability. Yet as far as agricultural activities are concerned, the practical problems confronting the farmer and the development planner alike result not so much from inadequate rainfall as from the *variation* in rainfall conditions and *uncertainty* over its time and place of occurrence. It is the fluctuating nature of the physical environment, with marked seasonal and annual variations in rainfall and streamflow, that make regions such as the Ghor extremely hazardous for human occupation.

Within this type of fluctuating physical environment, man has pursued a variety of strategies designed to overcome the scarcity and variability of moisture. When successful, these strategies have supported a growing population. All too frequently, however, man has failed to perceive that the balance between water, soil, and vegetation in the semi-arid lands is a delicate one, and that the modification of a single process—such as the removal of the natural forest cover for fuel or for lumber, the overgrazing of the vegetation cover, or the overdrawing of an aquifer—can initiate a disastrous sequence of events, resulting ultimately in a marked degradation of the resource base. In the southern Ghor, for example, the widespread adoption of modern pump-well technologies during the 1950s has led to a serious overdraft situation with rates of withdrawal greatly in excess of present rates of recharge. This depletion of groundwater resources has been aggravated by a serious deterioration in the quality of groundwater. Groundwater of good quality is essentially confined to the alluvial fans bordering the Ghor escarpment; these fans are underlain and bordered by formations of the Lisan Series which comprise interbedded clays and marls of lacustrine origin deposited beneath an enlarged predecessor of the Dead Sea. As a result of the chemical composition of the Lisan Series, however, groundwater in the formation is extremely saline. With the withdrawal of groundwater from the alluvial fan aquifers and the creation of marked depression cones in the areas of maximum withdrawal, saline groundwater from the Lisan Series has intruded into the alluvial fan aquifers. Between 1961 and 1967, some 250 wells in the southern Ghor were abandoned as a result of falling water tables and increasing salinity. The history of the Middle East is filled with similar examples of the way in which man's modification of the natural environment can generate positive feedback, disrupting the equilibrium that existed within the natural system prior to his intervention. All too frequently, the possibility of a low but sustained increase in productivity is sacrificed for spectacular but short-term gains.

It should be apparent therefore that within such regions as the Central Jordan Rift Valley, the development planner is confronted by a series of problems. The question thus arises as to what contribution the geographer can make towards the interpretation and solution of these problems. For, as White has pointed out, such analysis will not only permit a better understanding of the history of resource management in the semi-arid zone but allow for more intelligent use of resources in the future (White, 1966a).

In the first instance, it should be evident that the development process introduces new elements and factors into the landscape which must disrupt the existing ecological system. Development proposals cannot therefore be evaluated solely in terms of narrowly-focused cost-benefit analyses but must be considered in terms of their broader ecological impact. It is as critical to identify and predict the changes that will occur in hydrological régimes and in biogeochemical cycles as a result of the construction of a reservoir as it is to measure the direct economic impact of the project. Similarly, it is critical to recognise that, in physical and social terms as well as in economic terms, there will be a spatial dimension to development proposals with the effect of resource development being felt in adjacent and functionally related areas. In the Ghor, for example, the increased withdrawal of water from Lake Tiberias for the Israeli National Water Carrier and from the River Yarmuk for the Jordanian East Ghor Canal have resulted in a steady increase in the salinity of the lower Jordan, threatening the livelihood of those farmers in the Zor who are directly dependent upon the River Jordan for their irrigation supply. The identification of such linkages and the analysis of probable social, economic and ecological impacts, require multidisciplinary insight and expertise. However, given the nature of geographical training, which emphasises both the complexity of man-environment interactions and the interpretation of spatial relationships, the geographer should be able to make a major contribution to the development process. In particular, by defining the linkages and impacts to be examined the geographer can help to provide a framework for more comprehensive studies.

THE APPRAISAL OF WATER RESOURCES

In addition to the *general* contribution that geographers can make to the development process, there are more direct applications of geographical expertise. One of the most difficult problems confronting the planner in the developing world, for example, is the lack of information concerning the extent and quality of such basic resources as land and water. The establishment of adequate data-gathering networks and the completion of detailed inventories may take years, and, in such circumstances, the geographer can contribute to a generalised appraisal of the distribution of the major resources through space and through time. Figure 5.2, for example, showing the probability of places within the Ghor receiving more or less than 200 millimetres (mm) of rainfall during the winter months, emphasises the significance of spatial and temporal differences in rainfall conditions in the semi-arid zone.[1, 2] It will be noted that even within the comparatively small area of the Ghor, the precise nature of the risk as far as agricultural activities are concerned is immensely varied and the range of uncertainty great. In relation to the management of the region's water resources, it is of significance that in no part of the Central Jordan Rift Valley (indeed in no part of the entire Hashemite Kingdom of Jordan) is a winter's rainfall of less than 200 mm an impossibility. In general, however, winter rainfall in the northern Ghor is usually adequate to permit the cultivation of a cereal crop using rainfed techniques. Rainfed agriculture is still possible in the central Ghor where there is a 70–90 per cent probability of receiving 200 mm during the winter season, but conditions must be considered more marginal. Rainfall can be expected to fail in up to three years out of ten and, although there may be more good years than bad, there is no guarantee that two or three drought years will not occur in succession. In the southern Ghor, where there is less than a 70 per cent probability of receiving 200 mm, the chances of drought occurrence are too great for regular rainfed cultivation. Without a supplementary irrigation supply, agriculture, even in the winter months, becomes a distinct gamble. Rainfall is likely to fall below the minimum requirement in at least three years out of ten, while the probability of a succession of deficit years is correspondingly higher. It should perhaps be emphasised that while the probability of experiencing drought conditions during the winter cropping season can be calculated in this manner, and while the farmer, based on his personal experiences, may have a rough idea of the risks over a decade, in any particular year the farmer must make his decision as to which crops he will plant with no guarantee as to whether that year's rainfall will be adequate.

It should also be noted that these threshold values must not be considered as clear-cut boundaries, since factors other than the amount of rainfall influence the possibility of rainfed agriculture. In the marginal areas, for example, soil characteristics—in particular the ability of the soil to retain moisture—take on added significance for farming. The adequacy of 200 mm for rainfed cultivation in the Middle East also assumes reasonable distribution through the growing season. In many years this is clearly not the case

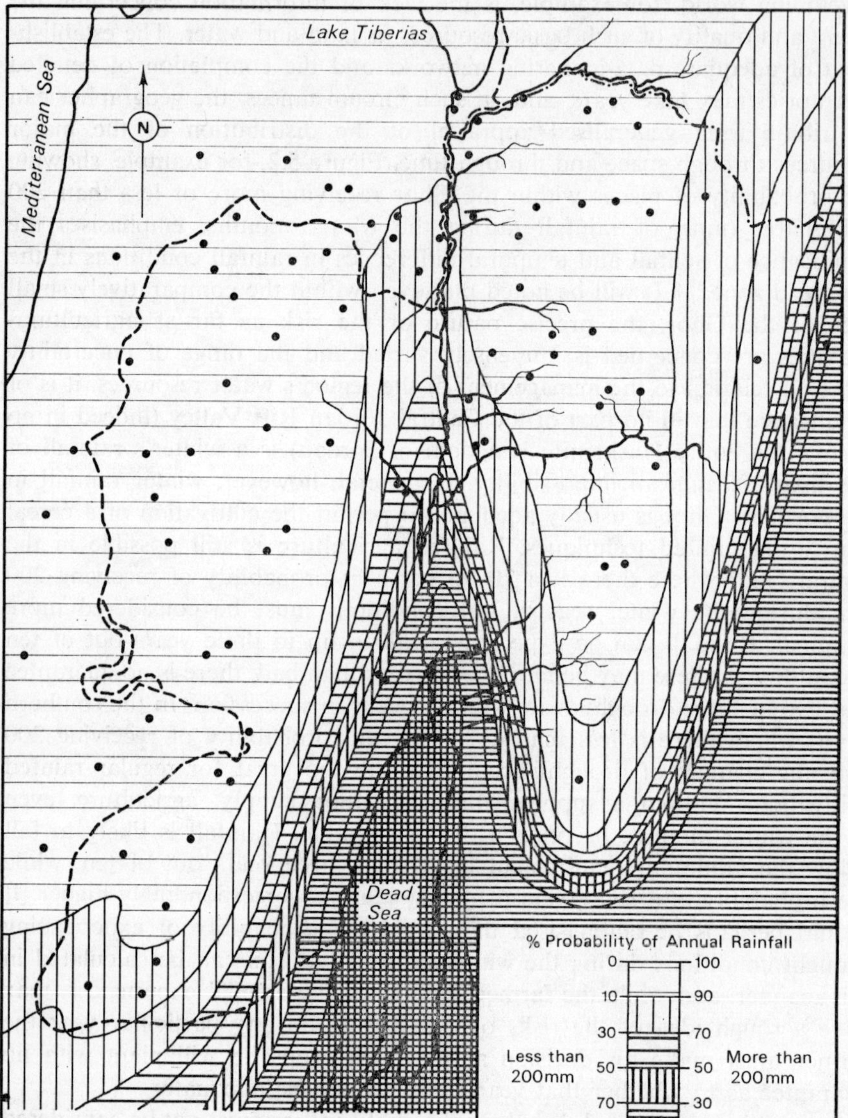

Figure 5.2 Rainfall probability in the central Jordan rift valley and environs: probability of receiving more or less than 200 mm of rainfall annually. Based on rainfall data for the period 1936–37 to 1965–66.

As a result, even in those areas in which rainfed agriculture is regularly possible, poor seasonal distribution may result in crop failure or reduction in yields. In this respect, it is equally important to analyse the variability of rainfall conditions during those periods of the cropping season when crops are most sensitive to any deficiency in moisture. In Jordan, for example, the extreme variability of rainfall conditions during October/November and again during April/May, at the beginning and end of the winter rainy season, has a considerable influence on both agricultural practices and crop yields.[3] Thus in Jordan, sowing does not usually take place until December, a practice which is in marked contrast to the situation in Syria and in the northern part of the fertile crescent where sowing is completed by early November. The later start to the season, however, means that the other phases of crop growth are correspondingly delayed. In particular, the heading period when cereal crops are sensitive to moisture deficiency is delayed until mid-April, by which time the reliable rainfall season has ended (Wallen and Perrin de Brichambaut, 1962). Thus the coincidence of the active growth period, when an assured moisture supply is critical, with the end of the rain season, when rainfall is particularly unreliable, means that agricultural conditions in the Jordan Valley are even more marginal than is apparent at first sight. In particular, unless a supplementary irrigation supply is applied, yields will be significantly lower than in areas where the heading period falls within the reliable rainfall season.

Although an understanding of general rainfall conditions is valuable, actual local conditions affecting the availability of water for plant growth are likely to be of greater practical interest as far as the irrigation planner is concerned. In this respect, water balance analysis as developed by Thornthwaite represents a major geographical contribution to the problem of resource assessment (Thornthwaite, 1948; Thornthwaite and Mather, 1955). Figure 5.3 shows water balance diagrams for stations in the Central Jordan Rift Valley constructed according to the Thornthwaite–Mather model (Thornthwaite and Mather, 1957). The water-balance diagrams, by providing information as to periods of moisture deficit and surplus, illustrate aspects of the hydrological cycle not immediately revealed by streamflow and rainfall data. In the Ghor, the available moisture from precipitation exceeds potential evapotranspiration rates only during the winter months. At this time, recharge of soil moisture occurs. As soon as potential evapotranspiration again exceeds precipitation, the moisture stored in the soil is drawn upon by the vegetation cover. The initial rate of moisture loss from the soil is large but a rapid decrease takes place as the soil becomes drier and as soil moisture tension affects the rate of depletion. At some stage, the *actual* rate of evapotranspiration will fall far below the potential rate; unless a supplementary irrigation supply is made available during the period

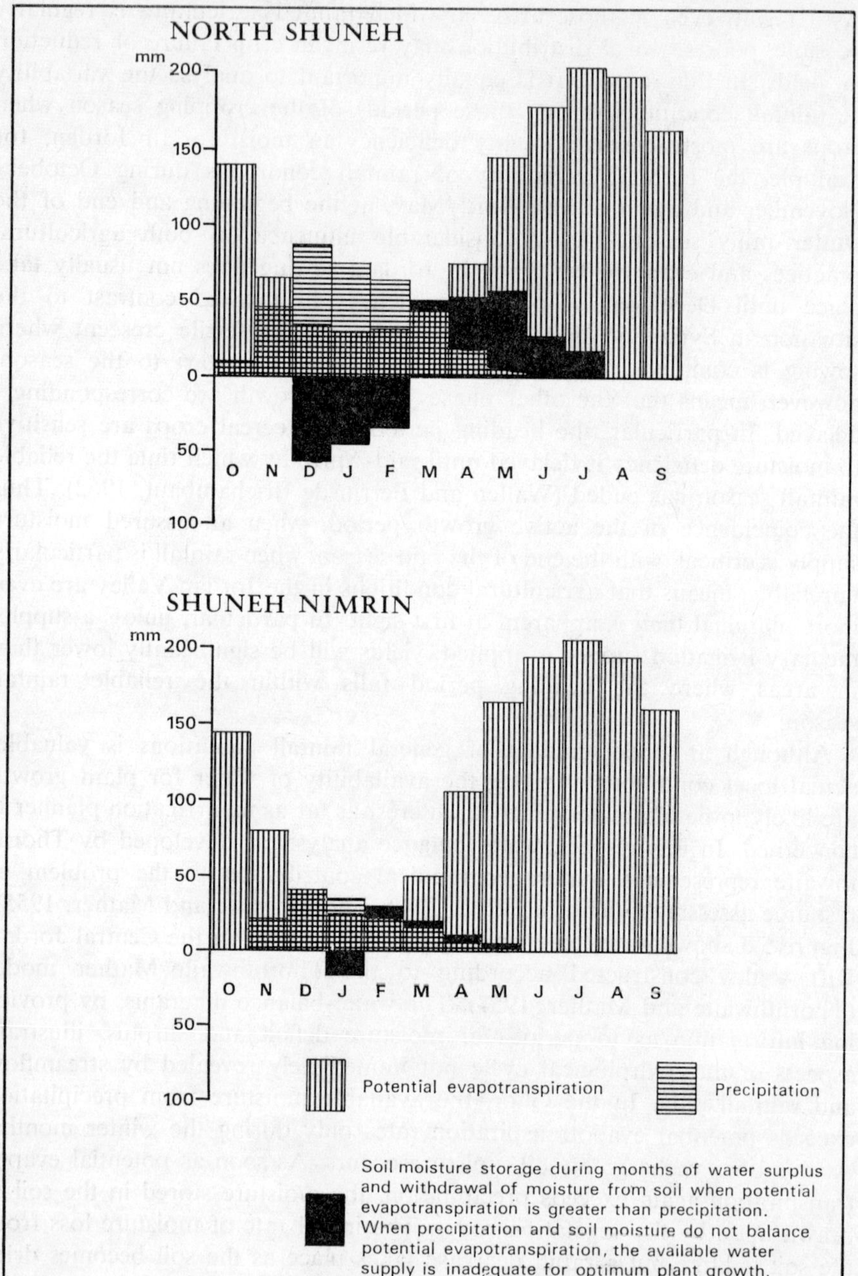

Figure 5.3 Average annual water balance for North Shuneh and Shuneh Nimrin in the central Jordan rift valley

of moisture deficiency, plant growth will be retarded and crop yields will be less than optimum.

Water-balance analysis not only provides insight into the factors of evapotranspiration, soil moisture and precipitation that directly affect crop growth but has practical value in computing crop water needs and irrigation diversion requirements during the periods of moisture deficit. The use of water balance analysis in this manner is complicated by the need for accurate data on both soil characteristics and evapotranspiration rates. Actual evapotranspiration rates are difficult to determine in view of varying moisture conditions and characteristics of soils, and while potential evapotranspiration rates can be determined directly by using lysimeters, the high cost of instrumentation prohibits this approach when water-balance conditions need to be computed over a wide area. However, a number of empirical formula have been developed by Thornthwaite (Thornthwaite, 1948; Thornthwaite and Mather, 1955), and Blaney and Criddle (Blaney and Criddle, 1950; Blaney, 1955) amongst several others, that compute potential evapotranspiration and water need from measurements of common meteorological elements, and these are widely used in the planning stage of irrigation projects. The Blaney–Criddle formula, which incorporates a crop coefficient, thereby recognising that the rate of potential evapotranspiration will vary according to the nature of the vegetation cover, is perhaps the most widely used procedure.

Once crop water requirements for a particular area have been determined, the planners can compute actual irrigation diversion requirements by making allowance for the water losses which will inevitably occur both on the farm and in the course of conveying water from the diversion point to the farm headgate. In the United States, it has been estimated that 'about one-third of all water diverted for irrigation is lost in conveying it to the land and another third percolates too deeply or runs off during the process of application' (Lauritzen, 1955, p. 311). In estimating likely water losses, the planner must deal with another dimension of the development process for he must make certain decisions as to the kinds and types of technology that can be applied to the use of water. In the Central Jordan Rift Valley, as in most developing regions of the world, there exists a tremendous gap between the technological skills that are available and those that are actually applied in the field by individual farmers. Various techniques are available, for example, that can substantially reduce wastage and loss of water during the actual field application of irrigation water, yet many farmers continue to use traditional 'wild-flooding' methods of applying water to their fields. The irrigation planner, recognising the existence of this gap, must make certain assumptions as to the extent to which this gap can be narrowed through the use of socio-economic incentives and restraints and the willingness of farmers to adopt more efficient techniques.

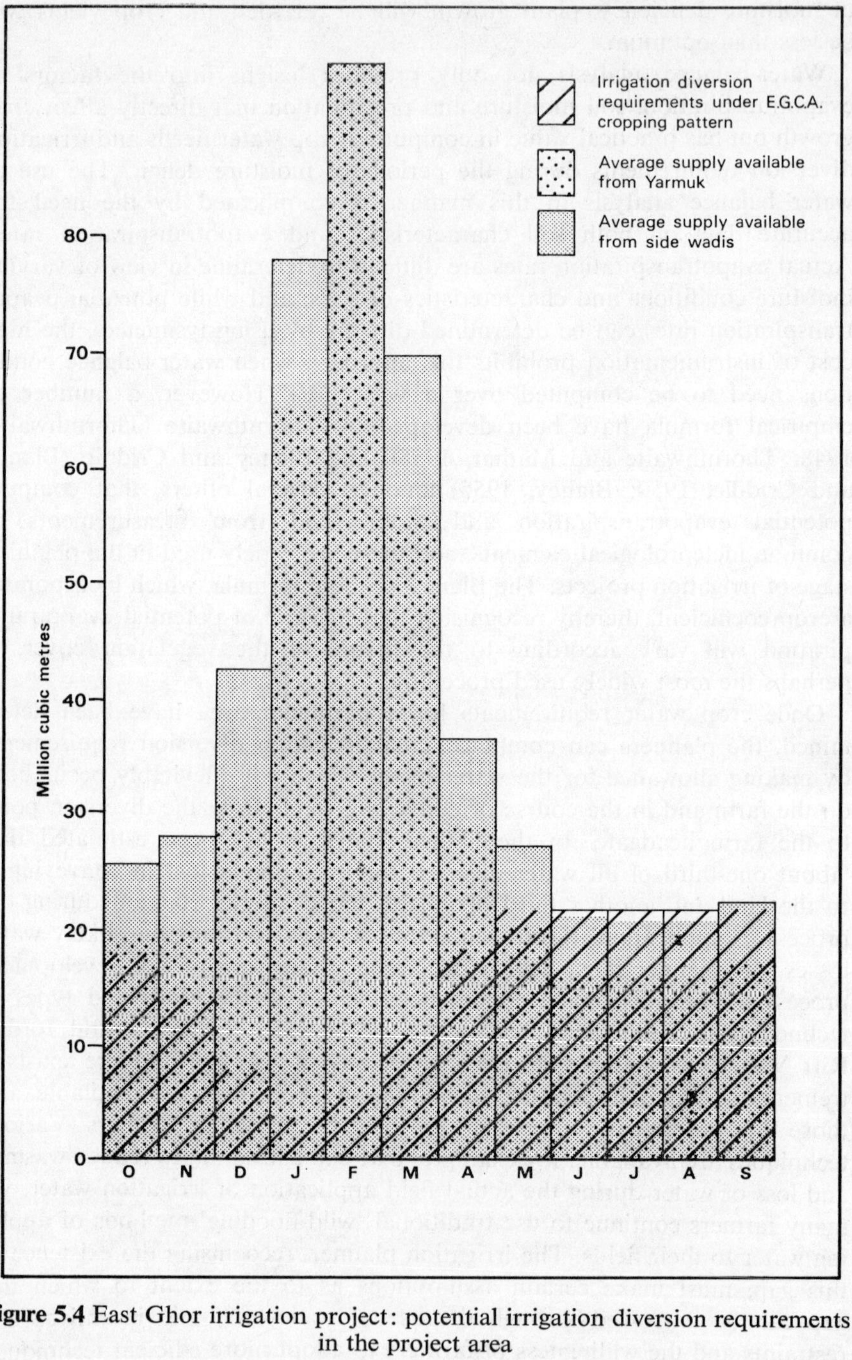

Figure 5.4 East Ghor irrigation project: potential irrigation diversion requirements in the project area

As will be discussed, problems may well result when the farmer does not evaluate the new techniques in the same way as the planner.

Estimates of irrigation diversion requirements computed through the use of water-balance analysis will vary according to: (1) the method used to determine crop water needs, (2) the cropping pattern used in the calculations, and (3) the anticipated standard of water management. Figure 5.4 shows the potential irrigation diversion requirements for the 125 000 dunums in the northern Ghor currently irrigated from the East Ghor Canal,[4] computed according to the Blaney–Criddle procedure and using the same cropping pattern employed by the East Ghor Canal Authority for planning and operational purposes. Annual water duty amounts to 1486 cubic metres per dunum or an irrigation diversion requirement of 185·75 million cubic metres of water per annum for the project area. The average annual supply available to the project area from (1) the River Yarmuk, after allowance has been made for a minimum flow of 1 cumec to pass below the canal intake works for use in the Israeli Yarmuk triangle and (2) the side wadis, after allowance has been made for the irrigation requirements of the areas served from existing diversion weirs, is 510·5 million cubic metres (Manners, 1969). However, computing water needs on an annual basis is misleading, and, as is evident from Figure 5.4, shortages can be anticipated within the project area in the period June to September due to the seasonal nature of the water supply. As this analysis clearly reveals, therefore, in an average year the water supply is inadequate to meet the potential irrigation diversion requirements of the area presently commanded from the East Ghor Canal. In view of this, it should also be noted that the calculation of irrigation diversion requirements assumed that the East Ghor Canal Authority's goal of an overall irrigation efficiency of 65 per cent can be achieved. In view of past experience in the United States, this figure would seem unduly optimistic.

However, as indicated earlier, the potential irrigation diversion requirements were calculated on the basis of the cropping pattern suggested by the East Ghor Canal Authority. This cropping pattern assumes that a large proportion of the project area will be devoted to the cultivation of vegetable, fruit and fodder crops—crops which have comparatively high water requirements. In part, this reflects the tremendous increase in fruit and vegetable acreage that has occurred in the last decade in response to market opportunities. At the same time, however, a land redistribution programme completed within the area during project construction has created a medium-sized proprietory with average farm holdings of less than 50 dunums (Manners, 1970). As a result, each farmer must devote a large part of his holding to the cultivation of high value crops and make maximum use of double cropping if he is to achieve a reasonable standard of living *and* repay his share of project construction and operation costs as specified in the East Ghor Canal Law. The cropping pattern used by the East Ghor Canal Authority for planning purposes takes into account the shift in cropping patterns that will have to be achieved if these goals are to be realised. The economic advantages of high value crops, however, are partially offset by their high water requirements.

Of course, the *present* cropping pattern in this part of the Ghor does not yet achieve either the degree of double cropping or the crop allocation envisaged under the Authority's cropping pattern. In particular, a much greater proportion of the area is still devoted to cereal crops with low water requirements. The net result of the analysis, however, is to demonstrate that if, as seems likely in response to market opportunities and to various social and economic incentives offered by the East Ghor Canal Authority, there is a further shift in the land-use pattern with increased emphasis on the cultivation of high value crops with high water requirements, *without* any significant increase in the efficiency of irrigation operations, then the risk of periodic water shortages in the project area will become appreciably greater.

A final note of caution must be added with regard to the use of water-balance analysis for agricultural and irrigation purposes. The empirical formulae developed by Thornthwaite and others are derived from observed correlations between measured potential evapotranspiration and macro-climatic data. The Thornthwaite formula, for example, is based on lysimeter readings of potential evapotranspiration from the central and eastern United States and data on monthly air temperature. A daylength adjustment factor is then applied to correct for season and latitude. But since such relationships are likely to vary from one climatic region to another, there is some doubt as to the validity of applying such formulae on a world-wide basis (Chang, 1968). Stanhill, for example, has shown that both the Thornthwaite and the Blaney–Criddle formulae underestimate potential evapotranspiration values in the Middle East (Stanhill, 1961a and 1961b). Penman has developed a more satisfactory approach to calculating potential evapotranspiration using the energy budget concept (Penman, 1948). Although the Penman equation is based on sounder physical principles, its solution requires data on radiation conditions, vapour pressure and wind speed. Thus, while studies indicate that the Penman approach is more accurate, data limitations prevent its wider application in regions such as the Ghor. In conclusion therefore, it should be apparent that the one important need in relation to water-balance analysis is further research into methods to facilitate the accurate computing of evapotranspiration.

In relation to resource appraisal, it should be emphasised that the absence of adequate data can have far-reaching repercussions as far as resource management is concerned. The Jordan Basin provides an excellent illustration of such a situation, where differing estimates of the amount of water available not only served to exacerbate pre-existing tensions between Israel and the Arab world but affected the design of water development projects. The early proposals for the development of the water resources of the Jordan Basin, for example, assumed that the available water was sufficient to meet the entire irrigation requirements of both Palestine and Transjordan (Lowdermilk, 1944; Hays, 1948). Accordingly, an integral feature of the Lowdermilk and Hays proposals was the diversion of the flow of the Jordan surplus to the irrigation requirements of the Ghor for use in the Negev. However, as a government study undertaken for the Anglo-American Committee of Inquiry commented: '. . . the desire to substantiate the contention that Palestine can support a very much larger population has led to the conception of vast projects, based, it would seem, not entirely upon authoritative and reliable hydrological data, but also upon conjecture and assumption' (Government of Palestine, 1946, p. 411).

Subsequent investigations have confirmed that the Lowdermilk and Hays proposals significantly overestimated the availability of water, yet their early proposals have continued to influence the thinking of Israeli planners (Stevens, 1956, 1965; Doherty, 1965). The National Water Carrier, the first stage of which was completed in 1964, retains the basic idea of the diversion of Jordan waters (from Lake Tiberias) for use in the Negev. It should be noted, however, that the National Water Carrier scheme has not fully realised early expectations. In particular, it appears that little water from the Jordan is reaching the Negev, most of the diverted water being used to recharge groundwater aquifers in Israel's coastal plain that are seriously overdrawn and subject to increasing contamination due to salt water intrusions (Smith, 1970).

Early Jordanian development proposals were based on a much more conservative estimate of available water resources and on the principle that 'the waters in a catchment should not be diverted outside the area unless the requirements of all those who use or genuinely intend to use the waters within the area have been satisfied' (MacDonald and Partners, 1951, p. i). In the absence of any political agreement over integrated river basin development, Jordan, like Israel, has proceeded unilaterally with its own diversion scheme, the Yarmuk Project. The first phase of the Yarmuk Project, involving the diversion of the base flow of the River Yarmuk through the East Ghor Canal to provide a perennial irrigation supply for approximately 35 000 acres of cultivable land in the North East Ghor, was completed in 1963. However, the failure to reach agreement for an integrated development project has had serious repercussions for the Jordanian planners. In the first place, the use of Lake Tiberias as a natural storage reservoir for Yarmuk waters became impractical and the Jordanians have been forced to proceed with the costly alternative of constructing storage dams on the Yarmuk itself. Work on the first of two proposed dams, the

Mukheibeh Dam, began in 1963. However, as a result of the occupation of the north bank of the Yarmuk in the vicinity of Mukheibeh by Israeli forces in June 1967, work has been indefinitely suspended. Of equal significance, disagreement between the various riparian states has effectively denied Jordan the use of water from the River Jordan itself. In this respect, the construction of low-level canals offtaking from the River Jordan within Jordanian territory is uneconomic due to the small area of land that can be commanded. The net result has been a considerable reduction in the amount of water that Jordan, as a major riparian state, had always assumed would be available for development. This 'deprivation' was one of the arguments used to justify the Arab League Plan to divert the Hasbani and Banias headwaters of the River Jordan along the western edge of the Gaulan Plateau to a storage reservoir on the River Yarmuk, since in the prevailing political situation this represented the only way in which Jordan could obtain its share of the flow of the main Jordan River (Smith, 1966). In turn, the Arab League Plan influenced the location of the Mukheibeh Dam: Mukheibeh, a far from ideal reservoir site, being located at a sufficiently low altitude (300 metres below sea level) to receive water from the Arab League diversion canal. In this respect, the Jordan Basin provides a very clear illustration of the way in which basic differences in perception—in this case of the amount of water available for development—have influenced the whole strategy of water resource development within the basin. Such unresolved differences effectively narrowed the range of development strategies open to planners in the various government agencies, resulting in the construction of 'rival' irrigation projects and in the inefficient and suboptimal use of available resources. These developments, however, have established patterns for resource use in the area for an indefinite period to come.

CONFLICTS IN WATER RESOURCE MANAGEMENT

In terms of developing the resources of such regions as the Central Jordan Rift Valley to their optimal productivity, there clearly exist resources that have been under-utilised through the failure to apply modern technologies to the management of land and water resources. Major increases in the productivity of the Ghor can be obtained, for example, through applying known technologies to transporting, distributing and applying water. This approach has become particularly critical in the Ghor since, in the short term at least, political circumstances have narrowed the range of options available to the planner. Even assuming the eventual completion of the Mukheibeh and Maqarin Dams as called for in the Yarmuk Project, water will continue to be a critical factor limiting the extension and intensification of irrigation agriculture in the southern Ghor.

In such circumstances, it is tempting for the planner to decide that improved techniques of water management can be adopted and to assume that, given the right incentives, farmers will be prepared to practise those methods of land preparation and water application best suited to achieving maximum physical efficiency. However, realistic planning must take into account the decision-making behaviour of individual farmers. Given the rapidly changing social and technological circumstances and the expanding economic opportunities that are likely to be found within the area of a development project, what criteria will farmers apply when making decisions about future agricultural operations? Given their perception of the situation and the means that they have at their disposal, how will individual farmers evaluate new technologies of land and water management? Such questions raise a number of issues that are fundamental to the development process, for unless the farmer perceives the situation in dimensions similar to those of the planner, he cannot be expected to share the planner's conclusions as to the best way of responding. In this respect, as White has pointed out, the actual process of development is broadly similar for both the farmer and the planner. Although one operates at the local level and the other at the regional or national level, both must make similar decisions about the same set of resources (White, 1963 and 1966b). Their final decision will reflect their appraisal of the available resources, the kinds of technologies and capital at their disposal, and their estimate of prospective returns (White, 1966b). It is probable, however, that quite different criteria will be applied in the decision-making process. The farmer, for example, may apply criteria that range from purely economic to largely behavioural. He may decide upon a strategy that produces a 'satisfactory' return, providing him with reasonable leisure time and other social returns as well as an adequate income. Alternatively, in the face of risk or uncertainty, the farmer may opt for a strategy that minimises the maximum loss that he can incur. The planner, on the other hand, may base his decisions on the assumption that farmers will act on rational economic grounds, seeking an 'optimum' solution, aimed at maximising returns. There will clearly be several different but equally good solutions for any given set of circumstances depending upon the criteria that are applied (White, 1966b). Unfortunately, it is all too frequently the planner who specifies the goals, assumptions and criteria underlying development programmes without recognition that those for whom they are doing the planning may apply very different criteria. Unless these differences are recognised and accommodated during the initial stages of planning, conflict is likely to occur whether one is concerned with an irrigation project in Jordan, the siting of a low-income housing project in a high-class residential area such as Forest Hills, New York, or the location of London's third airport.

The introduction of new technologies of water management in the area of the East Ghor Irrigation Project has been characterised by such a conflict. As already suggested, concern had been expressed at the outset as to whether the available water supply was adequate to meet the potential diversion requirements of the project area. This concern re-emerged in 1966 when, in anticipation of the availability of water from the Mukheibeh storage reservoir, work began on a 20 kilometre extension to the main canal and on a Zor Canal, the latter designed to provide those farmers in the Zor threatened by the increasing salinity of the lower Jordan with an alternative water supply. Thus, following the cessation of work on the Mukheibeh Dam in June 1967, the East Ghor Canal Authority was confronted with a situation whereby a vastly expanded area would have to be provided, at least temporarily, with an irrigation supply from existing sources. And inherent in this situation was the heightened danger of periodic water shortages within the project area at a time when farmers were beginning to adjust their cropping patterns to the availability of a dependable irrigation supply. The net result has been a tremendous emphasis on improving irrigation technologies; in particular, the entire thrust of the Agricultural Extension Service programme has been on the promotion of improved methods of land preparation and water application.

The Agricultural Extension Service recommends a bench–border method of applying irrigation water as being best suited to the soil and topographic conditions of the Ghor. This method involves the construction of parallel border levees across the slope of an irrigation unit, with laterals to distribute water to each carefully graded inter-border area. In this way, the head of water applied can be carefully controlled, while careful levelling ensures that the water is spread evenly. In addition to achieving a high degree of physical efficiency, the bench–border method involves a considerable saving in terms of the amount of labour expended in the actual supervision of irrigation operations. To promote the use of the bench–border method of application, various incentives have been provided in the form of cash grants and low-interest loans. Each farmer who accepts the advice and assistance of the Extension Service in the preparation of their unit receives a cash grant amounting to 50 per cent of the cost of development.

Despite the intensive efforts of the Extension Service, however, the overwhelming majority of farmers in the project area continue to use traditional methods of applying water to the soil. In the course of a farm survey undertaken in the project area in 1967, nearly 80 per cent of the farmers interviewed used traditional flood and furrow methods of irrigation on their farms (Manners, 1969). The majority of these used 'wild-flooding' techniques where water is allowed to flow freely across the field without any form of control. Unless levelling has been undertaken very carefully, many areas of the field will remain untouched during irrigation by this method of application while other areas receive too much water. At the same time, the uncontrolled flow of water down the field in the direction of greatest slope means that the water has little chance to infiltrate the soil—the net result is a very low application efficiency with wastage of most of the water and serious underwatering of the crops. If the farmer attempts to compensate for the latter by increasing the head of water applied, the danger of soil erosion and the development of gullying is considerably increased.

In the course of fieldwork in the Ghor, it became increasingly apparent that the farmers' 'priorities' were very different from those of the project planner and that, when viewed from the point of view of the individual farmer, the dimensions of the problem of water management are rather different. In particular, farmers within the project area were having to adjust to a situation in which there was no longer any limit to the water supply available but where water had now to be paid for—hence the cost of water was now a major factor in their calculations. It was, therefore, common to hear complaints about the cost of water and at the same time to see wasteful and inefficient use of water. The interviews revealed that farmers were not unaware of the Extension Service programme. A majority of farmers had received at least one visit from an extension agent in the preceeding year, while all were in constant communication with the water master. Most farmers were also aware of the cash grants and loans available for land improvement. However, it was also apparent that even those farmers who were best informed had a rather imperfect understanding of the advantages to be derived from the adoption of new techniques of water application. In particular, a significant number of the farmers commented on having seen the demonstration fields of the Extension Service lying 'under water', while others expressed apprehension that the suggested methods could lead to an increase in their irrigation needs. This response indicated a real concern over water consumption and reflected the fact that farmers were having to adjust to a situation whereby they had to pay for their water supply.

CONCLUSION

This brief account of some of the problems that have confronted planners within the East Ghor Canal Authority suggests another broad area in which geographical analysis can contribute to the development process. In this particular situation, more modern technologies could provide for a more efficient and productive use of available resources. Yet 'conflicts' have occurred (both at the national level and at the local level) due to differences in aims and perceptions amongst those involved in the development process. At the national level, such conflicts could have been avoided through a more realistic appraisal of available water resources. At the local level, such conflicts could be reduced given a better understanding of pre-existing patterns and techniques of resource use. Analysis of spatial differences in management practices, which is essentially geographical, together with the investigation of the social and economic criteria affecting management decisions and the the networks through which the decision-makers obtain their information, both of which profoundly influence the types of technologies that are adopted, would make it possible to predict more accurately the likely alterations in management practices in response to a changing environment. While such considerations are not solely geographical, geographers can make a major contribution to this type of analysis, thereby creating a much sounder basis for development programmes.

NOTES

1. It should be noted that in Jordan rainfall is measured for the period October to September: thus all the rainfall falling during the winter cropping season is included in the total for that 'water year'.

2. Rainfall probability was calculated from the expression:

$$d = \frac{x - \bar{x}}{\sigma}$$

where x = the critical value (200 mm); $\bar{x}$ = the mean value; and σ = the standard deviation. The derived values of probability, expressed as a percentage, were then plotted and an isopleth map prepared. The key has been prepared so that the probability of receiving both more than and less than 200 mm is indicated. The value of 200 mm was selected as critical as it represents approximately the minimum amount of rainfall necessary for the cultivation of a winter cereal crop in the Middle East by rainfed techniques (Wallen and Perrin de Brichambaut, 1962).

3. The length of the rainy season in the Middle East appears to be closely related to the northwards and southwards displacements of the Subtropical and Circumpolar Jet Streams in response to the changing pattern of solar heating of the earth (Wallen and Perrin de Brichambaut, 1962).

4. 1 dunum = approximately $\frac{1}{4}$ acre. The East Ghor Irrigation Project involves the gravity-flow diversion of the River Yarmuk. With the exception of a small dam on the Wadi Ziqlab (Figure 5.1) which has a live storage capacity of only 4·3 million cubic metres, no facilities have yet been completed within the project area for the storage of winter flood flow from the Yarmuk or the side wadis.

REFERENCES

Blaney, H. F. and W. D. Criddle (1950) *Determining Water Requirements in Irrigated Areas from Climatological and Irrigation Data* (US Department of Agriculture, Soil Conservation Service, Technical Paper No. 96; Washington DC: US Government Printing Office).

Blaney, H. F. (1955) 'Climate as an index of irrigation needs', *Water, The Yearbook of Agriculture, 1955* (Washington DC: US Government Printing Office), 341–345.

Chang, Jen-Hu (1968) *Climate and Agriculture: An Ecological Survey* (Chicago: Aldine).

Doherty, K. B. (1965) *Jordan Waters Conflict* (International Conciliation No. 553; New York: Carnegie Endowment for International Peace).

Government of Palestine (1946) *A Survey of Palestine* (Jerusalem: Government Printer).

Hays, J. B. (1948) *T. V. A. on the Jordan* (Washington DC: Public Affairs Press).

Lauritzen, C. W. (1955) 'Ways to control losses from seepage', *Water, The Yearbook of Agriculture, 1955* (Washington DC: US Government Printing Office), 311–320.

Lowdermilk, W. C. (1944) *Palestine: Land of Promise* (New York: Harper).

MacDonald and Partners, Sir M. (1951) *Report on the Proposed Extension of Irrigation in the Jordan Valley* (Westminster: Cook, Hammond and Keel).

Manners, I. R. (1969) 'The development of irrigation agriculture in the Hashemite Kingdom of Jordan with particular reference to the Jordan valley' (unpublished D. Phil. dissertation submitted to the Board of the Faculty of Anthropology and Geography, Oxford University).

Manners, I. R. (1970) 'The East Ghor Irrigation Project', *Focus* (American Geographical Society) **20,** No. 8, 8–12.

Penman, H. L. (1948) 'Natural evaporation from open water, bare soil, and grass', *Proceedings of the Royal Society* **A 193,** 120–145.

Smith, C. G. (1966) 'The disputed waters of the Jordan', *Transactions of the Institute of British Geographers* **40,** 111–128.

Smith, C. G. (1970) 'Water resources and irrigation development in the Middle East', *Geography* **55,** No. 249, 407–425.

Stanhill, G. (1961a) 'A comparison of methods calculating potential evapotranspiration', *Israel Journal of Agricultural Research* **11,** 159–171.

Stanhill, G. (1961b) 'The accuracy of meterological estimates of evapotranspiration in arid climates', *Journal of the Institute of Water Engineers* **15,** 7–20.

Stevens, G. C. (1956) *The Jordan River Valley* (International Conciliation 506; New York: Carnegie Endowment for International Peace).

Stevens, G. C. (1965) *Jordan River Partition* (Hoover Institute Studies 6; Stanford: The Hoover Institute on War, Revolution and Peace).

Thornthwaite, C. W. (1948) 'An approach towards a rational classification of climate', *Geographical Review* **38,** 55–94.

Thornthwaite, C. W. and J. R. Mather (1955) *The Water Balance* (Publications in Climatology, Vol. VIII, No. 1; Centerton, NJ: Drexel Institute of Technology, Laboratory of Climatology).

Thornthwaite, C. W. and J. R. Mather (1957) *Instructions and Tables for Computing Potential Evapotranspiration and the Water Balance* (Publications in Climatology Vol. X, No. 3; Centerton NJ: Drexel Institute of Technology, Laboratory of Climatology).

Wallen, C. C. and G. Perrin de Brichambaut (1962) *A Study of Agroclimatology in Semi-Arid and Arid Zones of the Near East* (Rome: FAO, UNESCO, WMO).

White, G. W. (1963) 'Contributions of geographical analysis to river basin development', *Geographical Journal* **129,** 412–435.

White, G. W. (1966a) 'The world's arid areas', *Arid Lands; A Geographical Appraisal*, ed. E. S. Hills (London: Methuen), 15–31.

White, G. W. (1966b) 'Deserts as producing regions today', *Arid Lands; A Geographical Appraisal*, ed. E. S. Hills (London: Methuen), 421–438.

Chapter 6

Rainfall Reliability and Agriculture in the Semi-Arid Tropics: The Case of the Jaguaribe River Basin, Ceará, North-East Brazil

Newton Rose

Introduction

The subject of this chapter is an inquiry into one aspect of the relationship between rainfall reliability and agriculture in a part of the semi-arid interior of north-east Brazil. In spite of rapid economic development on a national scale and the use of funds diverted from the more prosperous southern parts of the country to build an industrial sector and infrastructural elements, north-east Brazil remains one of the most poverty-stricken areas of the Americas. Chief among the physical problems of the region is uncertainty with regard to the amount and sequence of occurrence of rainfall. An attempt is made here to quantify the reliability of rainfall by testing whether the rainfall régime of the Jaguaribe basin can support the growth of a drought-tolerant crop, grain sorghum. The underlying question is: can this crop, not now grown in the study area (and little exploited within the larger region) be added to the mix of crops already utilised to provide an additional measure of security to the people?

The study area, the Jaguaribe river basin, is located in the interior and along the coast of Ceará state, occupying approximately one-half of the state's territory. The population of the basin (1·7 million in 1970) is overwhelmingly rural (71·5 per cent rural in 1970).[1] The growing of a drought-resistant variety of cotton and subsistence crops and the raising of cattle are the main economic activities.[2]

Mr. Rose is Lecturer on Geography and Geology in the Department of Geology, Washington Square College of Arts and Science, New York University.

Rainfall patterns

Within the basin, as throughout most of the interior of the north-east, rainfall sufficient to support agriculture falls only for part of the year. Depending on location within the basin, the long dry season begins in May or June and ends between December and February. During this season the landscape has a parched appearance and rainfall, though it does occur, is minimal. Besides this major irregularity in the distribution of rainfall, there are potentially two others with more serious consequences for agriculture. First, the rains may be unevenly distributed within the wet season (December, January or February to May or June), resulting in drought stress in certain plants and lower yields. Secondly, there are years with very low rainfall, which have brought a degree of notoriety to the region. A farmer born in the mid-1940s would have experienced the severe drought of 1958 as a teenager and, had he chosen to remain in the basin, the severe drought of 1970 would have found him as a young adult. If this hypothetical farmer then decided to spend the remainder of his life in the basin, or if he returned when the rains resumed, he could be almost certain to experience one or more such dry spells later in his lifetime.[3] In past years these droughts had tragic consequences in terms of human survival. The great drought of 1877–79 is estimated to have been directly or indirectly responsible for the deaths of half a million people (Hirschman, 1965). The lessons of the latest drought indicate, however, that modern communications, transportation, medicine and government aid programmes have, for the most part, eliminated the most severe of the consequences of drought.

BASIC CONSIDERATIONS

The data base

Rainfall data for the basin are available from several sources. The older and more extensive monthly totals (collected by the agency which was established to make the region less susceptible to drought, the Departamento Nacional de Obras Contra as Sêcas, or DNOCS) are now housed in the offices of the north-east development agency, the Superintendência do Desenvolvimento do Nordeste (SUDENE), in Recife, Pernambuco.[4] Long periods of daily data are available for only four stations within the basin. These data, collected by the Serviço de Meteorologia do Ministério da Agricultura (SM), are scattered between the agency's offices in Recife and Rio de Janeiro. Data from these sources may be regarded as a reasonably reliable basis from which to draw general conclusions.

The rainfall station chosen for the analysis presented here is Iguatu. The DNOCS monthly data series for this station covers a period of fifty years

(1912–1961) and there is also a thirty-year (1932–1961) daily series collected by SM.[5] In addition, long-term average data are available for temperature, insolation, relative humidity and wind speed. The municipality of Iguatu is centrally located within the basin and is an important agricultural producer. Further, its mean annual rainfall (718·0 mm according to the DNOCS data) is intermediate between the low and high mean totals for other stations within the basin.[6]

Grain sorghum

The attributes of sorghum which are most often exploited are its ability to grow with less water than a maize crop and its ability to survive a period of drought and to resume growth thereafter. In parts of China, India, and Africa the grain is a staple element in human diets. Almost all of the large crop produced annually in the United States is fed to livestock. Grain sorghum is believed to have originated in the north-eastern parts of Africa and from there to have spread to other parts of Africa, to India, the Middle East and China (Doggett, 1965). The crop has been produced continuously in the United States for over a century. Other leading producers in the Americas are Argentina and Mexico, as well as several Central American and Caribbean countries.

Sorghum has not been planted to any extent in the semi-arid interior of north-east Brazil. The author has seen the crop growing at an experimental farm within the Jaguaribe basin, but there is only one report in the available literature of the crop's being grown commercially in the basin.[7] Several persons and organisations concerned have made the suggestion that the crop be introduced into the region, but, so far as the Jaguaribe basin is concerned, little had been done toward this end by 1968.[8]

Although grain sorghum is usually thought of as being at the low end of the water-requirement spectrum, the crop grows well where rainfall is adequate as well as under irrigation. In addition, grain sorghum is able to withstand greater extremes of heat than most other grains. It can be grown on all types of soils and can tolerate considerable quantities of alkali or salt. 'Grain sorghums when flooded are not so subject to drowning out as corn (maize). However, grain sorghums habitually do not grow best under extremely wet conditions' (Ross and Webster, 1960, p. 2).

Temperature and water requirements of sorghum

As a rule grain sorghum cannot be planted where the soil temperature is below 21·1 °C. (Ross and Webster, 1960). While there are no published soil-temperature measurements for the Jaguaribe basin, Table 6.1 indicates that soil temperature will be adequate throughout the growing period.

Table 6.1 Mean, mean maximum and mean minimum air temperatures (in °C.) before and during the growing season at Iguatu, Brazil[9]

Month	$\overline{T}$	$\overline{T}$ max.	$\overline{T}$ min.
December	28·5	35·3	23·3
January	28·0	34·1	23·2
February	26·6	32·5	22·8
March	25·9	31·6	22·5
April	25·6	31·0	22·3
May	25·1	31·1	21·4
June	24·7	31·6	20·3

Setting a rainfall value below which full development of a sorghum crop cannot take place is somewhat arbitrary (Glover and Robinson, 1953). For the purposes of this analysis, a value of 304·8 mm has been accepted. Glover and Robinson cite 12 inches (304·8 mm) as the lower limit for the growth of grain sorghum and flint maize in East Africa. Hubbell also accepts 12 inches (304·8 mm) as the general lower limit for producing a grain sorghum crop (Hubbell, 1965). In both sources, 304·8 mm is given as the total water requirement for the growing season rather than for the entire year. The water demand for grain sorghum reaches a peak 'roughly in the 4- to 5-week period before and during flowering' (Glover and Robinson, 1953, p. 276) and falls off following this event.

STATISTICAL ANALYSIS

The rainfall régime at Iguatu

Figure 6.1 shows the histograms of (1) the mean monthly rainfall and (2) the monthly Penman estimates of the potential evapotranspiration at the Iguatu station.[10] The plot of the rainfall clearly shows the year divided into a wet part (January through May) and a dry part (June through December). The Penman values reach a high in October and are at their lowest in June.

Following a scheme contained in a United Nations study (United Nations, 1967), the data for precipitation and potential evapotranspiration were smoothed into curves for comparison and interpolation. (To enhance the clarity of the diagram these curves have not been reproduced in Figure 6.1.) The point at which the precipitation is equal to one-tenth of the potential evapotranspiration ($P=E_t/10$) is defined as the beginning of the period of preparation for the coming wet season. This point occurs at Iguatu on November 12.[11] Rains which fall during the preparatory period are con-

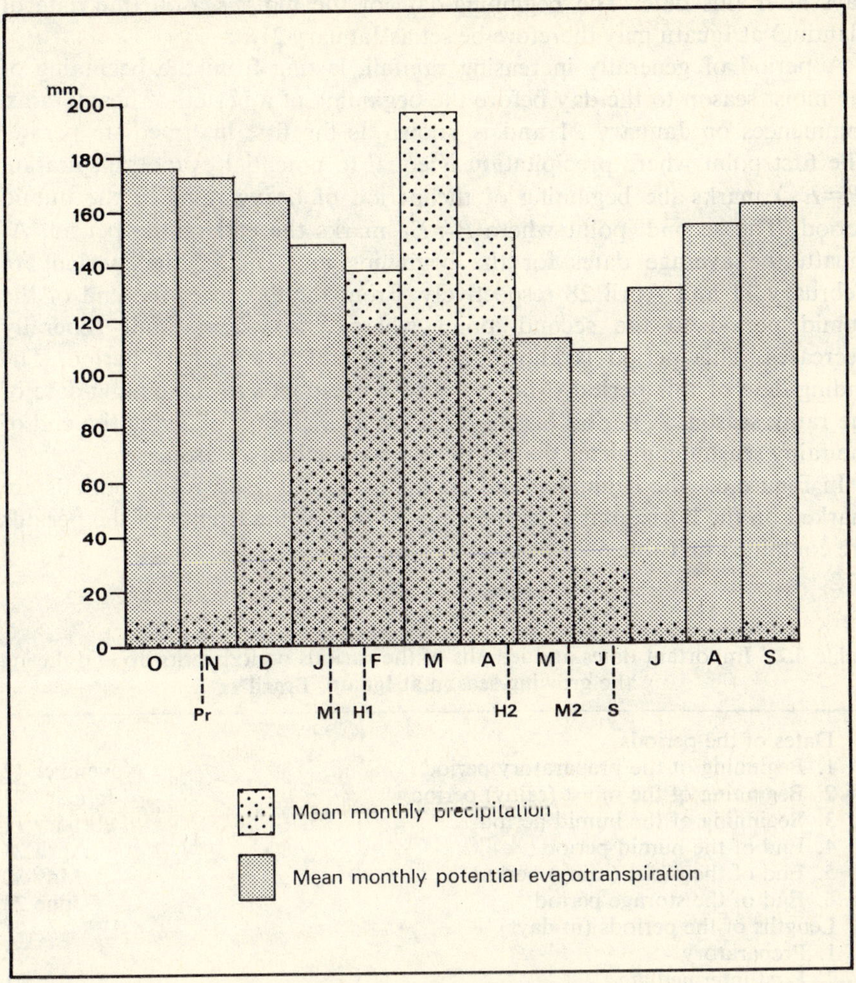

Figure 6.1 Mean monthly precipitation and mean monthly potential evapotranspiration at Iguatu, Brazil

sidered to evaporate and to be unavailable for the growth of annual food crops.

In the United Nations scheme, the first point where precipitation is equal to one-half of the potential evapotranspiration ($P=E_t/2$) is said to be the average beginning-date of the moist season, or the date on which planting of crops can take place. The point where $P=E_t/2$ at Iguatu is January 13, but the rainfall data show a considerable number of zero values for some days after this date. The beginning-date of the moist season (the date of planting) at Iguatu may therefore be set as January 21.

A period of generally increasing rainfall, lasting from the beginning of the moist season to the day before the beginning of a period of heavy rains, commences on January 21 and is known as the first intermediate period. The first point where precipitation is equal to potential evapotranspiration ($P=E_t$) marks the beginning of the period of heavy rains or the humid period. The second point where $P=E_t$ marks the end of the period. At Iguatu the average dates for the beginning and end of this period are February 11 and April 28 respectively. From the day after the end of the humid period to the second point where $P=E_t/2$, rainfall generally decreases. This period is known as the second intermediate period. The ending-date of this period (May 30) also corresponds to the ending-date of the rainy season. A period of soil-moisture availability following the end of the rainy season completes the entire water-availability season.

In Figure 6.1 the beginning and ending dates of these various periods are marked on the horizontal axis; the dates as well as the lengths of the periods are contained in Table 6.2.

Table 6.2 Important dates and lengths of the various periods prior to and during the growing season at Iguatu, Brazil

A. Dates of the periods	
1. Beginning of the preparatory period	November 12
2. Beginning of the moist (rainy) period	January 21
3. Beginning of the humid period	February 11
4. End of the humid period	April 28
5. End of the moist (rainy) period	May 30
6. End of the storage period	June 23
B. Lengths of the periods (in days)	
1. Preparatory	70
2. First intermediate	21
3. Humid	77
4. Second intermediate	32
5. Storage	24
Total for the rainy season (2, 3 and 4)	130

From the foregoing analysis it should be noted that the rainfall régime at Iguatu is non-homogeneous; that is, there are distinct periods of wet weather and dry weather and, moreover, during the wet season there are times of pre-heavy rain, heavy rain and post-heavy rain. This non-homogeneity in the rainfall régime has been an important consideration in selecting the statistical method employed.

The growing periods of sorghum

The growing periods of the sorghum crop are not directly equivalent to those defined on the basis of precipitation and potential evapotranspiration. Rather, the periods are those which occur during the vegetative cycle of the sorghum plant. There is, however, a basic comparability between the nature of the rainfall régime at Iguatu and the water needs of sorghum; sorghum requires a period of moderate moisture, followed by a time of abundant moisture and terminated by an abatement in moisture supply. To arrive at an acceptable estimate of the lengths of the different phases of sorghum growth, the author has relied especially upon the work of Glover and Robinson in East Africa. In the model constructed by these authors, the planting of grain sorghum and flint maize takes place during the early part of December, the peak water demand coming sometime in early February. Thus, the first period of growth lasts approximately 60 days, followed by a period of approximately 4–5 weeks of greatest water demand during flowering. Glover and Robinson imply that these lengths have some general validity and the author concurs that these periods (60 and 30 days) are acceptable for preliminary analyses. A period of approximately 50 days (arbitrary but corresponding generally to the known ripening time required by the crop) is accepted for the time from completed flowering to maturation. This gives a total growing cycle of approximately 140 days (extending beyond the 'second intermediate' period previously defined, to include 10 days in the dry season when stored soil moisture can be used by the plant). The 140-day cycle accepted for Iguatu falls midway between the lengths of the cycles recorded for Africa south of the Sahara, from 80 to more than 200 days (United Nations, 1967).

Moisture requirements within each period

The next step in the analysis is to set a definition of rainfall requirements within each of the three growing periods. Again it must be admitted that these definitions are arbitrary but that they were arrived at with regard to what seem to be reasonable arguments. Once established, these definitions should not only permit one to judge the possibility of growing sorghum in the basin; they will also provide a crude index of rainfall reliability.

In the author's preliminary work, the SM daily data were grouped into five-day periods or pentads. These pentads were then regrouped into periods corresponding to the lengths of the three growth periods, that is, periods of 60, 30 and 50 days. The problem was to determine critical values, within each period, below which sorghum could not properly develop.

The critical value in the flowering period was taken from Glover and Robinson: 'At this period grain sorghum can consume water at a rate equivalent to some 3 in (76·2 mm) of rain per month' (Glover and Robinson, 1953, p. 276). All values greater than 76·2 mm in the rainfall records were given the designation 'very wet' and all values equal to or lower than 76·2 mm were designated 'dry'.

The requirements for the first and third growth periods were determined by making certain assumptions. First, crop growth was considered possible in all years having a total of 304·8 mm per season, provided that no growth period had an extremely low total. Second, it was assumed that a moderate amount of rainfall in either the first or third periods could be made up in the other two periods. As a guide to these critical values, the average rainfall over the first 60 days of the growing season at Iguatu is 308·6 mm, and over the last 50 days, 121·6 mm. (The mean amount of rainfall over the 30-day second growth period is 190·5 mm.) The admittedly low values of 100·0 mm and 25·4 mm were accepted as the critical values in growth-periods 1 and 3 respectively. Thus, in the first period all values greater than 100·0 mm were given the designation 'wet', and any value lower than or equal to 100·0 mm was designated 'dry'. In the third growth period all values greater than 25·4 mm were designated 'wet' and all values equal to or lower than 25·4 mm were given the designation 'dry'.

It must be noted that a value below the critical value in any period automatically eliminates the possibility of growing sorghum in that year, as the model is constructed. However, 'moderate' rainfall (above the critical value but below the mean for that period) was accepted, provided that the total rainfall for the growing season was still equal to or greater than 304·8 mm. In actuality, the critical values as they have been defined eliminated successful crop growth in all years with a total seasonal rainfall of less than 365·5 mm.

The Markov chain method [12]

The non-homogeneity of rainfall within the growing periods at Iguatu provides the justification for the statistical method to be used below. The statistical device is a finite Markov process, which by definition is: 'a multi-stage stochastic or probabilistic process such that the conditional probability of being in any one of a finite number of states in stage $k+1$ (given the past history of the process), depends only on the state the process is in in stage k for any integer k' (Burford, 1967, p. 50).

A Markov process ('chain'), in its non-homogeneous form, is used to answer the question: what is the probability of successfully growing a sorghum crop at Iguatu? In other words, what is the probability for the occurrence of the sequence of events: 'dry' to 'wet' to 'very wet' to 'wet'?

The state of the process prior to the beginning of the moist season is always dry. As the model is constructed, the possible states of the process within the first 60 days and the last 50 days of the growing cycle are 'wet' and 'dry' (although the definitions of 'wet' and 'dry' differ from one period to the other). The possible states during the middle or second growing period are 'very wet' and 'dry'.

The application of the method

The thirty-year daily rainfall data were grouped into the 60, 30 and 50-day periods and arrayed in the following way (the values in the growth periods, or stages 2, 3 and 4, are in millimetres and stage 1 is always dry):

Table 6.3 Rainfall in the growth periods of a grain sorghum crop

	1	(60 days) 2	(30 days) 3	(50 days) 4	(140 days) sum
Year	mm	mm	mm	mm	mm
1932	—	245·6	15·6	72·2	333·4
1933	—	241·8	216·6	39·0	497·4
·	·	·	·	·	·
·	·	·	·	·	·
·	·	·	·	·	·
1961	—	144·3	224·2	147·5	516·0

A count was then made to determine how many times there was a transition from 'dry' in stage 1 to (1) 'wet' and (2) 'dry' in stage 2. Next, a count was made of how many times a transition was made from 'wet' and 'dry' in stage 2 to (1) 'very wet' and (2) 'dry' in stage 3. Finally, a count was made of how many times the transition was from 'very wet' and 'dry' in stage 3 to (1) 'wet' and (2) 'dry' in stage 4.

Once the count was made, the conditional or transition probabilities were calculated. The transition probability of going from 'dry' in stage 1 to 'wet' in stage 2 was found to be 0·93. The transition probability of going from 'wet' in stage 2 to 'very wet' in stage 3 was 0·86; and the transition probability of going from 'very wet' in stage 3 to 'wet' in stage 4 was 0·96. The joint probability of achieving the sequence of events leading to the growth of the crop is the product of the three transition probabilities, or: $0{\cdot}93 \times 0{\cdot}86 \times 0{\cdot}96$ which equals 0·77. The entire process is illustrated by means of a tree diagram (Fig. 6.2). The desired path is the uppermost 'branch' of the 'tree'; the other states and transition probabilities are shown as the other 'branches' of the 'tree'.

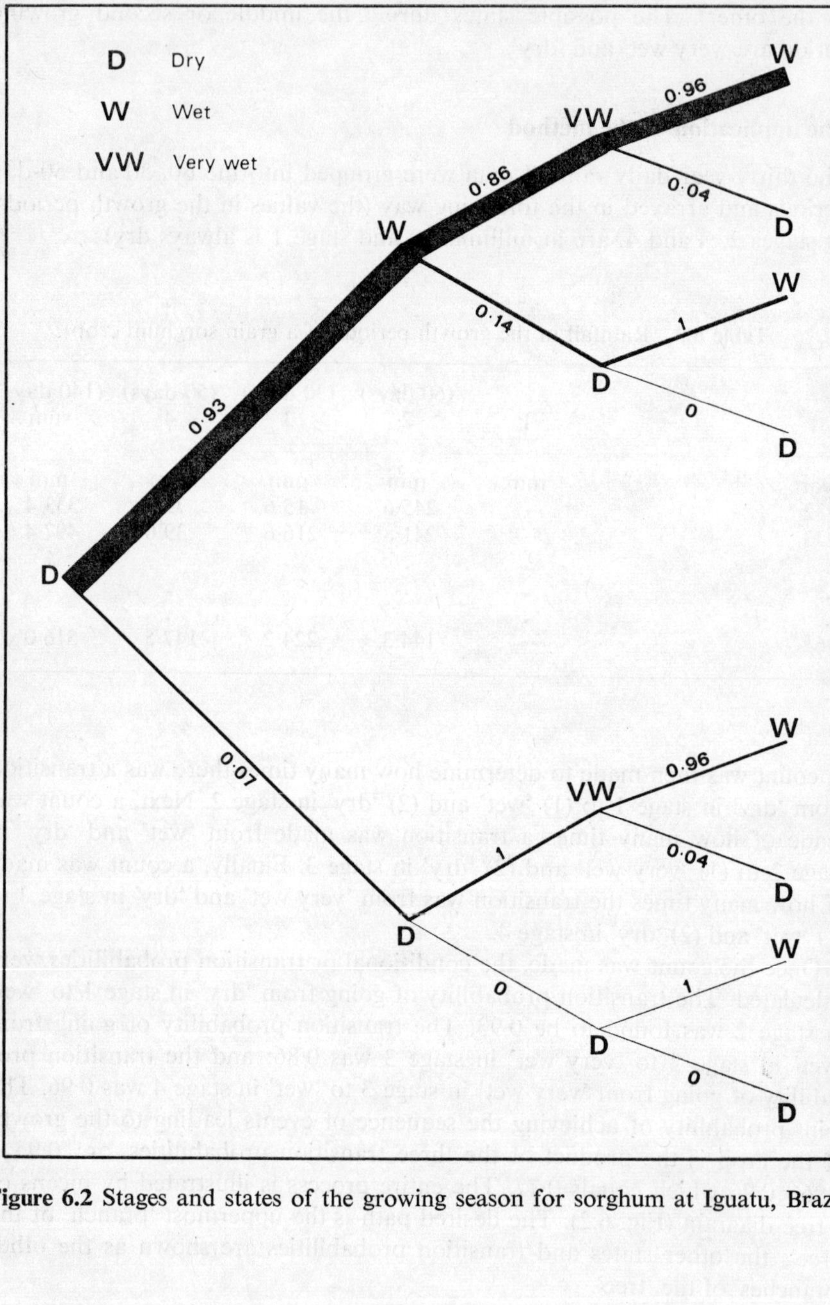

Figure 6.2 Stages and states of the growing season for sorghum at Iguatu, Brazil

DISCUSSION OF THE RESULTS

The results of the analysis indicate that a grain sorghum crop planted at the beginning of the moist season at Iguatu should be successful in 77 years out of 100. The probability is high enough to suggest that grain sorghum might be a crop to introduce into the area. The author suggests that, at this point, research be undertaken to ascertain the proper varieties to be grown, the best date of planting, the proper spacing of the plants, etc. The many varieties of sorghum allow one to select the variety best suited to actual field conditions. The failure of a crop planted at the beginning of the moist season might be compensated for by the planting of an earlier-maturing variety later in the growing season. Furthermore, 'the grain of most types of sorghum weathers badly and is often covered by a fungus growth when the weather is rainy or humid during ripening' (Ross and Webster, 1960, p. 1). If experiments prove this to be a problem, then the crop could be planted in the wet season in order for it to mature under more ideal conditions.

The social and economic problems of sorghum's introduction are, perhaps, more difficult to grapple with. The nutritional value of sorghum compares very favourably to that of other subsistence crops now grown in the basin, including manioc and maize (Miracle, 1966). However, whether it is possible to change the dietary habits of the people of the basin is a difficult question to answer. From the author's experience it appears that the crop would find its greatest use in the feeding of cattle and, perhaps, in the future, as the base of a broiler (poultry) industry.[13] With the advent of satellite communications, television sets in the public squares of the towns and steady supplies of electricity, governmental agencies have the opportunity at least to inform the people of ways in which their situation might be improved by the crop's introduction. Experimentation with sorghum and conveying these results to the general farming public would be the responsibility of federal, state and local agencies.

CONCLUSION

Using a set of 'low' definitions the author has calculated a measure of rainfall reliability and has demonstrated that a grain sorghum crop has a good chance of being grown in the Jaguaribe river basin. In the drier parts of the basin, a crop planted at the beginning of the moist season appears to be an additional hedge against the uncertainty in amount and sequence of occurrence of rainfall. It also appears that a crop planted during the moist season has a good chance of success in the wetter parts of the basin. Grain sorghum might also be used as an 'insurance crop' during years of low rainfall when the growth of other subsistence crops is impaired. The author suggests the need for further research into both the technical requirements and the social and economic acceptability of this crop.

NOTES

1. Calculated by the author from data contained in the preliminary results of the 1970 census.

2. The main subsistence crops are maize, beans, rice and manioc (cassava); sugar cane and bananas are also grown in the basin.

3. The author is preparing a paper on the problems of periodicity in the occurrence of drought years and the supposed relationship between these dry years and the sunspot cycle. At this point, he can report that there does not appear to be any direct relationship between drought in north-east Brazil and the sunspots.

4. SUDENE has been collecting its own data, on a daily basis, since 1962. In time, these data series may become the most accurate ones, since observers are trained in measurement and recording techniques. At the time of the research for this paper (1968–69) the series for Iguatu was too short to be useful.

5. The arithmetic mean of the 50-year DNOCS series is 718·0 mm and the standard deviation is 251·9 mm. The mean and standard deviation of the 30-year SM series are 774·6 mm and 204·8 mm respectively.

6. Of the DNOCS stations within the basin which were analysed by the author, mean annual rainfall ranged from a low of 507·3 mm (the standard deviation was 171·4 mm) to a high of 1054·5 mm (the standard deviation was 325·8 mm).

7. It was reported in Volume 6 (*Agricultura*) of the *Estudo geral de base do vale do Jaguaribe* that sorghum was being grown during the dry season on the 'more or less saline margins of temporary lakes in the lower part of the basin'.

8. In 1963 and 1964 the Grupo de Estudos do Vale do Jaguaribe experimented with grain sorghum at Iguatu. In the earlier year, planting did not take place until sometime during March, the reason apparently being lack of seeds. The following year, planting of several varieties took place at the beginning of March. The results indicated that sorghum could be successfully grown at Iguatu, but the two years were unusually wet 1175·1 mm and 1339·2 mm respectively) and cannot be taken as typical for predictive purposes.

9. These temperature data were taken from *Normais climatológicas da área da SUDENE.* The same publication contains data on insolation, relative humidity and wind speed.

10. Using the Penman method (Frère, 1972), estimates were calculated for both the potential evaporation from an open water surface (E_0) and the potential evapotranspiration (E_t); however, both estimates should be considered preliminary. In this regard, the author's calculation of the Penman estimate of the annual E_0 (1973·6 mm) falls within the limits (1890 mm to 2310) of another 'preliminary estimate'. The latter was made from Piche-evaporometer data; details can be found in Volume 5 (*Monografia hidrológica*) of the *Estudo geral de base do vale do Jaguaribe.*

11. All these dates are average dates for the beginning or ending of the various periods.

12. Several researchers have used Markov chain probability models to portray the behaviour of spells of rainy and dry weather. Fitzpatrick and Krishnan (1967) provide a short review of these usages—as well as of attempts which have been made to modify the method to achieve greater accuracy between estimated and observed sequences. Fiering (1967) has suggested that in certain cases the multiple-lag model may yield significantly better results than the Markovian models.

13. A broiler industry using sorghum as feed has been developed in the southern part of Brazil, according to a personal communication from Dr. Paul I. Mandell of the Food Research Institute, Stanford University.

REFERENCES

Bond, J. J., T. J. Army and O. R. Lehman (1964) 'Row spacing, plant populations and moisture supply as factors in dryland grain sorghum production', *Agronomy Journal* **56,** 3–6.

Brazil, Superintendência do desenvolvimento do Nordeste (1963) *Normais climatológicas da área da SUDENE* (Recife, Pernambuco), 15.

Brazil, Superintendência do desenvolvimento do Nordeste; and France, Association pour l'organisation des missions de cooperation technique (1967) *Estudo geral de base do vale do Jaguaribe* (10 vols.; Recife, Pernambuco: Grupo de estudos do vale do Jaguaribe), vol. 6, *Agricultura,* 101–103, 198–199; vol. 5, *Monografia hidrológica,* 45.

Burford, R. L. (1967) *Introduction to finite probability* (Merrill's mathematics and quantitative methods series; Columbus, Ohio: Merrill Books), 44–61.

Doggett, H. (1956) 'The development of the cultivated sorghums', *Essays on crop plant evolution,* ed. J. Hutchinson (Cambridge, England: Cambridge University Press), 50–69.

Fiering, M. B. (1967) *Streamflow synthesis* (Cambridge, Massachusetts: Harvard University Press), 87.

Fitzpatrick, E. A., and A. Krishnan (1967) 'A first-order Markov model for assessing rainfall discontinuity in central Australia', *Archiv für Meteorologie, geophysik und bioklimatologie,* **B 15,** no. 3, 242–259.

Frère, M. (1972) *A method for the practical application of the Penman formula for the estimation of potential evapotranspiration and evaporation from a free water surface* (Rome: Food and Agriculture Organisation of the United Nations), 1–20.

Glover, J. and P. Robinson (1953) 'A simple method for assessing the reliability of rainfall', *The Journal of Agricultural Science* **43,** 275–280.

Hirschman, A. O. (1965) *Journeys toward progress; Studies of economic policy-making in Latin America* (Garden City, New York: Anchor Books), 42.

Hubbell, D. S. (1965) *Tropical agriculture—an abridged field guide* (Kansas City, Missouri: Howard W. Sams International Corporation), 139.

Miracle, M. (1966) *Maize in tropical Africa* (Madison, Wisconsin: University of Wisconsin Press), 10.

Robbins, W. W. (1924) *The botany of crop plants* (2nd edition; Philadelphia: P. Blakiston's Sons), 193–203.

Ross, W. M., and O. J. Webster (1960) *Culture and utilisation of grain sorghum* (Agriculture information bulletin; Washington, DC: Agriculture research service, US Department of Agriculture), no. 218, 1–18.
United Nations (1967) *A study of the agroclimatology of the semi-arid area south of the Sahara in West Africa* (Rome: Food and Agriculture Organisation of the United Nations), 203–222.

Chapter 7

The Demographic Variable in Regional Planning in Kano State, Nigeria

Michael Mortimore

Introduction

Kano State in Nigeria contains one of the larger concentrations of population in tropical Africa. About 5·6 million people live within the state, but the greatest rural densities are found in the Kano close-settled zone, an area about 160 km long and 80 km wide surrounding the city of Kano. This zone (Figure 7.1) had 2·4 million inhabitants in 1962, living at more than 132 per km^2 and rising to over 264 per km^2 in the centre. The city or urban area of Kano had an additional 250 000 at that time and its present population exceeds 400 000. The significance of the close-settled zone may be measured against the fact that, unlike many better favoured areas of high density in tropical Africa, this region lies in the Sudan savanna zone and has a mean annual rainfall of only 750 mm to 900 mm which is available during a rainy season of five months or less. Standards of living are threatened by a continued growth of the rural population, at an annual rate which presently exceeds 2·5 per cent, because nearly all land that can be utilised with traditional techniques is fully occupied. Rising expectations flowing from the rapid extension of the money economy set the problem of development in bold relief. This chapter aims to survey the nature of the problem confronting regional planning in Kano State, the development strategies available and the role of research into the spatial aspects of the situation.

THE DIMENSIONS OF THE PROBLEM

Rural population growth

The present distribution and density of population must be seen as part

---

Mr. Mortimore is Senior Lecturer in the Department of Geography, Ahmadu Bello University, Zaria.

of a process. The lack of statistical information tends to discourage analysis in diachronic terms, especially before the beginning of the colonial administration in 1903. However, it is clear that quite slow rates of natural increase must have created a changing situation. Recently there have been improvements in public health and diet and much faster natural increase. Some indication of the changes which have occurred may be seen in Figure 7.2, which brings together an early estimate of population density in 1913 (Gowers, 1913) and the increasingly reliable results of the censuses of 1931, 1952 and 1962. These maps show the dimensions of the process in question during the present century, when, however, the context of the changes has been significantly affected by the colonial presence and aftermath, together with all the innovations that resulted. In order to account for the origin and early development of this process it is necessary to look back much farther. At this point the lack of data makes itself felt. A backward projection may be possible, but has not been attempted yet. Until it has been, we must be content with an explanation in broad outline only.

It will be noticed from Figure 7.2 that the growth of the close-settled zone during the present century displays two aspects. Demographic expansion by natural increase, and by immigration in some areas, has raised population densities progressively while simultaneously there has been an outward geographical extension of the central core area of highest density. The link between the two is provided in the nature of the subsistence economy. Nearly all families aim to grow their own staple food requirements. Since soil conditions are remarkably homogeneous, a fairly constant ratio between land and population emerged. Once all the available land in the centre had been occupied, the landless had to go to the periphery where a surplus still existed. Inside this migrating frontier of population pressure the more slowly rising population had to intensify its use of the land by progressively abolishing fallow, and applying greater quantities of manure.

Figure 7.2c shows three concentric growth zones that existed between 1931 and 1952. Inner and outer zones (I and III) of rapid growth (2·2–2·3 per cent per annum) were separated by an intervening zone (II) of slower growth (1·4 per cent). Zone III appears to have contained the frontier where land could still be obtained by migrant farmers, while Zone I benefited from promixity to Kano city (see below). In the period 1952–62 (Figure 7.2d), however, this pattern gave way to a broad regional division between north and south. The mean annual district growth rate in the north was 4·6 per cent and in the south 1·6 per cent. The frontier had by then moved out of the state southwards and the symmetry of the pattern was distorted as different ecological conditions were encountered (Mortimore, 1968).[1]

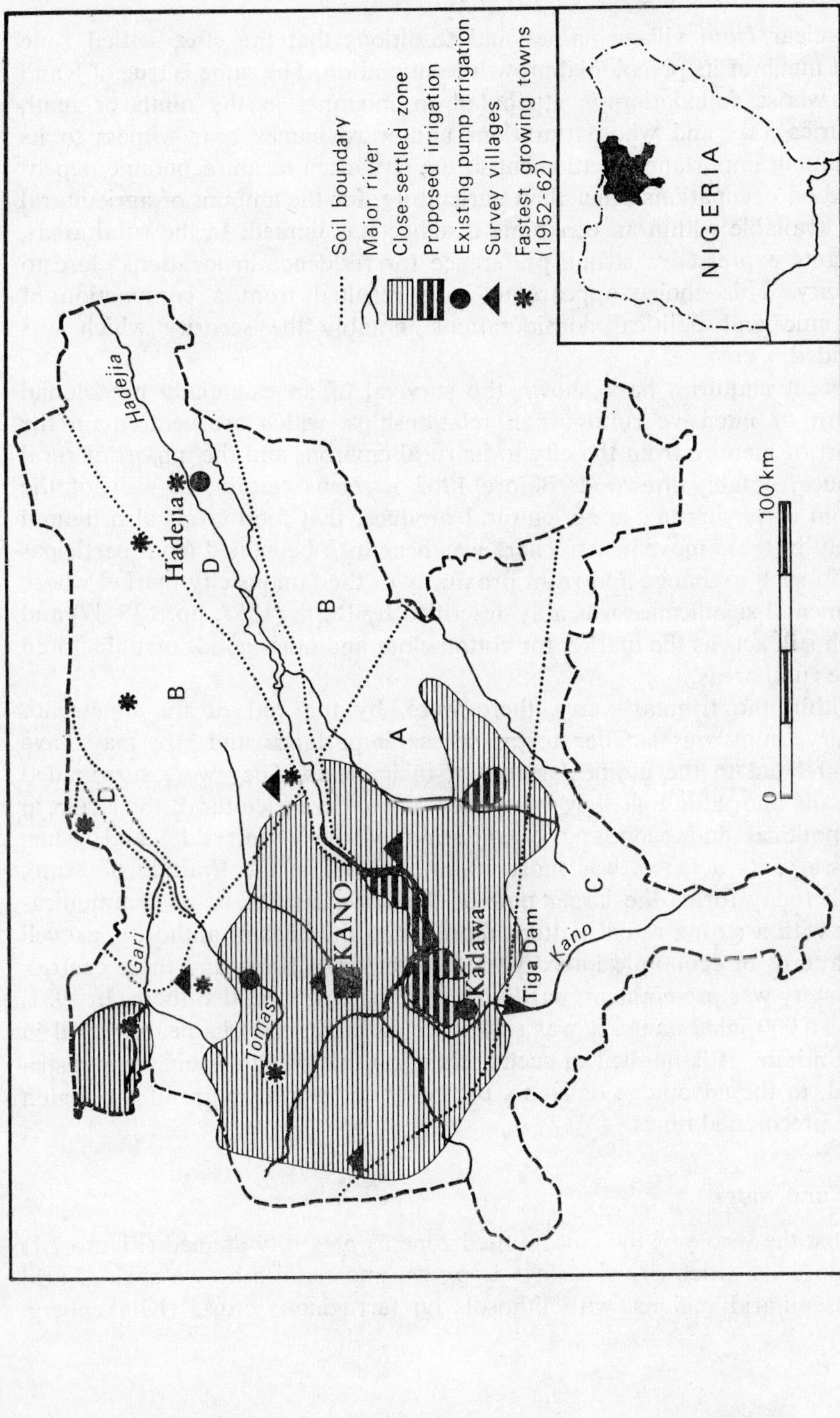

Figure 7.1 Kano State, Nigeria. Soil groups: A, brown and reddish-brown soils with lithosols; B, brown and reddish-brown soils on dune sands; C, ferruginous tropical soils; D, halomorphic and hydromorphic soils

Urbanisation and migration

It is clear from village names and traditions that the close-settled zone owed much of its precolonial growth to migration. The same is true of Kano city, whose foundation is attributed to migrants in the ninth or tenth centuries A.D., and whose traditions and ward names bear witness to its continuing importance. Settlement in the city often meant economic dependence on occupations other than agriculture, for the amount of agricultural land available within an economic distance was limited. In the rural areas, migrants expressed a strong preference for residence in locations close to the city. This choice appears to have resulted from a conjunction of economic and political considerations, notably the security which was provided.

Recent enquiries have shown the survival of an essentially precolonial pattern of intensive rural–urban relationships, which are centred on the export of manure from the city to its rural environs and the import of rural produce, notably firewood. Before 1903 it seems certain, in view of the custom of paying tax in agricultural produce, that food crops also figured largely in these movements. There was benefit to be gained from participation in such exchange and from proximity to the famous city market whose continental significance was ably described by Barth (1857, pp. 125–47) and which still acts as the market for cotton cloth and other goods manufactured in the rural areas.

Within the tributary area there were, by the end of the nineteenth century, numerous smaller towns whose importance and size may have been related to the political status of their rulers. They were surrounded by walls and, although they were predominantly agricultural, their role in the political and economic organisation of the countryside, and, when necessary, its defence, was most important. Within the Emirate of Kano, which today forms the larger part of the state, a network of communications with a strong radial pattern represented the lines of authority, as well as patterns of economic interdependence, between Kano and these centres. Kano city was pre-eminent, partly because the chiefs resided there. In 1931, with 96 000 inhabitants, it was sixteen times larger than its nearest rival in the Emirate. It is implicit in such a situation that wealth is unevenly distributed, to the advantage of Kano, the economic growth point of the region since precolonial times.

Soils and water

Almost the whole of the close-settled zone as presently defined (Figure 7.1) lies on soils which are classified as brown and reddish-brown soils of arid and semi-arid regions, with lithosols on ferruginous crusts (Klinkenberg,

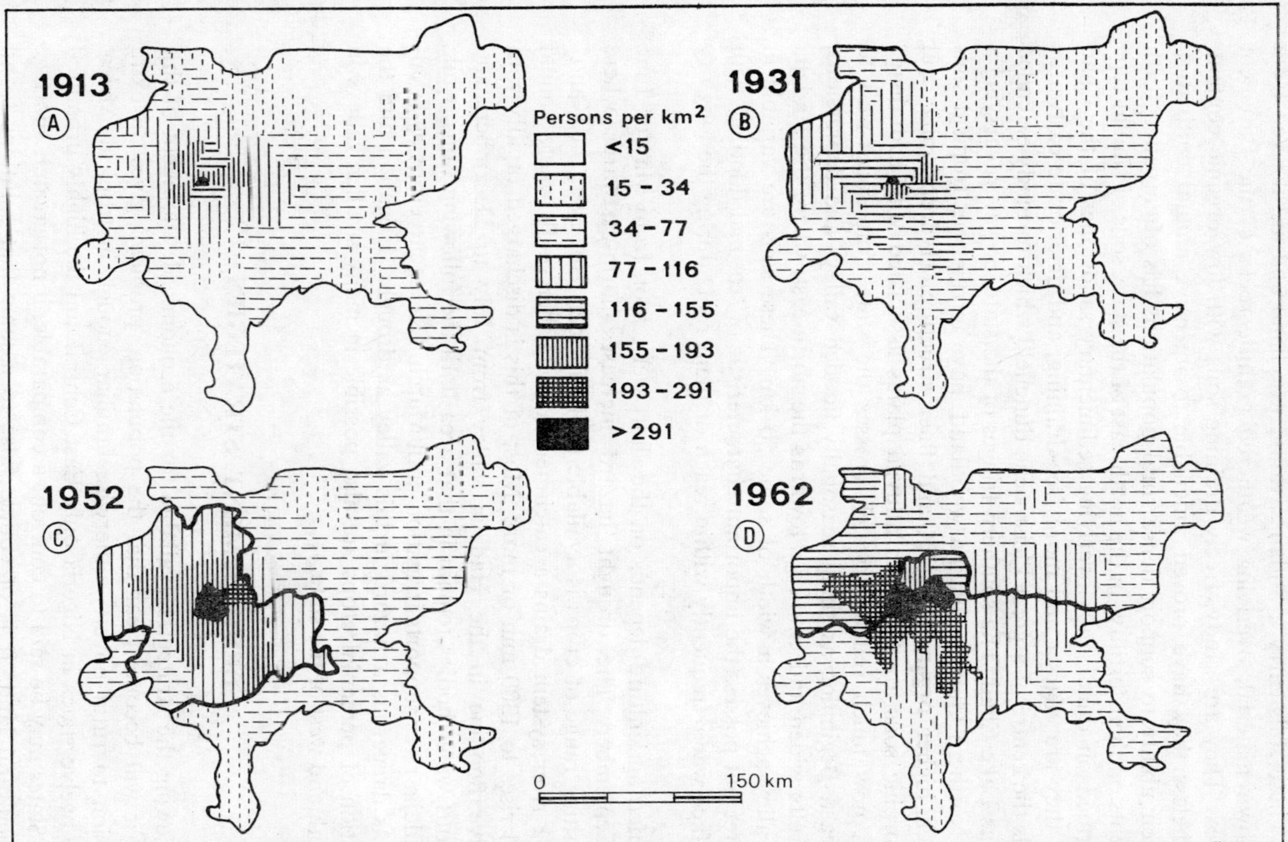

Figure 7.2 Density of population in Kano State, Nigeria. The growth zones of 1931–52 are shown on C where Zone I is solid black and the heavy line separates Zones II and III; those of 1952–62 are shown on D where the heavy line separates northern and southern zones (see text)

1970). These soils carry a Sudan savanna woodland and are derived from windblown material, overlying weathered granitic rocks of the Basement Complex. They are sandy, freely draining soils with low organic content. Nevertheless they have proved amenable to intensive annual cultivation and their ability to support dense rural populations has often called forth comments on their fertility. In the north-east of the state, soils of similar type are formed on dune sands overlying sedimentary sandstones of the Chad Series; they are coarser and more free-draining, and deteriorate eastwards towards the centre of a great Pleistocene dunefield which formerly extended as far as Lake Chad. The southern fringes of the state, however, carrying a richer Northern Guinea savanna woodland, have ferruginous tropical soils of much higher organic content and finer structure. Hydromorphic and halomorphic soils are found in alluvial plains associated with a drainage system, now much reduced, which crosses the state from south-west to north-east. Beginning as small seasonally flooded valleys and depressions, these belts widen and converge towards the north-east, where the Hadejia river valley achieves a width of some 50 km. These soils are variable in character but possess the important characteristic of seasonal flooding, with subsurface water frequently within reach of plant roots during the long dry season.

Rainfed agriculture depends on the short wet season from May or June until September. The drought hazard increases north-eastwards where diminishing range of crops is available to the farmer. However, the Kano–Hadejia river system drains a catchment area farther south where annual rainfall rises to 1300 mm or more. Most of this is discharged rapidly with extensive flooding in the Hadejia valley from July to December, and numerous water bodies remain in that area until the following rains, while the Hadejia river supports irrigation until March. Damage to crops by flooding is as dangerous a hazard in this valley as drought. Irrigation and the cultivation of perennial crops are also possible in many smaller valleys in the south and west of the state.

DEVELOPMENT STRATEGIES

In reviewing the strategies available for the economic development of this region it will become clear that the population 'problem' gives an extra dimension to rural poverty otherwise familiar enough in tropical Africa. Of the twelve states in Nigeria only East Central and the highly urbanised Lagos States may be said to confront a comparable, if not greater, shortage of agricultural land. It is of course false to assume that solutions to all economic problems can be found within the boundaries of the state. But again the size of the population is of overriding significance. There is no assurance yet in Nigeria that massive interregional migration will be politically acceptable, even on the scale allowed before the civil war of 1967–70, or that national urban centres can hope to absorb successfully in the near future a manifold increase in the number of migrants. So it will be assumed in the present discussion that while migration out of the state (see below) will continue in a relatively small scale, the basic objective of regional planning must be the more productive use of existing resources.

Urban development

The rate of increase in the population of the state between 1952 and 1962 was 2·6 per cent (Table 7.1), although a higher estimate of 2·8–2·9 per cent

Table 7.1 Population projections for Kano State

	(1) Population 1962	(2) Annual increase 1952–62	(3) Estimated population 1972	(4) Projected population 1982	(5) Projected population 1992
		%			
(a) Kano State	4 329 300	2·6	5 612 000	7 279 000	9 436 000
(b) Kano city	249 000	6·7	436 133	1 034 000	2 452 000
(c) Other towns	212 000	2·4	272 300	349 900	449 500
(d) Fastest growing towns	52 800	8·9	124 100	290 700	682 000
(b) and (c) as a percentage of (a)	10·5		13·5	17·5	23

is available (Green, 1970). If the lower of these is maintained (a conservative assumption) the population will approach ten million in the next twenty years, an increase of about four million. It is of some value to estimate the proportion of this increase which could be absorbed if urban development within the state were to continue at its present rate.

Kano city has grown faster than the average for the state since 1931. However, the rate of growth observed for the period 1952–62 was not maintained, owing to the communal rioting in 1966 which led to considerable loss of life and the exodus of at least 35 000 Ibos and others from the town. However, annual growth now appears to exceed 6·7 per cent and is probably nearer 9 per cent (Green, 1970). The population may therefore reach a million before 1982 and (if this rate is maintained) two and a half million by 1992. Four alternative projections made before the events of 1966 all anticipated a population of at least one million by 1982 (Trevallion, 1966).

Although Kano has become one of the major industrial growth nodes of Nigeria, the capital cost of creating industrial employment is high and is certain to remain limiting. An unknown proportion of immigrants in the town fail (or do not seek) to obtain wage employment, but the risk is considered justified by the difference between rural and urban incomes and by the low status of agriculture. Many live instead by exploiting the tertiary sector to the full. Small-scale business, trading, manufacturing and services support more people than is generally realised. Considerable new wage employment (and thus patronage for the tertiary sector) has resulted from public investment, on a scale hitherto unknown, following from the establishment of the capital of the new state in 1967. Such developments are able to employ workers from a zone of rural settlement within commuting distance by foot or cycle from Kano. Many of the commercial benefits must also be diffused to the villages close to Kano by the mechanisms mentioned above. But beyond this zone, participation in the fruits of urban growth can only be obtained by migration.

It is evident from the projections in Table 7.1 that even if its present high rate of growth is maintained, Kano city cannot hope to accommodate more than a part of the anticipated increase in the population of the state, let alone contribute towards a net reduction in pressure on the land. Furthermore, like most other major African cities, Kano is experiencing considerable social and infrastructural strain, quite apart from problems of employment, in maintaining its present rate of growth. In view of the size of these problems, nobody believes that Kano should be allowed to continue growing at such a rate.

A vigorous programme to develop smaller urban centres in the metropolitan region has been recommended (Trevallion, 1966; Green, 1970), and some such development, coupled with the devolution of local government to new headquarters, is part of official policy in the state. Thus far too little evidence is available to show whether smaller towns can compete effectively with Kano for migrants or private investment. The primacy of Kano city in the hierarchy of settlements reflects a strong tendency towards polarisation in the region, implicit in the political basis of precolonial settlement. The construction of the railway (1911), the attraction of industry (since World War 2) and the growth of the administration have all served to consolidate its pre-eminence. Therefore if such a policy is to succeed strong governmental incentives will be required.

Between 1952 and 1962 the thirty-five larger settlements in the state, excluding Kano city, grew at a rate not significantly different from that of the state as a whole to comprise a total population of 212 000. Many of these settlements are small old towns without modern sectors, growing mainly by natural increase, and dependent to a large extent on agriculture. Table 7.1 shows that, unless this rate is exceeded, these towns will not reach half a million inhabitants by 1992 and that they will be responsible for a diminishing proportion of the population living in towns. However, the growth of some towns was much faster. The eight fastest growing towns, all in northern areas (Fig. 7.1), grew rapidly between 1952 and 1962 and, although their present size is small, they and others like them that have also succeeded in attracting investment and migrants may add considerably to the scale of urbanisation in the future.

It should be remembered that a large component in the growth of Kano city, and in the future probably of smaller towns also, consists of immigrants from outside the state. Governmental action to encourage the growth of smaller towns is seen as a means of slowing immigration to the metropolis. If this fails, and both metropolis and other towns continue to grow at the 1952–62 rates, the proportion of the state's population living in them will more than double to 23 per cent (Table 7.1). This represents a minimum, for it does not take account of smaller settlements growing to the urban threshold nor of the increasing part which will be played by the fastest-growing towns. But, even if these projections are exceeded, it is unlikely that during the next twenty years urbanisation will provide a means of stopping the growth of the rural population, and the deterioration in the supply of agricultural land.

Rural development

We conclude, therefore, that if living standards are to be raised, rural development must take place in order to benefit the majority of the population (perhaps three in four) who will still be living outside the towns in twenty years' time. The four main sectors which offer possibilities are agriculture, stockraising, forestry and rural manufacturing. The last will not be considered here since little thought seems yet to have been given to it. Stockraising and forestry offer possibilities as subsidiary activities integrated with agriculture, but, except in certain localities unsuited to cultivation, they are unlikely to take the centre of the stage because of their extensive land requirements. Rural development in Kano State therefore resolves itself mainly into a problem of raising farm incomes, for which there are several well-known strategies available, including:

(1) Raising efficiency under existing techniques;

(2) Pricing and subsidy policies designed to make agriculture more profitable;

(3) Infrastructural improvements, especially new roads to remote areas;

(4) Cultivating more land; and

(5) Intensifying the use of existing land.

The first two, although they are likely to operate unevenly, do not require spatial location decisions to be made at the policy stage and therefore will not be discussed here. The last three are of greater significance from the viewpoint of regional planning. However, they must operate within the general economic climate of agriculture in northern Nigeria, and three observations are relevant here. First, it has been shown that farmers respond to economic incentives by increasing output or adopting new techniques provided that they do not conflict with their basic objective of securing their subsistence requirements (Norman, 1972). Second, it is generally recognised that the government fails to create sufficient incentives for agricultural producers. Third, the resources hitherto given to research and extension have not been adequate. Although successes have been achieved, especially in the cash production sector, they have failed to revolutionise the structure of agriculture. Indeed it is doubtful whether sufficient proven innovations for such a revolution are yet available. The need for better road communications is recognised and new roads are being constructed. At the present stage almost any selection of routes can be justified and this strategy does not present many problems.

Cultivating more land may be achieved in two ways. The first is the enlargement of existing holdings to permit animal or powered mechanisation. The most successful move towards such larger-scale methods has been the growth of ox-ploughing, with strong governmental assistance in some areas. But this only serves to emphasize the inadequacy of holdings, which at 1·5 ha or less in the central close-settled zone, are far too small to justify mechanisation. Customary tenure permits the enlargement of holdings by purchase, but this requires relatively large sums of capital, which is scarce at economic rates of interest. It can only be achieved by dispossessing smaller farmers, which will increase unemployment.

The second way to cultivate more land is to colonise unoccupied areas. Although relatively low population densities are still to be found in peripheral areas of the state, these reflect a lower proportion than in the close-settled zone. Forest reserves occupy extensive areas, formerly uninhabited, in the south. Thus although some vacant land undoubtedly exists, it is available in quantities too small to provide much hope for large-scale colonisation. Rural migration into southern districts has slowed noticeably since 1952, while in northern districts the very rapid immigration which has developed since then (from Niger) may be expected soon to slow as land ceases to be available. In any case this is very poor land and was largely uninhabited during the nineteenth century. Kano farmers now travel in search of land far beyond the state boundary to such areas as Gombe in North-Eastern State, 300 km or more from Kano. No information is available on the numbers involved, but it serves the interests of the state that such migration should continue to provide an outlet for some farmers without adequate land.

Colonisation thus far has consisted simply of the transfer of traditional techniques to new areas, freed from the constraint of land shortage. Government participation in the process is required if it is to achieve a major increase in incomes. The largest area of potential agricultural land still awaiting effective development within the state is in the Hadejia valley. Flood hazards have prevented its full use under indigenous techniques, notwithstanding the widespread use of rice. Complex geomorphology and soils raise the costs of levelling for irrigation schemes and necessitate detailed study before land-use recommendations can be put forward. Existing land use is not understood and the area provides important dry-season grazing grounds for Fulani cattle. Nevertheless its potential may be considerable.

Intensification of land use has been the solution sought historically in answer to the problem of an increasing shortage of land. Apart from the intercropping of grains and legumes and the raising of labour inputs in weeding, the principal means of achieving this has been the use of manure derived from animal droppings, ash and compound sweepings. Average applications of manure reach 3·7 tons per hectare in central areas, although they drop considerably at greater distances from Kano city, where urban supplies are not available and holdings are somewhat larger. Such intensification makes use of a stock of knowledge within the indigenous system, and appears to result directly from the growth of population density. Today chemical fertilisers are also used, but they cost more. With their help it remains to be seen whether continued intensification on this pattern can increase productivity fast enough to cope with both population increase and the additional demands made by rising expectations among the rural population. Increased extension inputs will be of critical importance.

Alternatively, the climatic constraint can be partially overcome by irrigation, by which means it is possible to grow two crops a year instead of one. Irrigated land has been confined until now to the valleys of perennial streams, perhaps 5 per cent of the total area, where the shaduf is employed for lifting water 2 to 4·5 m on to smallholdings. Such *fadama* agriculture is supplementary to the cultivation of rainfed land in the wet season and is fully commercial, being concerned with the production of vegetables, wheat and fruit for urban markets. *Fadama* land beyond the reach of such markets remains unexploited agriculturally, except where it can be used for perennial crops of sugar (which grows without irrigation) and sometimes for cassava.

Pilot irrigation schemes on the Tomas, Gari and Hadejia rivers are currently being extended under the direction of the Ministry of Agriculture, especially at Hadejia where pump-irrigated wheat now occupies over 300 ha. The Hadejia valley presents problems for this type of development, as noted earlier, but irrigation of existing upland farms (i.e. outside the seasonally flooded *fadama* lands) is an attractive possibility arising from a recent study of the Kano river basin (United States Agency for International Development, 1968). The state government has embarked on a pilot project at Kadawa, which is irrigated from a small dam on a nearby tributary of the Kano river. Without awaiting conclusive results, work has commenced on a major dam at Tiga which will store and divert sufficient water to irrigate 26 000 ha in Kura District and convert them to perennial cultivation. The preliminary survey suggests that this figure can be more than doubled by extending irrigation to other localities (Figure 7.1), with supplementary storage and diversion dams. Although the creation of large dams necessitates some resettlement and the loss of valuable land, these heady proposals for doubling the use made of large areas of existing land are a logical outcome of the conjunction of cumulative demographic growth and an inelastic supply of agricultural land.

RESEARCH FOR REGIONAL PLANNING

In such a context, research may contribute in one or more of four ways: description, explanation, selection of developmental objectives and determination of policies. These have been termed the positive, hypothesis-testing, normative and policy-making phases of applied research (Norman, 1971). Early published research on spatial aspects of development in Kano emphasises the first two, and was concerned with population distribution (Prothero, 1959), the relations between population distribution patterns and systems of agriculture (Grove, 1961; McDonnell, 1964) and the geomorphology, soils and vegetation of north-eastern Kano (Grove, 1958; Pullan, 1962; Clayton, 1957). Some consideration has also been given to problems of soil erosion (Dennison, 1956). Most normative and policy-oriented work was undertaken by government departments, such as agriculture and forestry, and tends to be recorded, if at all, in the files of the departments concerned.

In 1962 a Greater Kano Planning Authority was set up with instructions to prepare a twenty-year development plan for an area which includes Kano City and the two adjacent districts of Ungogo and Kumbotso. The preparation of this plan (Trevallion, 1966) by a small professional staff initiated socio-economic research of a more specific nature than had previously been undertaken in the Kano area. In the same year Ahmadu Bello University was established at Zaria, 175 km away, with a department of geography. The conjunction of these two developments, otherwise unrelated, made it possible for geographers to undertake research projects on behalf of the Board, and although the Board ceased work shortly afterwards, the initial objective of providing information inputs for urban and regional planning continued to guide the interest of the department in Kano.[2]

The most urgent priority was to establish the distribution of the population from the preliminary returns of the census of 1962, at a scale larger than Prothero had been able to undertake for that of 1952. Standard procedures for mapping population were employed, insofar as the nature of the information allowed. This census was subsequently annulled and followed in 1963 by another, whose results were inflated for political purposes and must therefore be discarded, so the 1962 map remains the best approximation available until the results of the 1973 census become available (Mortimore, 1965).

Standard mapping procedures were also applied to the preparation of urban and rural land-use maps for metropolitan Kano, based on air photography of 1962. Such photography taken during the dry season provides only limited information about agricultural land use, but nevertheless the first fully reliable figures for the inner close-settled zone are now to hand for the main categories of land use. It is from such basic information that the impact of rapid urban expansion into the agricultural environs of Kano city can be anticipated, for 85 per cent of the land is under annual or continuous cultivation. Dislocation and resettlement on the peri-urban fringe can now be quantified, in terms of land-use changes over a short period, because the area was photographed again in 1969.

Land-use mapping requires complementary analysis of the rural economy in the field and information was urgently required about the nature of the rural economy within Ungogo and Kumbotso Districts. In such exploratory situations the case-study approach has a good deal to recommend it. It must assume, without conclusive evidence, some degree of typicality for its findings and it may fail to detect important area differentiations. But its power to detect intersectoral and other types of relationship in the context of the family or village is strong. A study was therefore made of three villages some 8 to 16 km north of Kano city, with the aim of uncovering as many aspects of the village economy as possible in the time available, but laying emphasis on systems of resource use (Mortimore and Wilson, 1965). This technique has subsequently been applied successfully and with much greater intensity by an interdisciplinary team working elsewhere in northern Nigeria (Norman, 1972; Goddard, Fine and Norman, 1971).

In attempting to move from basic analysis toward explanation of the Kano close-settled zone it becomes necessary to test hypotheses arising from the case studies by means of sampling procedures in a larger area. This work is focused on two aspects: the graduated changes which occur between centre and periphery and the functions of the centre itself. In studying the first, a preliminary comparison between the case study and some outlying villages revealed a number of contrasts relating to the intensity of land use, which help to confirm the existence of a relationship between population density and land use (Mortimore, 1970). This suggests that profitable use may be made of a model of agricultural growth in which population increase acts as an independent variable and causes progressive changes in the intensity of land use (Boserup, 1965). The utility of such a model has been argued for northern Nigeria as a whole (Mortimore, 1971) and it can be tested in two ways: by the examination of evidence relating to the historical antecedents of the contemporary situation and by the extensive sampling of land use and related functions in the wider region. This work is in progress and no conclusions are yet available.

The second aspect concerns the identification of the functions of the centre. Without Kano city, the close-settled zone in its present form is inconceivable. The spatial relationship between town and country is manifested in various forms and further simple investigation into one of these, the movement of pedestrians and donkeys carrying goods into and out from the sixteen gates of the walled city of Kano, has thrown some light on the precolonial economic relationship and the nature of the changes brought about by the growth of modern Kano (Mortimore, 1972).

Some research oriented towards specific planning objectives is presently directed towards three problems arising from the developments discussed earlier. The irrigation developments planned for Kura District are being pursued with very little information about the ability of farmers in the area to adapt their social and economic habits and in particular their inherited system of land use. This system is now under study in the neighbourhood of the Kadawa pilot project, based on interviews with a stratified sample of the farming population.[3] Adaptation to change is facilitated by a universally accepted heirarchical structure of authority extending from the Emir to the farmer, but nevertheless there are sufficient examples of agricultural development projects both in Nigeria and elsewhere (e.g. Baldwin, 1957), which have experienced difficulties through neglecting this issue, to lend some urgency to such studies. In the Hadejia valley a second project aims to assist with the provision of a classified evaluation of land resources in this little-understood area. Basic geomorphological and soil classifications already exist and land use is being intensively classified and mapped, using these units, in a small area near Hadejia, with a view to projecting the conclusions to the whole valley, subject to extensive testing.[4] Since future land-use recommendations are likely to be made within the framework of the geomorphological–soil units, existing land use in terms of these units forms the most satisfactory basis for evaluating the disruption likely to be caused by, and the cost-effectiveness of, such recommendations. A third and smaller project is concerned with gathering data on the farming and landholding interest in Kano city, which is indicating signs of stress in the forms of speculative purchases of land, mounting prices and irregular practices in anticipation of government compensation, all consequent on the increasingly rapid expansion of the built up area of the city.

Studies directed primarily towards spatial aspects form only a part of research which is relevant in such a developmental context. No mention has been made of other work in Kano in the physical and social sciences.[5] Within the context of regional planning, several deficiencies may be noted which are not unique to this situation: first, insufficient resources are available for research, especially in the social sciences, either under government or independent (usually university) sponsorship. Second, research is inadequately integrated into the decision-making procedure, especially when conducted by independent workers. The development projects under way in the state have been selected intuitively by various ministries on the basis of their technical feasibility.[6] Third, regional planning is weakly conceived and little attention has yet been given to its spatial aspects. It comes within the sphere of the state's Ministry of Economic Planning, but urban planning in Kano city is supervised now by the Kano Metropolitan Planning and Development Board. Planning powers and procedures are inadequate and expertise is scarce.

CONCLUSION

In the absence of much of the basic information required for regional planning,[7] it has been suggested that the time-honoured techniques of mapping and case studies may be appropriately followed by more extensive sampling techniques, in order to build up a body of descriptive data and test explanatory hypotheses. Frequent alternation between these methods, rather than sequential operation, will be necessary as the questions being asked shift from distributional to analytical aspects. As population has defined the nature of the problem in Kano, so it will continue to be the most urgent research priority. The census of 1973 will provide a starting point for further analysis of rates of growth, migration and urbanisation, while further studies in the field must establish the dimensions of land-use changes and identify areas of threatened ecological imbalance. The geographer's interest in spatial relationships is a safeguard against the common and dangerous assumptions of homogeneous problems and universalistic solutions in rural Africa, while his traditional occupation with ecological or man–environment relationships has value where the cultural apparatus interposed between the two is severely restricted by poverty.

NOTES

1. The census of 1931 was a sample census and subject to considerable inaccuracy, probably negative, in places. But the regularity of the spatial patterns which emerge seems to justify the simple analysis made here.

2. The initiation of this link was due to the work of H. A. Moisley, formerly Professor of Geography at Ahmadu Bello University.

3. This work has been undertaken as a postgraduate research project by J. M. Baba.

4. This work has been undertaken as a postgraduate research project by J. U. Uba.

5. For example, J. N. Paden's continuing publications on the history and society of Kano (Paden, 1970); a current study of industrial employment by P. Lubeck; and a team project on the Kurmi market by C. J. Greenhill and N. Perchonock. Becker (1969) has given a coordinated account of the urban geography of Kano.

6. In the northern states of Nigeria, at the time of writing, some effort is being made to incorporate the University into more planning projects at state level.

7. The extent of this problem in northern Nigeria is underlined by the fact that Kano State is alone among the six northern states in possessing population data of usable reliability and none of these states had (in 1972) complete cover of contoured topographical mapping.

REFERENCES

Baldwin, K. D. S. (1957) *The Niger Agricultural Project* (Oxford: Blackwell).

Barth, H. (1857) *Travels and discoveries in North and Central Africa* (5 vols.; London: Longmans), Vol. 3.

Becker, C. (1969) *Kano, eine Afrikanische grosstadt* (Hamburger Beitrage zur Afrikakunde Band 10; Pfaffenhoven: Afrika Verlag).

Boserup, E. (1965) *The conditions of agricultural growth* (London: Allen and Unwin).

Clayton, W. D. (1957) 'The swamps and sand dunes of Hadejia', *The Nigerian Geographical Journal* **1,** 31–37.

Dennison, E. B. (1956) 'Problems of reclaiming certain infertile areas in the Sudan zone of Northern Nigeria (Kano Province)', *World Crops* **8,** 131–136.

Goddard, A. D., J. C. Fine and D. W. Norman (1971) 'A socio-economic study of three villages in the Sokoto close-settled zone. 1. Land and people; 2. Maps', *Samaru Miscellaneous Papers* **33, 34** (Samaru: Institute for Agricultural Research, Ahmadu Bello University).

Gowers, W. F. (1913) *Kano Province Annual Report for 1913* (Nigerian National Archives, Kaduna).

Green, L. (1970) *Northern Nigeria; population models for national and regional planning 1952–67* (Physical Planning Research Division Interim Report no. 3; Ibadan: Nigerian Institute for Social and Economic Research), mimeo.

Grove, A. T. (1958) 'The ancient erg of Hausaland, and similar formations on the south side of the Sahara', *Geographical Journal* **124,** 528–533.

Grove, A. T. (1961) 'Population densities and agriculture in northern Nigeria', *Essays on African population*, ed. K. M. Barbour and R. M. Prothero (London: Routledge and Kegan Paul), 115–136.

Klinkenberg, K. (1970) *Soils map of northern Nigeria* (Samaru: Soil Survey Section, Institute for Agricultural Research, Ahmadu Bello University).

McDonnell, G. (1964) 'The dynamics of geographic change: the case of Kano', *Annals of the Association of American Geographers* **54,** 355–371.

Mortimore, M. J. (1965) 'Distribution of population in Kano Province, 1962'. *Occasional Paper* 2 (Zaria: Department of Geography, Ahmadu Bello University), map.

Mortimore, M. J. (1968) 'Population distribution, settlement and soils in Kano Province, Northern Nigeria, 1931–62', *The population of tropical Africa*, ed. J. C. Caldwell and C. Okonjo (London: Longmans), 298–306.

Mortimore, M. J. (1970) 'Population densities and rural economies in the Kano close-settled zone, Nigeria', *Geography and a crowding world*, ed. W. Zelinsky, L. A. Kosínski and R. M. Prothero (New York: Oxford University Press), 380–391.

Mortimore, M. J. (1971) 'Population densities and systems of agricultural land use in northern Nigeria', *The Nigerian Geographical Journal* **14**, 3–15.

Mortimore, M. J. (1972) 'Some aspects of rural–urban relations in Kano, Nigeria', *La croissance urbaine en Afrique noire et à Madagascar* (Paris: Centre National de la Recherche Scientifique), Vol. 2, 871–888.

Mortimore, M. J. and J. Wilson (1965) 'Land and people in the Kano close-settled zone', *Occasional Paper* 1 (Zaria: Department of Geography, Ahmadu Bello University).

Norman, D. W. (1971) 'Economic and non-economic variables in village surveys', *Samaru Agricultural Newsletter* **13**, 61–64.

Norman, D. W. (1972) 'An economic survey of three villages in Zaria Province. 1. Land and Labour relationships; 2. Input–output study; 3. Maps', *Samaru Miscellaneous Papers* **19, 37, 23** (Samaru: Institute for Agricultural Research, Ahmadu Bello University). (Parts 1 and 3 published in 1967).

Paden, J. N. (1970) 'Urban pluralism, integration and adaptation of communal identity in Kano, Nigeria', *From tribe to nation in Africa; studies in incorporation processes*, ed. R. Cohen and J. Middleton (Scranton: Chandler), 242–270.

Prothero, R. M. (1959) *Northern Region, Nigeria 1:1 000 000 Population distribution* (London: Directorate of Overseas Surveys).

Pullan, R. A. (1962) 'A report on the reconnaissance soil survey of the Nguru-Hadejia-Gumel area', *Bulletin* **18** (Samaru: Soil Survey Section, Institute for Agricultural Research, Ahmadu Bello University), mimeo.

Trevallion, B. A. W. (1966) *Metropolitan Kano. Twenty-year development plan, 1963–1983* (London: Newman Neame).

United States Agency for International Development (1968) *Report on the hydrological survey of the Chad basin.*

ACKNOWLEDGEMENTS

Dr. G. J. Williams and Dr. D. W. Norman kindly consented to read and comment on a draft of this paper and their assistance is gratefully acknowledged.

Chapter 8

Changing Patterns of Pastoral Nomadism in North Africa

Douglas L. Johnson

Introduction

Pastoral nomadism is a specialised livelihood form that is ecologically adjusted to use the resources of rainfall deficit environments. In most instances these resources are too marginal to be exploited by sedentary agricultural populations on a secure and stable basis. Using the mobility inherent in their flocks, nomads are able to adapt to the environmental instability of recurrent drought by shifting their activities to seasonally more favoured locales. In regions where periodic drought or general conditions of aridity impose strict limits on the practicality of areally extensive cereal agriculture, nomadic groups have thrived because the available pastoral technology, when viewed in terms of the existing known alternatives, was able to provide an internally satisfying and viable way of life. Thus, it is along the arid margins of dry farming that pastoral nomadism operates most successfully. In these zones it is the nomad's superior ability to bring into production the fodder resources represented by semi-arid grassland environments which represents pastoralism's *raison d'être*. Within the context of this understanding of the ecological role of pastoral nomadism, this chapter analyses the traditional pastoral ecology of Mediterranean Africa and suggests some of the consequences of acculturation which result from the impact of external stimuli upon nomadic society.

TRADITIONAL PATTERNS OF NOMADIC MIGRATION AND LAND USE

Ecological setting

The pastoral land-use systems traditionally prominent in North Africa

Dr. Johnson is Assistant Professor in the Graduate School of Geography, Clark University.

were based on regular seasonal patterns of movement to gain access to grazing resources that varied in occurrence in both time and space. Not aimless wanderers pursuing sporadic and ephemeral rainshowers, the pastoralists of North Africa relied instead upon regular, periodic, predictable and ecologically astute movements to gain access to essential resources. These movements are closely tied to the seasonal nature of the Mediterranean climatic regime. Rhythmic alternation of winter–spring rain with summer drought ensures repetitive patterns of surpluses and deficits in the availability of the region's grazing and water resources. Coastal and upland districts receive relatively copious precipitation which, in turn, provides suitable conditions for dry-farm agriculture in the winter and adequate water storage and fodder reserves for animals in the dry summer months. Interior districts receive sufficient winter–spring rain to provide seasonal pasturage, but these areas lack the water reserves (except for locally favourable oasis settings) to sustain men and animals throughout the entire year. To utilise these grazing resources effectively, pastoral nomads oscillate between summer pastures in the coastal and upland districts and winter rainy-season pastures on the fringes of the Sahara.

The essential ingredient in the human land-use system is the existence of an unexploited niche. Sedentary agriculturalists, by virtue of their restricted mobility and primary interest in agricultural pursuits, are unable fully to exploit all of the grazing resources potentially available in both the better-watered and the semi-arid zones. Limited pasture in regions primarily devoted to field crops reduces the number of animals that can be supported; usually the numbers are less than the potential of the available stored water to support animals in the dry season. Similarly, concentration on agricultural pursuits reduces the ability of the denizens of the cultivated districts to maintain large herds, since these herds cannot be moved to the seasonally available pasture of the interior. To do so would require a corresponding reduction in the attention paid to cultivated crops and in the availability of labour to care for them. At the same time, the seasonal pasture resource represented by the ephemeral pastures of the interior is capable of supporting large numbers of animals only if they can be moved to the pasture at the appropriate season and as long as sufficient water can be found for the dry season.

It is as an attempt to balance these deficits and surpluses that the nomadic lifestyle can best be understood. By reducing or abandoning altogether his interest in agriculture, the nomad is able to develop the mobility required to achieve access to pasture and water throughout the year. Thus, mobility permits the nomad to utilise the distant rainy season pastures that are unexploitable by settled communities. The small numbers of animals that can be maintained by the agriculturalist reduces the pressure on water resources in the summer months. With surplus water available in cisterns and wells near the coast, the nomad is usually assured of enough water to meet his flock's needs. In this sense the nomad exploits seasonal pastures that cannot be reached by the coastal farmers while making use of the dry season surplus water that the settled agriculturalist does not need for his own animals. In this way the seasonal rhythm of nomadic movement is ideally designed to utilise unexploited winter pasture and summer water reserves.

Historical conditions

In North Africa pastoral nomadism is a time-sanctioned way of life. Herodotus mentioned tribes whose patterns of pastoral movement were analogous to those followed by contemporary nomadic groups. The Nasamones, a tribe dwelling along the Gulf of Sirte, were one such group (Herodotus, 1954, p. 300). Annually they moved from the coast to the interior oasis of Awjila where they acquired the dates and other agricultural produce that were essential components of their food supply. This pattern of movement, and the importance of oasis agricultural production in the economy of the pastoral tribe, corresponds to the livelihood system of many of the nomadic clans of that area in the recent past (Cauneille, 1954). Seasonal nomadic migrations are known for a large number of North African provinces in classical antiquity (Gautier, 1942) and pastoral motifs are a common occurrence in the art of the Roman provinces (Précheur-Canonge, 1962).

Although well established in antiquity as an economic system, the nomadic tribes of North Africa were dominated politically and militarily by the agricultural populations of the coastal and upland districts. Despite periodic episodes of conflict between farmer and nomad, it was not until the invasion of the Bani Hilalian tribes in the twelfth century that pastoralism became the dominant lifestyle of the region. This change is not directly attributable to the appearance of the Bani Hilal. Rather, for various reasons, the political and military integrity of the settled community went into eclipse and in the insecure and unstable conditions that followed large tracts of land were abandoned by agriculturalists and converted to pastoral pursuits (Poncet, 1954). The occupants of agricultural areas who once practised only supplemental animal husbandry now began to concentrate on the mobile capital represented by sheep and goats with consequent reduced emphasis on agriculture. Only the most productive coastal districts and the refuge regions of the higher mountain ranges preserved an essentially agricultural character.

Basic types of North African pastoral adjustments

Arising out of these conditions, and building upon pre-existing pastoral traditions, two basic nomadic adjustments emerged. The first of these is found in or near the coastal and upland districts and involves utilising pastures which occur at differing elevations. The onset of the winter rains finds the nomadic tribes dispersed, or in the process of dispersing, in the interior lowlands. The areas grazed are often far from settled agricultural centres and are usually located along the fringes of the interior deserts. Here the animals graze on the annual grasses that appear after rainfall and on the more permanent vegetation of depressions and seasonal stream courses. These depressions and seasonal grasses are found in a broad arc south of the Saharan Atlas and in the interior areas south of the Jabal Nafusa and Jabal al-Akhḍar massifs in Libya. The animals herded and the peculiarities of the seasonal régime usually determine the length of the annual displacements, although the same general zones are visited repeatedly. Thus, the Arbaa (Figure 8.1) are regularly found in the *daya* regions north and west of Ghardaia (Capot-Rey, 1939; Nicolaisen, 1950) while the Ouled Djellal and Arab Cheraga habitually return to areas south and west of Biskra (Despois, 1964).

Once pastures begin to wither at the onset of the summer season, the nomads trek back toward the coastal and upland districts. Generally, this involves a movement over considerable distances, along prescribed routes, and through mountain districts where topographic features give definition to the migration pattern. The movement of the Ouled Djellal and the tribes of the Biskra region through the Aurès mountains to the high plateau areas south and east of Constantine are typical of the migration pattern of nomadic groups throughout the highland zones of North Africa (Despois, 1964). Again, this movement is to clearly defined territories, usually those in which the nomads have traditional rights to the use of pasture on the stubble of the agricultural harvest and access to assured water supplies. The dry season is spent in the uplands in relative security until the inception of the winter rains prompts a return to the pastures of the south. In every case this movement involves the use of winter pastures in the south and summer pastures in the northern uplands with an oscillating movement over traditionally sanctioned routes between the two zones.

Eastern Libya offers a good example of this type of movement. For illustrative purposes the following discussion is focused on a cross-section drawn from Ra's al-Hilāl on the Mediterranean to the edge of the interior sand sea (Figure 8.2). In the northern portions of the cross-section a low upland close to the Mediterranean coast, located on the southern edge of winter cyclonic storm tracks, creates locally favourable environmental conditions. While this dissected plateau country, called the Jabal al-Akhḍar

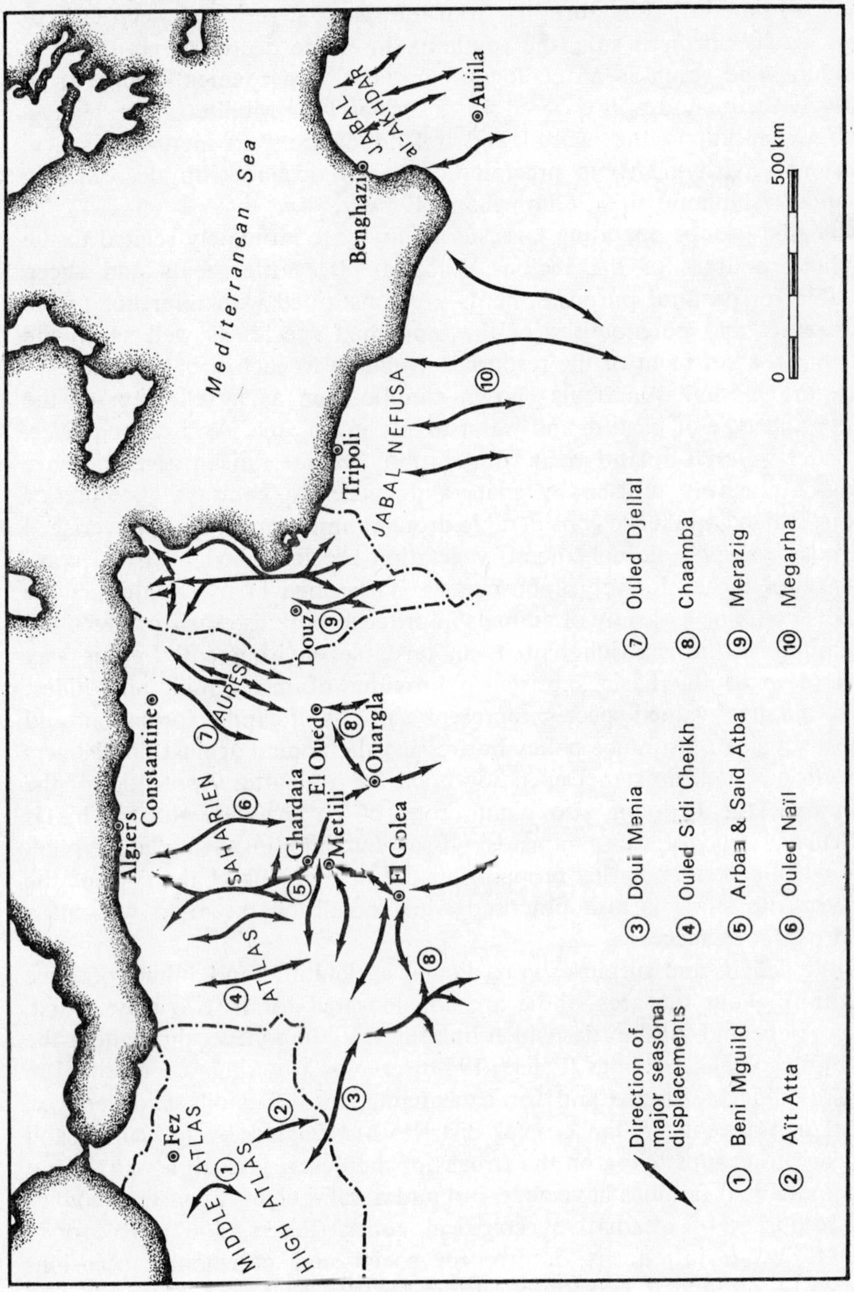

Figure 8.1 Selected pastoral migration patterns in North Africa

(Green Mountain), rises abruptly from the sea in a series of steep escarpments on its northern side, the southern dip slope decreases gradually in elevation. The result is an ecological gradient. Environments range from relatively secure and well-watered zones supporting a Mediterranean maquis shrub vegetation in the coastal and upland districts, to increasingly arid grassland environments as precipitation totals decline with decreases in elevation (Ahlmann, 1928; Gimingham, 1954).

Nomadic groups operating in eastern Libya are intimately related to the variable resources of the region. With camels, cattle, goats and sheep available for pastoral purposes, herds are constituted with reference to the peculiarities and potentialities of the individual species as well as on the basis of an assessment of the resources available to each tribal component. Thus, the particular animals chosen can be seen as a reflection of the quality and type of pasture and water in any given area. As a consequence, the better-watered upland areas with shrubby Mediterranean vegetation are almost exclusively utilised by goats and cattle. In contrast, the interior regions which experience considerable drought and support an impoverished perennial, or sporadic ephemeral, vegetation are inhabited by tribal components relying exclusively upon camels. Tribesmen in more intermediate districts combine a variety of animals in order to optimise their productivity and minimise the risk inherent in an unstable environment. In this way goats serve as the basic subsistence provider of meat, milk and hides; sheep, the most valued species, represent a means of capital formation; and camels act as an insurance policy buttressing the nomad against the dangers of environmental disaster. Depending upon the agricultural potential of the local area, the bedouin also plant crops of barley and wheat. This is particularly true of those nomads possessing land in the upland zones, although more risky barley cropping in the depressions at the foot of the southern dip slope is also practised when conditions seem to warrant a fair prospect of success.

While deficits and surpluses in particular agricultural and animal produce exist throughout the area, these are compensated for in two ways. First, the marriage and kinship ties which link lineage components throughout the range of ecological settings (Peters, 1960) serve as a non-market mechanism for providing assistance and for exchanging desired goods and services. Tribal components in the coastal districts are related by genealogy and marriage to groups living on the fringes of the desert in such a way that at need individual families have access to and a valid claim upon the produce and resources of alternative ecological zones (Peters, 1967). Resort to markets is an infrequent occurrence; social and ceremonial occasions frequently provide a substitute setting for the exchange of goods and services. Although this particular method of balancing agricultural and pastoral activities is not encountered uniformly among bedouin, attempts

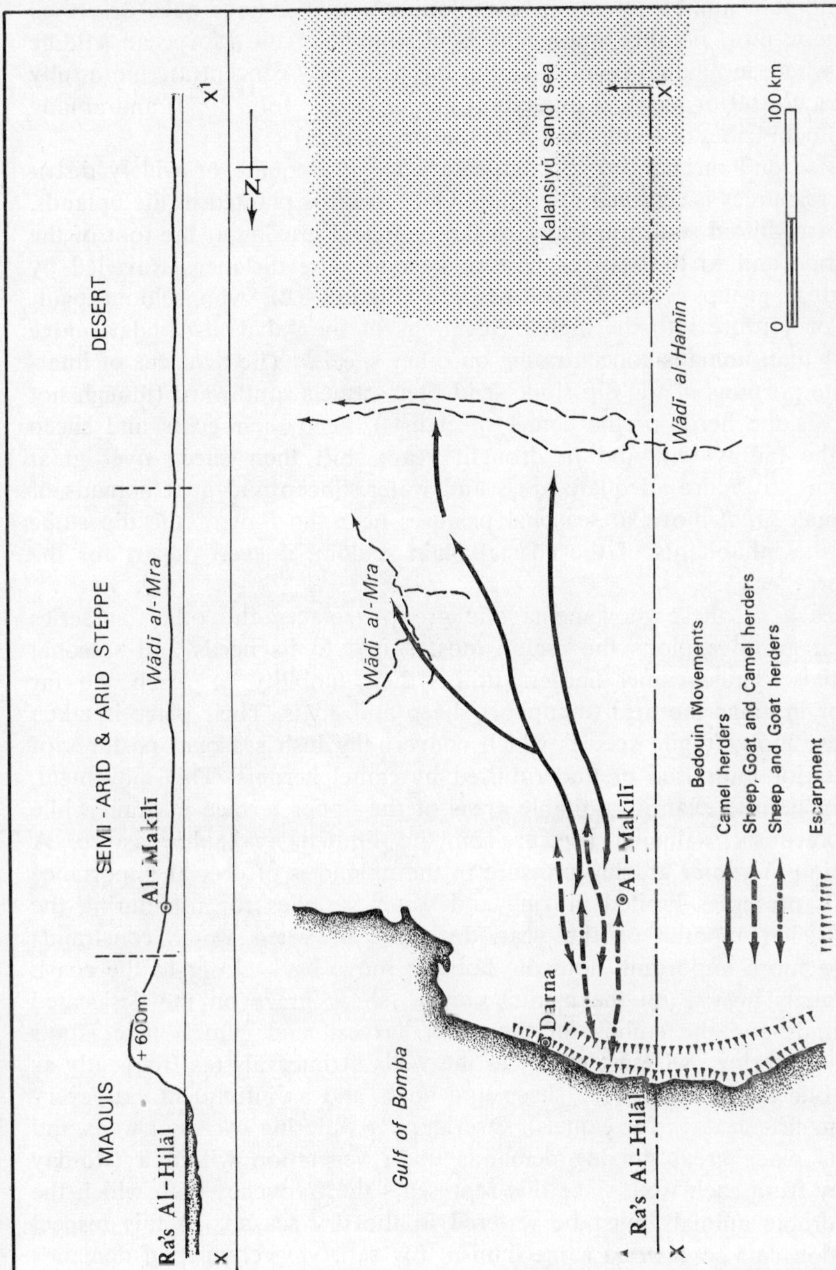

Figure 8.2 Pastoral migration systems in the Jabal al-Akhḏar, Cyrenaica

to combine a number of resource-exploiting activities have been described elsewhere, most notably among the Beni Mguild of the Moroccan Middle Atlas where individual families within the tribe may concentrate more fully on agricultural or pastoral pursuits (Célérier, 1927; Joly, 1954) and among the Baluchi tribes of south-west Asia (Salzman, 1971).

The second mechanism to compensate for inadequate or widely distributed resources is seasonal migration. Once grain is planted in the uplands, herds are shifted southward toward the semi-arid grazing at the foot of the dip slope and to the sparser pasture beyond. The distances travelled by individual groups vary. Camel herders (Figure 8.2), who seldom plant grain or venture into the higher elevations of the Jabal al-Akhḍar, move further than nomads concentrating on other species. The denizens of intermediate portions of the dip slope send their camels southward (though not as far as the herds of the camel specialists), keep their goats and sheep near the family tent and in drought years shift their sheep over great distances to secure adequate grass and water. Sheep and goat nomads of the upper *jabal* move to seasonal pastures near the foot of the dip slope while the inhabitants of the coastal plain seldom, if ever, depart for the interior.

In each of these movements one group replaces the other in series as each group exploits the region most suited to its needs and seasonal potentials. Thus, camel herders utilise their mobility to reach the far interior in areas too arid to support sheep and goats. Their place is taken by these more fragile species which convert the lush seasonal pastures of the interior unutilised or underutilised by camel herders. This movement, in turn, permits planting suitable areas of the upper terrace in grain while simultaneously reducing pressure on the limited available pasture. A reduction in winter grazing pressure in the upland is of crucial importance since it preserves limited grazing and water supplies for use during the water-deficit months of the year. In summer, when water constraints become more important, nomadic families move back closer to the coast. Here goats browse on the maquis shrubs, sheep graze on the desiccated grasslands and the stubble of the grain harvest, and camels take jaunts into the interior. All must return to the wells at intervals (as frequently as every one to three days for sheep and goats and as infrequently as every eight to fifteen days for camels). Overlapping watering cycles at wells and cisterns place great grazing demands upon vegetation within a two-day journey from each well, since this represents the frequency with which the less durable animals must be watered in the dry season. In this respect migration can be seen as a mechanism for satisfying competing demands for pasture and water at different times in the seasonal cycle. The combination of available physical and animal resources, seasonal migration, land tenure and water ownership rights, and genealogical and kinship relationships can be regarded, when viewed through the medium of a culture possessing pastoral values and technology, as an integrated resource-use system having a considerable measure of ecological astuteness and as being indicative of similar adjustments elsewhere in North Africa.

The second type of traditional pastoral migration is based upon oases scattered throughout the Sahara. Located far from major sources of atmospheric water in zones where precipitation totals are negligible, relief is slight, and perennial vegetation is scanty or non-existent, attention of necessity focuses upon those infrequent sites where ground water exists close enough to the surface to be exploited by relatively unsophisticated technologies. These oasis sites play a crucial rôle in the desert nomad's lifestyle, for not only is he dependent, as are most nomads, for most of his agricultural requirements and equipment on the output of sedentary agriculturalists (de Planhol, 1966) but also the date grove serves as a fixed point in the nomadic yearly cycle. The rôle of the oasis as a provider of agricultural products to nomadic groups in an area that generally is not conducive to agricultural activities is of paramount importance. Although in some instances components of a nomadic family may be left in the oasis to engage directly in agriculture, a more common technique, at least before the imposition of European colonial authority, was to employ slaves to work the oasis gardens or to extract a forced tribute from the settled population. In many instances nomads own large numbers of palms in the oases that they frequent. Martel (1965) reports that the 1200 nomadic families of the Merazig tribe in southern Tunisia own approximately 3000 palms in their oasis centres and the incidence of palm ownership among other Saharan tribes is equally widespread. Without control over access to the agricultural resources of the oases nomadism in extremely arid areas would be non-viable.

The dry season usually finds the desert tribes such as the Chaamba camped in the vicinity of the oasis, for both the date harvest and the scarcity of water encourage spatial concentration. Conversely, in the rainy season, long-distance migration across the desert to zones of ephemeral annual grassy vegetation constitute the migration pattern. Less regular in terms of the areas visited, desert nomadism none the less exhibits a seasonal pattern analogous to the migrations of the mountain districts, although more broadly shaped pulsation from the oasis to pastures in the surrounding zones is characteristic. Thus, movement is more a direction than a journey to a precisely delimited zone (Johnson, 1969) so that distances covered as well as the areas visited often vary considerably from year to year. The precise distribution of rainfall in any given year has a decided impact on the areas being grazed in that, although the nomads traditionally return to the same general area year after year, the specific area being used and the distance travelled into the interior can vary considerably. Thus, in seven out of every eight years, Chaamba based on Metlili south of Ghardaia usually visit the depressions and seasonal streams on the northern fringes of the great western sand sea, while in drought years they frequent the western fringes of the sand sea closer to the Saharan Atlas (Cauneille, 1968). Once the date harvest approaches and the pasture and surface water produced by the winter rains has dissipated, most of the desert nomads will then return to the oasis to collect their share of the date harvest and to await the beginning of the next rainy season.

THE COLLAPSE OF TRADITIONAL PASTORAL LIFESTYLES

Pressures brought to bear upon the pastoral tribes of North Africa during the last one hundred years have altered traditional livelihood modes in profound, and perhaps irreversible, ways. Pastoral nomadism can be regarded both as a resource-use system that functions at a given technological level in adjustment to precarious and unstable environments and as a cultural system with both explicit and implied values about how certain patterns of resources should be utilised to support life. Both the pastoralist's perception of optimal resource use and the technological and political systems that support nomadic existence have been challenged. Changes in the scale within which social and economic interaction takes place have had a detrimental impact on pastoral life, for forces originating completely outside the control of the nomadic community have posed a serious threat to the underpinnings of bedouin life. In many parts of North Africa the basic rationale for pastoral nomadism has been destroyed.

Colonial governments as agents of change

Colonial governments, and today the new independent states, have had a profound impact on nomadic society. Many of the pressures emanating from the governmental sector are associated with the imposition of a centralised colonial régime. In many instances long and bloody military campaigns were required before the bedouin were forced to acknowledge the authority of the central government. In Libya in general and in Cyrenaica in particular, numerous campaigns and a protracted guerrilla war were required before the Italians were able to overcome the hostility of the bedouin population and extend their sway into the interior portions of the country (Evans-Pritchard, 1949). Similar difficulties, if not analogous instances of protracted warfare, were encountered by the French elsewhere in the Maghreb. The ultimate consequence of European colonialism in

North Africa was the loss of the privileged social and economic position formerly held by the bedouin.

In large measure the success of nomadic pastoralism was directly proportional to the political power possessed by nomadic tribal confederations. The success of much of the oasis-oriented desert nomadism was dependent on an ability to dominate politically and militarily the settled agriculturalists in their midst. In many instances the farming population was relegated to a servile status and was forced to surrender much of its agricultural produce because physical isolation placed farmers at the mercy of their nomadic overlords. Once the tyranny of distance had been bridged by external political authorities possessing the inclination and the military power to impose their will upon the nomadic tribesmen, economic conditions were often altered to the nomad's disadvantage. Stagnation and decline in the nomadic sector and growth in the sedentary community was often the result (Josse, 1971). No longer able to exploit the servile populations of the oases and to extract agricultural commodities as tribute for providing protection from the depredations of other tribes, nomads have frequently been placed at an economically competitive disadvantage. The Aït Atta of Morocco is a tribe that has been adversely affected by these changes. Formerly able to dominate the sedentary farming population, whose fortified granaries were of crucial importance as a focus of the exchange of grain for animal products (Spillmann, 1931) and whose grain supplies were essential to the survival of the pastoral community, the Aït Atta, never noted for their long-range migrations, have become increasingly sedentary (Jongmans, 1956). Unable to subsist successfully solely on the produce of their herds or to enter economically into a free trade market vis-à-vis their sedentary contacts, a drift towards sedentary life has been inevitable in most nomadic communities. Not only has the privileged position of the nomad been lost in his economic relations with the indigenous sedentary population but his freedom from governmental restraint has also been greatly reduced. In the colonial era this was exemplified by the expansion of European agriculture at the expense of nomadic uses of the same territories. In Eastern Libya the bedouin were denied access to the upper portions of the Jabal al-Akhḍar where a planned programme of Italian peasant demographic settlement was undertaken by Mussolini's Fascist government (Istituto Agricolo Coloniale, 1947). Although the inception of World War 2 prevented these plans from attaining either their full areal potential or a position of economic profitability, they represented a very real and present threat to the survival of the nomadic régime. Prevented from planting barley and wheat in the best-watered portions of the *jabal* in the winter and from utilising the water reserves of the upland zones in the summer, the viability of their agricultural activities and survival of their flocks would have been seriously jeopardised. Since the only areas with a potential barley cultivation remaining in bedouin hands were those at the foot of the southern dip slope (the very areas most susceptible to climatic variability and repeated and prolonged drought), the opportunity for disastrous crop failures and the resulting removal of one of the bedrock components in the Cyrenaican bedouin economy could only make the nomadic economy more precarious. Similarly, the loss of the water resources of the upper terrace would have exposed the herds to analogous risk. Equally serious was the decimation of the bedouin population and its herds as a consequence of forced incarceration in concentration camps as part of Italian efforts to establish full military control over the indigenous society. Even after fighting ended, Italian attempts to disrupt the very fabric of bedouin society continued. Encouraging a proliferation of leaders in an attempt to destroy the authority of tribal spokesmen and reassigning well-ownership in order to weaken patron–client relationships based on water rights (Peters, 1968) exemplified the type of measures undertaken to reduce the cohesion of nomadic society. Ultimately the result of the Italian colonial policy, if allowed to continue unchecked, would have been to reduce the already demoralised and impoverished bedouin to a state of incipient servitude and compelled them to eke out a precarious existence on the marginal resources of the interior.

Although colonial policy elsewhere in North Africa was less severe, the colonial régimes did have an unsympathetic attitude toward the nomadic tribes (De Fraquier, 1953) and this attitude caused the erosion of many of the foundations of nomadic society. In many areas agriculture, particularly arboriculture, was consciously favoured. In coastal Tunisia large tracts of grazing land were converted to olive orchards and this conversion was accompanied by a drift of nomadic tribesmen into agricultural labour (Despois, 1940). Contemporary movements of migrant agricultural labourers through the coastal plain is associated spatially with the routes followed by the small nomadic groups still practising the traditional migration from the Jefara Plain south of Gabès to the highlands of central and northern Tunisia. Throughout the Maghreb, conditions of increased security, coupled with a policy encouraging colonial agricultural settlement, favoured the expansion of agriculture into areas that had formerly been important summer and late spring pastures for the nomads. A host of laws controlling migration routes, watering points, rates of movement, areas permitted for pastoral utilisation, all designed to eliminate conflict between nomad and sedentary, served to constrict and interfere with the traditional operation of the nomadic migration régime.

Certainly this is the case in Algeria, where nomadic access to the uplands north of the Atlas Saharien is prevented completely or prohibited until the completion of the agricultural harvest. As a result the Ouled Naïl, the Ouled Sidi Cheikh and other tribes are now largely confined south of the Atlas along the fringes of the Sahara (Bisson, 1962). The imposition of artificial political boundaries also had an adverse effect on the migration régimes of certain tribes. The Doui Menia formerly migrated annually across the Algerian–Moroccan frontier, a pattern of movement that is blocked today (Capot-Rey, 1952). The traditional outlet of the Cyrenaican bedouin to animal markets in Alexandria (Evans-Pritchard, 1949) was cut off by the Italian–Sanusi hostilities and the economic interchanges associated with this exchange pattern have never been resurrected. The consequence everywhere was interference in the normal phasing of the pastoral migration, greatly restricting the scope of pastoral movements and, at least in the case of herders of sheep and goats, forcing them to utilise, for extended periods of time, resources of a marginal nature. Decimation of the herds in drought years with resulting impoverishment of the nomadic way of life was inevitable.

Changes in technology

Equal in importance to the nomad's loss of his dominant political and economic position have been massive changes in technology. Transportation is one important area where the introduction of alien technology has undercut the nomad. Specifically, the rapid spread of motor vehicles has eliminated the caravan trade. In the past many nomadic tribes derived much income from the trans-Saharan caravan traffic both by providing guides and pack animals to merchants and by extorting protection money as a guarantee of safe passage. Today this source of revenue has disappeared. Camels now are far less valuable, since they can no longer produce significant amounts of cash income. This is reflected in a gradual decline in the reported number of camels in eastern Libya where the total fell from 58 163 in 1965 to 47 807 in 1968 (Libyan Arab Republic, 1969, p. 91). No longer useful in caravan traffic, often replaced by a truck or jeep in the movement of the bedouin family's household baggage, the camel is being abandoned as a primary subsistence animal by all except the inhabitants of the most marginal areas.

Recent discovery of huge oil and gas reserves in several North African countries dramatically altered their resource base. Although oriented towards the production of petroleum products for export to European markets, and despite being heavily staffed by foreign technicians, the oil industry has often stimulated spectacular economic growth and development. Despite the oil reserves being located far from the centres of population, oil developments have served as growth points in interior areas (Lerat, 1971) and infused considerable amounts of capital into the budgets of both bedouin and sedentary families. Exposure to contact with western products and values has contributed to a considerable amount of acculturation, has often resulted in the uneasy juxtaposition of different value systems (Thomas, 1961), has stimulated demands for certain goods and services and has provided governments with some of the resources needed to meet the newly-generated material expectations and demands. Libya's financial condition is a particularly dramatic illustration of the impact of the discovery of oil. From a monetary deficit as late as 1958, with few prospects for substantial and spectacular economic development (Farley, 1971), governmental revenues swelled ten years later to £L 277 227 000, or more than $US 750 000 000 (Libya, 1969). Despite considerable waste and mismanagement, very considerable funds were devoted to development projects.

While large amounts of revenue have been channelled into building an infrastructure of roads, seaports, public buildings and hospitals, and agricultural development programmes, much of the new-found wealth eventually has had an effect on the individual citizen. Social service programmes in particular have had a marked impact on the non-urban population. In Libya the government has been active in building schools, hospitals, clinics, marketing facilities, electrification schemes, wells and cisterns, and public housing projects. All these facilities encourage nomadic settlement. In an attempt to induce increased stability in rural residence patterns and retard migration to the coastal cities, public housing is frequently provided in scattered fashion at well sites, Islamic shrines and other settlement foci. Similar development programmes have been launched elsewhere in North Africa, though larger total populations and more restricted funds have reduced the magnitude of their impact. Thus, not only are governments generally eager to bring the benefits of modernisation directly to their citizens but they are also convinced that selective adoption of western values and the conversion of nomads into farmers will make them better citizens and more productive contributors to the national economy (Abou-Zeid, 1959). In the present context North African governments have both the lure of financially rewarding employment and services outside the nomadic sector and the political and economic muscle to guide development into channels that suit national goals. The end-product has been the sedentarisation of an increasing proportion of nomadic society.

Patterns of sedentarisation

Once the conditions that formerly contributed to the success of nomadic pastoralism changed, abandonment of the traditional *genre de vie* in favour of an alternative lifestyle was rapid. Certainly this was true of Egypt, where in the space of only a few generations the separate identity of the bedouin tribes was lost as they were absorbed into the surrounding mass of peasant agriculturalists (Awad, 1954). Although the process is further advanced in Egypt than in most of the remainder of North Africa, here, too, the process of change has accelerated rapidly. Three main patterns of sedentarisation appear to be emerging throughout North Africa. These are: (1) settlement in rural areas in one portion of the tribal territory; (2) reoccupation of the abandoned colonial agricultural infrastructure; (3) migration from the rural areas to coastal urban centres accompanied by complete abandonment of the nomadic way of life.

Most of the nomadic tribes that have begun to become more stationary, or have abandoned the nomadic way of life completely, have settled in a particular part of their tribal territory. Two basic patterns can be discerned in this regard. In the desert the process takes place in the vicinity of the oasis. In the desert nomadic régime, the oasis has always played a crucial role, for it is in the oasis that the majority of craft products and agricultural goods must be obtained. Also, it is common for the nomad to own land or date palms in the oasis itself and the existence of ties between the nomad and the oasis are of long standing. That settlement for many of the nomads should thus take place in a setting that for them is a familiar one is not surprising. Certainly this is the process taking place in the vicinity of Ouargla, as well as at many of the other oases of the Algerian Sahara, where satellite villages have grown up around the edges of the main oasis communities (Rovillois-Brigol, 1963). Often encouraged by the investment of government capital into agricultural projects, many of the once proud protectors of caravans are becoming cultivators of dates.

Not all nomads shift so rapidly to complete involvement in agriculture. Some maintain a more mixed lifestyle, retaining their herds and engaging in short seasonal movements despite a developing interest in agricultural operations (Bataillon, 1955). Although, as Clark (1959) points out, this mixed agro-pastoralism can be a long-term adjustment, in most instances it is a preliminary to total renunciation of the pastoral lifestyle. As agriculture becomes more important, the difficulties involved in finding suitable shepherds at reasonable cost become insuperable. Since the migratory régime was designed to permit access to spatially and temporally variable grazing resources, as sedentarisation proceeds the increased difficulties associated with finding adequate pasture and water inevitably result in rapid decreases in herd size. This in turn, enhances the importance of and accelerates the shift to full-time agriculture. Yet the transition to a more settled livelihood mode is often fraught with difficulties and marred by imperfect results, if only because nomads are slow to adopt the planning horizons and attitudes more suited to their new pursuits (Bessis, 1956).

A second pattern of settlement occurs in the coastal districts fringing the Sahara. A variety of stimuli prompt sedentarisation, including planned government projects, loss of pasture to peasant cultivators, the attractiveness of employment in mining enterprises and the appeal of familiar watering and staging points, as is the case in the eastern Aurès Mountains of Algeria

(Wagner, 1971); yet it is clear that a common denominator is part of the process. This almost universal trait involves settlement in areas where the dry season pasture or winter agricultural operations once were located. These areas are usually claimed as the tribal homeland. For tribes in most of North Africa this means settlement in the upland areas that provided pasture, water and grain to support their pastoral livelihood. In eastern Libya the emerging settlement pattern characteristically is a dispersed one. While some tribesmen settle near tribal village centres or water points, many are widely scattered across the countryside. Most frequently the traditional location of an individual's barley patch determines where the family begins to stabilise. Gradually the tent is abandoned in favour of a tin *baraka* (hut); shepherds become more difficult to find and consequently more expensive as younger sons, who formerly would have tended the flocks, are sent to school or seek employment in the towns; seasonal movements become less frequent and of shorter duration; herds are sold or left with relatives; and the transition to a settled life is rapidly completed.

In some areas sedentarisation begins when nomads move into the farmhouses abandoned or sold by departing European colonists. Although indigenous peasant cultivators were the prime beneficiaries of this alteration in rural agricultural ownership patterns in most of North Africa, in eastern Libya the nomadic tribes reaped some of the rewards of Italian agricultural efforts. For if the Italians dispossessed the bedouin in order to build agricultural settlements, after 1943 the bedouin were in a position to reoccupy their former lands. The Italian farmhouses were valuable to the bedouin not as dwellings but as sites containing cisterns and cleared agricultural land. While the Libyan government claimed ownership of the Italian structures, rents charged were nominal and the farmhouses were distributed less on the basis of the potential ability and willingness of the occupant to perform farming activities than on the desire to distribute available dwellings on an equitable basis among the various branches of the tribal genealogy. For many years these houses were only occupied seasonally, as their owners continued to retain a primary interest in animal husbandry and the seasonal movements required to support it. Economic development based on the Italian agricultural infrastructure has been more difficult as a result (International Bank, 1960) and only with the massive infusion of oil revenues has it begun to show much progress. Nonetheless, occupation of the former Italian farmhouses has proved to be an important element in the modification of the bedouin régime.

Perhaps the major factor involved in sedentarisation has been the massive migration of rural dwellers to the major coastal cities. Between 1950 and 1971 Algiers, Casablanca, Rabat, Tripoli and Benghazi have approximately doubled in size while other cities have exhibited appreciable, if somewhat less spectacular, growth (United Nations, 1971). Although statistical data are seldom precise enough accurately to measure the components in this flow so that the relative percentage of nomads moving to the city can be determined, it is clear that participation by nomads has been considerable. Since the majority of the Cyrenaican population until recently was nomadic in some degree, much of the movement along the rural–urban gradient delineated by Hartley and Norris (1969) is undertaken by former nomads. This pattern, of course, shows considerable regional difference since Algerian nomads tend to move directly to the city whereas in Tunisia they generally are first transformed into rural peasants (Glauert, 1957). So spectacular has been the leap in urban population that immigrants far exceed the capacity of the local facilities to absorb them. *Bidonvilles*, or shanty towns, on the outskirts of the urban agglomeration, often occupied by migrants from a common region, are characteristic of the urban growth pattern of North Africa (Harrison, 1967). Only after a period of time does a shift from temporary materials to more durable vernacular architecture take place (Holz, 1970). Movement to the city represents a significant break from nomadic traditions. Although ties with relatives in the countryside remain strong and visits, money and assistance may be exchanged between extended family components in a continuation of the patterns of mutual self-help that typify nomadic society, for the migrant to the city a return to the values and lifestyles of the traditional society is usually impossible.

CONCLUSION

If in the past pastoral nomads utilised an unexploited niche, the viability of this niche has encountered serious difficulties. Although it was possible for especially talented and fortunate individuals to amass large amounts of capital in the form of animals, herds of five to six hundred sheep having been reported for wealthy nomads in Mauritania (Dubie, 1953, p. 151), most nomads operated on a subsistence basis. Even when surplus capital was accumulated during a succession of good years it was not consumed but rather preserved as insurance against anticipated future deficit years. Despite being a stable adjustment in terms of its survival for centuries, pastoral nomadism always remained a precarious way of life. Once political and military pre-eminence was lost to sedentary governments, increased security encouraged agricultural expansion into former grazing territories. Deprived of dry season pasture, uncompetitive in a free market economy, threatened by alternative transportation modes, confronted by often unsympathetic government officials, attracted by the prospect of employment in other sectors of the economy, and desirous of attaining some of the material comforts of western technology, it is not startling that nomadic pastoralists should display an accelerating drift toward settled life.

In at least one sense sedentarisation is unfortunate. While sedentarisation may well mean an improvement in educational standards, health benefits and general living conditions, it is almost invariably accompanied by an abandonment of marginal resources. These resources usually will not sustain an agricultural population, but are eminently suited to grazing and associated animal husbandry activities. Only nomads can exploit these grazing resources effectively and in many parts of the Near East animal products originating in the nomadic community are an important contributor to the national economy (Stauffer, 1965). In Libya, where sedentarisation and rural to urban migration has proceeded apace, demand for animal products has increased while domestic output has declined. The result has been to convert Libya from a net exporter to a net importer of animals and animal products. While a rich nation can afford this, a poor, even a developing one, cannot. Not only is it a drain on national resources, but it also wastes the talent and expertise of a substantial proportion of the local population. The crucial issue is how best to combine local traditions and knowledge with technological advances introduced from outside. To abandon completely the rational and successful ecological understandings developed through long periods by traditional bedouin societies unnecessarily restricts the range of options a developing economy should have available.

Yet, theoretically desirable as it might be to preserve some aspects of the pastoral nomadic lifestyle in order productively to utilise semi-arid and arid environments, the forces of modernisation operate in different directions. One does not require an exceptionally fertile imagination to envisage the demise of pastoral nomadism. At the present moment the traditional nomadic way of life, long successful on a subsistence level as an adjustment to drought deficit environments, seems destined to disappear from the North African scene.

REFERENCES

Abou-Zeid, A. M. (1959) 'The sedentarisation of nomads in the western desert of Egypt', *International Social Science Journal* **11,** 550–558.

Ahlman, H. W. (1928) 'La Libye septentrionale: études de géographie physique et humaine', *Geografiska Annaler* **10,** 1–118.

Awad, M. (1954) 'The assimilation of nomads in Egypt', *Geographical Review* **44,** 240–252.

Bataillon, C. (1955) *Le Souf; étude de géographie humaine* (Université d'Alger, Institut de Recherches Sahariennes, Mémoire No. 2; Algers: Institut de Recherches Sahariennes).

Bessis, A. *et al.* (1956) *Le territoire des Ouled Sidi Ali Ben Aoun; contribution à l'étude des problèmes humaines dans la steppe tunisienne* (Paris: Presses Universitaires de France).

Bisson, J. (1962) 'Les nomades des départements sahariennes', *Travaux de l'Institut de Recherches Sahariennes* **21,** 199–206.

Capot-Rey, R. (1939) 'Pays du Mzab et région des dayas: étude sur le relief de la dorsale saharienne', *Annales de Géographie* **48,** 41-62.

Capot-Rey, R. (1952) 'Transformation récentes dans une tribu du Sud-Oranais', *Annales de Géographie* **61,** 138–142.

Cauneille, A. (1954) 'Le nomadisme des Megarha (Fezzân),' *Travaux de l'Institut de Recherches Sahariennes* **12,** 41–67.

Cauneille, A. (1968) *Les Chaanba (leur nomadisme); évolution de la tribu durant l'administration française* (Paris: Editions du Centre National de la Recherche Scientifique).

Célérier, J. (1927) 'La transhumance dans le Moyen-Atlas', *Hespéris* **7,** 53–68.

Clarke, J. I. (1959) 'Studies of semi-nomadism in North Africa', *Economic Geography* **35,** 95–108.

De Fraquier, Cdt. (1953) 'La crise du nomadisme et de l'élevage sur les Haut Plateaux algériens', *Travaux de l'Institut de Recherches Sahariennes* **9,** 71–97.

Despois, J. (1940) *La Tunisie orientale; sahel et basse steppe* (Paris: Société d'Edition 'Les Belles Lettres').

Despois, J. (1964) *L'Afrique du Nord* (Paris: Presses Universitaires de France).

Dubie, P. (1953) 'La vie matérielle des Maures', *Mélanges Ethnologiques* (Memoires de l'Institut Français d'Afrique Noire, No. 23; Dakar: Institut Français d'Afrique Noire), 111–252.

Evans-Pritchard, E. E. (1949) *The Sanusi of Cyrenaica* (Oxford: Clarendon).

Farley, R. (1971) *Planning for development in Libya; the exceptional economy in the developing world* (New York: Praeger).

Gautier, E. F. (1942) *Le passé de l'Afrique du Nord* (Paris: Payot).

Gimingham, C. H., and K. Walton (1954) 'Environment and the structure of scrub communities on the Limestone Plateau of northern Cyrenaica', *Journal of Ecology* **42,** 505–520.

Glauert, G. (1957) 'Verämderungen in der Bevölkerungsstruktur Nordafrikas in den letzen Jahrzehnten', *Die Erde* **88,** 298-319.

Harrison, Robert S. (1967) 'Migrants in the city of Tripoli, Libya', *Geographical Review* **57,** 397-423.

Hartley, R. G. and J. M. Norris (1969) 'Demographic regions in Libya', *Tijdschrift voor Economische en Sociale Geografie* **60,** 221-227.

Herodotus (1954) *The Histories*, trans. Aubrey de Sélincourt (Baltimore, Maryland: Penguin Books).

Holz, R. (1970) 'Sandstone as a building material in Libya', *Pennsylvania Geographer* **7,** 1–10.

International Bank for Reconstruction and Development (1960) *The economic development of Libya* (Baltimore: Johns Hopkins).

Istituto Agricolo Coloniale, Firenze (1947) *La colonizzazione della Cirenaica* (Roma: Tipografia del Senato).

Johnson, D. L. (1969) *The nature of nomadism* (University of Chicago, Department of Geography, Research Paper No. 118).

Joly, F. (1954) 'Elevage: ovins et caprins', *Atlas du Maroc; notices explicatives* (Rabat: Comité de Géographie du Maroc), 25–26.

Jongmans, D. G., and J. H. J. Gerlings (1956) *Les Aït Atta; leur sédentarisation* (Institut Royal des Tropiques, Amsterdam, No. 115; Départment d'Anthropologie Culturelle et Physique, No. 50).

Josse, R. (1971) 'Problèmes de mise en valeur du Hoggar et de croissance urbaine à Tamanrasset', *Les Cahiers d'Outre-Mer* **95**, 245-293.

Libya, Kingdom of, Ministry of Planning and Development (1969) *Report of the Annual Survey of Petroleum Mining Industry 1969; Ref. Year, 1968* (Tripoli: Census and Statistical Department).

Libyan Arab Republic (1969) *Statistical Abstract; 1968* (Tripoli: Census and Statistical Department).

Lerat, S. (1971) 'Hassi Messaoud', *Les Cahiers d'Outre-Mer* **93**, 16–31.

Martel, A. (1965) *Les confins Saharo-Tripolitains de la Tunisie* (*1881–1911*) (2 vols.; Paris: Presses Universitaires de France).

Nicolaisen, J. (1950) 'Nomadismen i det centrale Algérie (Le nomadisme d'Algérie centrale)', *Geografisk Tidsskrift* **50**, 48–91.

Peters, E. L. (1960) 'The proliferation of segments in the lineage of the bedouin of Cyrenaica', *Journal of the Royal Anthropological Institute* **90**, 29–53.

Peters, E. L. (1967) 'Some structural aspects of the feud among the camel-herding bedouin of Cyrenaica', *Africa* **37**, 261–282.

Peters, E. L. (1968) 'The tied and the free: an account of a type of patron–client relationship among the bedouin pastoralists of Cyrenaica', *Contributions to Mediterranean Sociology*, ed. J.-G. Peristiany (Paris: Mouton), 167–188.

de Planhol, Xavier (1966) 'Small-scale industry and crafts in arid regions', *Arid Lands*, ed. E. S. Hills (London: Methuen; Paris: UNESCO), 273–285.

Poncet, J. (1954) 'L'évolution des genres de vie en Tunisie', *Les Cahiers de Tunisie* **2**, 315–323.

Précheur-Canonge, T. (1962) *La vie rurale en Afrique romaine d'après les mosaïques* (Paris: Presses Universitaires de France).

Rovillois-Brigol, M. (1963) 'La sédentarisation autour d'Ouargla', *Nomades et nomadisme au Sahara* (Recherches sur la zone aride, No. 19; Paris: UNESCO), 135–141.

Salzman, P. C. (1971) 'Movement and resource extraction among pastoral nomads: the case of the Shah Nawazi Baluch', *Anthropological Quarterly* **44**, 185–197.

Spillmann, G. (1931) *Districts et tribus de la haute vallée du Dra*' (Villes et Tribus du Maroc, vol. 9; Tribus Berbères, no. 2; Paris: Honoré Champion).

Stauffer, T. R. (1965) 'The economics of nomadism in Iran', *Middle East Journal* **19**, 284–302.

Thomas, F. C., Jr. (1961) 'The Libyan oil worker', *Middle East Journal* **15**, 264–276.

United Nations (1971) *Demographic Yearbook* (22nd edition; New York: Statistical Office of the United Nations).

Wagner, H.-G. (1971) 'Das Siedlungsgefüge im südlichen Ostalgerien (Nememcha)', *Erdkunde* **25**, 118–134.

Chapter 9

Spatial Aspects of Kenya's Rural Development Strategy

D. R. F. Taylor

Introduction

The importance of rural development in Kenya is paramount. The 1969 census (Republic of Kenya, 1970b) revealed that over 90 per cent of Kenya's eleven million people live in rural areas. Development is not the easiest term to define as it has many facets, but whatever criteria are used in the Kenya context to define and measure development these must be applied and justified in rural areas. The aims of national development have been set out quite clearly: 'Development of our Republic involves every man, woman and child. Prosperity is something in which all people have an equal stake' . . . 'The objective of national development is the social and economic transformation of all the people' (Republic of Kenya, 1969, pp. i and 166). The basic strategy of the Development Plan 1970–1974 (Republic of Kenya, 1969) is to achieve these aims through rural development: 'Rural development is the basic strategy of this Plan, for it is our aim that the fruits of development will be shared among the mass of the people as a whole, not just amongst a favoured few' (Republic of Kenya, 1969, p. iv).

Rural development is not defined in the Kenya Development Plan but it is generally accepted to be an holistic concept which recognises the complexity and interrelatedness of the many variables which influence the quality of life in rural areas. A definition accepted by the Working Committee on the Human Environment in Kenya reveals this: '. . . a series of quantitative and qualitative changes occurring among a given rural population and whose converging effects indicate in time a rise in the standard of living and favourable changes in the way of life' (Ominde *et al.*, 1971). Rural development and agricultural development are not synonymous,

---

Dr. Taylor is an Associate Professor in the Department of Geography and the School of International Affairs at Carleton University, Ottawa.

although the agricultural production base is of course a vital component of any rural development programme. In many African countries attention has been paid to agricultural development but until relatively recently there have been few attempts to develop a comprehensive rural development strategy. This is certainly true in Kenya where there has been considerable agricultural development especially over the last twenty-five years. The government itself, speaking of the development plans of 1964 and 1966, states, 'The first two plans paid insufficient attention to programmes and projects for the development of rural areas. Rural development was insufficiently integrated into the overall plan' (Republic of Kenya, 1969, p. 71).

Kenya's new rural development strategy contains, for the first time, an element of spatial planning. This chapter analyses the strategy with particular emphasis on these spatial aspects. It is argued that the spatial aspects, especially the growth centre concept, could be a vital element in rural development. This will however require a shift in emphasis so that the spatial aspect becomes a central element in planning rather than a peripheral one as it is at present. It will also require modification and development if it is to be successfully applied.

Development prior to 1970

The first attempt to transform rural Kenya which had any major impact was the Swynnerton Plan of 1954 (Swynnerton, 1955). The aim of this plan was 'to raise the productivity of the African lands, their human and stock-carrying capacity, the increased standards of living of the people, while at the same time effecting a substantial increase in the resources and economy of the colony (Swynnerton, 1955, p. i). Special attention was given to a number of points: the consolidation of fragments; security of land tenure; the introduction of cash crops; technical assistance to develop the land on sound lines; ready access to water; the introduction of marketing facilities; access to sources of agricultural credit; and an agricultural bias to education.

The plan dealt in theory with all of Kenya but in fact most of the estimated £11 million spent between 1954 and 1960 was concentrated in a relatively small, though very populous, area of the country—Kikuyuland. There is little doubt that the Swynnerton Plan initiated an agricultural revolution in Kikuyuland and brought about a great increase in the wealth of the people, but it was not without its drawbacks. The basis of the plan, land consolidation, meant that land ownership fell into the hands of fewer people than before and the plan did little to solve the problem of land hunger. Land consolidation intensified the problem of landlessness by driving many non-landowners off the land. The population of Kikuyuland in 1962 (Government of Kenya, 1964) was 1 005 800 of whom only 124 000 were registered landowners. The plan aimed essentially at increasing agricultural production, especially of cash crops; this aim was met, but there was a distinct lack of integrated rural development with little attention being paid to the quality of rural life. Not all the eight points in the plan were given equal emphasis and some were virtually ignored. A recent study (Taylor, 1971) has shown that in the important area of nutrition little progress was made, largely because of a lack of integrated planning. Rural education and social services also expanded very slowly. Kenya achieved independence on December 12, 1963, and the first Development Plan was presented on June 10, 1964, replaced by a revised version in mid-1966. These plans together with a fundamental document, Sessional Paper No. 10 of 1965 entitled 'African socialism and its application to planning in Kenya' (Republic of Kenya, 1965), formed the basis for development in Kenya prior to 1970. In terms of rural strategy these plans differed little from the approaches used by the Colonial Government in the late 1950s and early 1960s. The principal aims were an increase in agricultural productivity and the break up of the former 'White Highlands' by a process of land transfer and resettlement. By 1968, 47 000 African families were farming land previously owned by 2000 Europeans. The extent of this transfer is shown in Table 9.1 (Republic of Kenya, 1969, p. 24).

Table 9.1 Land transferred to African use up to 1968

Type of scheme	Acres transferred to Africans	No. of farms established
1. Subdivisional type settlement		
High density schemes	788 000	26 700
Low density schemes	188 000	5 200
Squatter settlement	86 000	14 000
Subtotal	1 062 000	45 900
2. Large-scale farms A.F.C., A.D.C.		
Assisted owners, etc.	953 600	1 192
Ol'Kalou Salient	121 000	19
3. Other cooperatives set up by Department of Settlement	170 000	15
Sub-total	1 244 600	1 226
Total	2 306 600	47 126

The expense involved in this land transfer was great and much of the capital available for agricultural development expenditure was used for this purpose. In 1963/64, 75 per cent of all agricultural development expenditure went into the land transfer programme and it was not until 1968/69 that this proportion fell to just under 50 per cent. The land transfer programme was necessary because of the socio-political situation prevailing in Kenya in the early 1960s, but, in the schemes, emphasis was laid on the economics of production and little attention was paid to the concept of rural development. The result has been that many more people have land but, with farmers growing more food crops, total agricultural production of cash crops has not noticeably increased.

Expenditure in these areas also meant that less money was available for development in other rural areas of Kenya. In these areas too there was a distinct lack of integrated rural development. This was due at least in part to the multiplicity of government departments and agencies working in the rural areas and the lack of effective cooperation between them. This was further complicated by the division of responsibility between the central government and the local government body—the County Council. Agricultural production in absolute terms increased but in relative terms, at least in the monetary sector, there was a decline. This was due to a variety of factors not least of which was a 6 per cent drop in the average price for agricultural exports between 1964 and 1968, sisal being particularly affected by the increasing use of synthetic fibres in the industrialised countries which caused sisal prices to drop by over 20 per cent. Coffee, one of Kenya's main exports, was seriously hit by coffee berry disease and production dropped from 41 407 tons in 1964 to 39 601 tons in 1968. The result was that despite increases in other crops such as maize, sugar and tea, 'no significant growth of agricultural exports at current prices had been achieved by 1968' (Republic of Kenya, 1969, p. 36).

Despite these difficulties financial returns to the mass of Kenya's farmers increased during the 1960s. Statistics indicate that the smallholders, predominantly African, did better than the large farmers and that there was increasing African participation in monetary agriculture. These figures were however influenced by the fact that over 400 000 hectares were converted from large into small farms over this period. Growth took place but, as

Table 9.2 Gross marketed production from large and small farms 1964–68

	Large farms		Small farms		Total		
Year	Value K£'000	Annual change %	Value K£'000	Annual change %	Value K£'000	Annual change %	Share of small farms %
1964	35·8	−4.0	24·6	16·6	60.4	3·4	40·7
1965	33·8	−7·0	23·8	−3.3	57·2	−5·3	41·6
1966	36.0	8·1	32·7	37·4	68.8	20·3	47·5
1967	33.0	−8·3	34.0	4·0	67.0	−2.6	50·7
1968	34·7	5·2	36·0	5·9	70·6	5·4	51·0

Table 9.2 (Republic of Kenya, 1969, p. 32) shows, the growth pattern was not steady. Growth rates were irregular but so also was the spatial pattern of growth with areas like Central Province, a predominantly Kikuyu area, advancing very much faster than the rest of the country.

This was the situation in 1969 and the assessment by the Government of Kenya of the progress made was '. . . that the overriding impression of Kenya over these years had been of steady progress towards greater prosperity. This does not mean that Kenya has become a rich country in four or five years, but there is no doubt that the people as a whole are better fed, better clothed, and to some extent better housed than they were five years ago' (Republic of Kenya, 1969, p. 20). There is however a different viewpoint amongst a significant minority of Kenya's peoples: 'Under the cloak of something called "African Socialism" Kenya is moving towards one of the more orthodox forms of capitalism to be found in the world today. . . . In only a few lush and favoured areas like Kiambu have peasant farmers increased their income. . . . However difficult it may be to organise and channel resources so that the mass of the people start enjoying a better fuller life, that is what the Government of Kenya must try to do because that is what it is put in power to do' (Kenya People's Union, 1969).

Between 1964 and 1970 the average economic growth rate was 6·6 per cent per annum and although this was offset by a population growth rate of 3·3 per cent per year the per capita income rose from £39·5 to over £50 (Republic of Kenya, 1971b). These figures would tend to support the government's view although they give no indication of the distributive element. However, unemployment currently stands at an estimated 250 000, over 25 per cent of all wage employment, and 125 000 school leavers attempt to enter the labour force each year (Ghai, 1971). There are also indications that '. . . present Kenyan politico-economic policies are basically benefiting a grand coalition of foreign interests and indigenous elite, at the expense of the mainly rural producers and workers of Kenya. It is felt that the present emphasis on production downgrades distribution issues and, indeed, that the type of economic strategy being pursued in Kenya leads almost automatically to ever widening gaps between income groups . . .' (Nellis, 1971, p. 2).

Kenya's rural development strategy

Rural development has been made a core element in the Kenya government's development strategy for the 1970s, outlined in the third Development Plan for the years 1970–74 (Republic of Kenya, 1969). Many current ideas on rural development in Kenya originated in a conference held at Kericho in 1966. Social scientists, government planners, administrators and some political leaders met to discuss related issues in education, employment and rural development; the conclusions were presented to the Government of Kenya and were summarised in a conference report published under the title, *Education, employment and rural development* (Sheffield, 1967). The overall recommendation was that the Government of Kenya should develop 'a comprehensive approach to rural transformation which would include short-term measures to promote agricultural productivity, a functional strengthening of the rural education system . . . and investments in the public sector and in private farming with long-term effects on agricultural productivity' (Sheffield, 1967, p. 5). The conference touched on many issues relating to rural development but proposed few concrete plans for action. Government planners, in formulating the new Development Plan, experienced some difficulty in implementing many of the suggestions made. As Nellis has pointed out, 'every involved person agreed on the urgency of the problem but no one proved able to state with sufficient precision the exact nature of a solution' (Nellis, 1972, p. 173).

The plan adopted by the government has a number of facets: increased agricultural production, a special rural development programme, and structures to allow a more effective local input to planning and the establishment of rural growth centres. These concepts were discussed at the Kericho conference. The concept of rural growth centres did not feature in the discussions, and it is interesting to note that it is this aspect of rural development strategy which contains a definite concept of spatial planning.

The basis of any rural development is of course increased agricultural production and the plan calls for a general rise in agricultural productivity by over 30 per cent between 1970 and 1974. In this area little new is planned and 'the path followed will not diverge markedly from that which has been established over the years since Independence' (Republic of Kenya, 1969, p. 166). Kenya has a diversified and productive agricultural base and output in this sector could be further increased. The plan intends to slow down the land transfer programme and to place emphasis on agricultural production especially in thc traditionally African agricultural areas. Expenditure on the land transfer programme is being drastically reduced and the main new expenditure is in the field of squatter settlement. It is likely that agricultural production in Kenya will meet the goals set out in the plan unless there are serious climatic problems. Failure of the rains can have a great impact on agricultural productivity. Apart from the weather the biggest factor affecting agricultural output is the price of Kenya's main exports on the world market. Between 1965 and 1970 the average price of these export crops dropped between 5 and 10 per cent. Kenya, like many other developing nations, has little power to affect this variable and can only hope that trade terms will move in her favour through favourable agreements with the main consumer countries.

A special Rural Development Programme grew out of suggestions made at the Kericho Conference that a series of small pilot projects should be established in which the concept of integrated rural development could be tested. This was followed up by a more detailed report which made eight specific suggestions as to how this was to be done:

(1) that a baseline survey of existing rural conditions be carried out so as to coordinate the many existing rural activities, and also so that new inputs could be evaluated;

(2) that all projects should be planned with a view to their 'replicability' so that projects found to be successful could be quickly and easily spread to other parts of the country;

(3) that a rural development national coordination committee be set up, possessing authority to integrate existing activity and to plan and implement new actions;

(4) that the first areas selected for any integrated effort should be drawn from various, considerably differing ecological and socio-economic areas within Kenya, the notion being that 'replicability' would be enhanced if the first few areas were each representative of a large number of other, similar areas;

(5) that in each project area a single government official should be placed in charge of all rural development efforts in order effectively to coordinate all activity;

(6) that an evaluation unit be built into the effort;

(7) that donors be utilised, but made to understand that this was an extra-ordinary programme demanding flexibility;

(8) that the programme be handled through the existing machinery of government and that one of the prime goals of the entire exercise be to enhance Kenyan capabilities in every respect of rural development planning and implementation.

These suggestions were adopted and work on the needed baseline surveys was begun in 1968. The size of the pilot unit was the subject of some discussion but it was eventually decided that 14 Districts in different areas of Kenya be chosen and that within each District a smaller administrative unit, known as a Division, be identified as the working unit. The Districts chosen were Kwale, Taita, Embu, Machakos, Meru, Murang'a, Nyeri, Baringo, Nandi, West Pokot, Kisii, South Nyanza, Busia and Kakamega. The baseline surveys took longer to carry out than had been anticipated and were not completed until early 1969. The full survey (Heyer, Ireri and Moris, 1971) led to the conclusion "that major economic breakthroughs of the "miracle" variety could not be expected, and that the problem of Kenyan rural development could only be solved by slow, steady incremental change on a broad front' (Nellis, 1972, p. 172); and it revealed the need for upgrading the efficiency of officials in the rural areas and for involving the local people in the planning process for their own development.

Many problems beset the programme between the completion of the baseline surveys and the implementation of specific plans for action; but eventually the programme moved into the planning phase and six of the fourteen Divisions had first phase plans developed. The remaining eight were to be developed in a second phase. Almost all of these pilot projects are being funded by external donors. Nellis (1972) argues that the major problem is lack of administrative coordination, both at the centre and in the field at the local level. He points out that 'for a multifaceted rural development operation to have any chance of success, it was essential that there be effective horizontal cooperation on all administrative levels, from Nairobi to the Division' (Nellis, 1972, p. 173). This, despite the existence of a National Rural Development Committee composed of the permanent secretaries of two of Kenya's most powerful ministries, Finance and Economic Planning, and the Office of the President, is most definitely lacking. Another major problem is the lack of effective Kenyan involvement in the Rural Development Programme. This programme is funded by external donors and the expatriate input in planning the programme is large. The problems arising from this special programme provide illustrations of what are felt to be much wider problems. One of these is Kenya's apparent dependence on overseas donors for much of its development. Another is the difficulty of creating structures which will ensure local involvement in developmental plans and strategies. The need for this involvement was stressed both by the Kericho conference and by further studies relating to the Rural Development Programme.

The ultimate key to success of any plan is the participation of the people. The 'Wananchi Declaration' (KPU, 1969) sums up the problem fairly well: 'Capital plays a large part in the rapid development of modern society. But in Africa above all, we must remind ourselves of the truth that it is and always will be man, working with his fellow men to apply his energies and skills to the land and materials around him—who creates development'. President Kenyatta adopted the national motto of *Harambee*—let us all work together—and significant local involvement in the form of self-help projects has taken place in Kenya. These are found throughout the country as a whole but are much more significant in some areas than in others. Nor are such efforts always the most effective that can be made; between 1964 and 1968, for example, 336 *Harambee* secondary schools were built; the effort of the people however declined after the construction of the school buildings and the government was expected to staff and run the schools which was by far the major financial effort required; no control of self help was in evidence and much effort was dissipated; the government had promised to support the self-help schools but found that it did not have the resources to do so; only seventy-three of the schools mentioned above were taken over by the government.

The colonial government of Kenya adopted a very paternalistic attitude towards Kenya's peoples. In some areas of Kenya many of the people still continue to regard the government as the provider, which weakens effective self-help effort at the grass-roots level. Participation has also been influenced by the political history of Kenya. Many of the smaller tribes were, and still are, distrustful of Kikuyu domination and are not particularly enthusiastic about plans which emanate from the Kikuyu dominated central government. Recent tensions between the Kikuyu and Luo have intensified this problem. The constitution under which Kenya was granted independence contained safeguards for minority groups in the form of local and regional governmental structures distinct from the central government. The principal manifestations of this at the local level were the powers vested in the elected County Councils. The government has steadily eroded and weakened these powers in a variety of ways; on January 1, 1970, the government removed from the County Councils the responsibilities for minor roads, primary education and health centres leaving the Councils as virtually powerless bodies. The reasons for centralising control were primarily political but there is ample evidence to show that many of the County Councils were not meeting their responsibilities effectively. This process has, however, meant that in many cases the local people felt they had no effective voice in their own affairs and in some areas has certainly made local involvement much more difficult.

The new plan establishes structures to facilitate local involvement both at the Provincial and the District level. Each District has a District Development Committee. This committee is the 'nerve ending' of government planning and is the lowest-level planning body in the rural areas to which the people can relate. The composition of this committee is interesting; it consists of the District Commissioner as chairman, the Provincial Planning Officer, the District Agriculture Officer, the Medical Officer of Health, the County Education Officer, the District Representative of the Ministry of Works, the Clerk to the County Council and the District Cooperative Officer. All of these, with the exception of the Clerk to the County Council, are government employees and in many cases they are not local people. There are no elected members and no provisions for local people to be part of this decision-making committee. Each of these committees is advised by a District Development Advising Committee but this is again dominated by central government employees. The chairman and secretary are the same and all D.D.C. members are also members of the D.D.A.C. The committee is supplemented by all Members of Parliament from the District, three members of the County Council, the chairman of the K.A.N.U. District Branch and two 'eminent citizens' nominated by the District Commissioner. Even in areas where the people are ardent supporters of the government it would appear that development is still controlled from above rather than coming from the grass roots level. The Districts are fairly large units rarely containing fewer than 200 000 people. Kenya has yet to create institutions which will encourage local participation on the scale which exists in Tanzania at the village level. Without this participation the chances of success of any plan are lessened.

The District Development Committees are vital links in the rural planning process. They represent the desire to decentralise the planning process and to take cognisance of local conditions. They are responsible for developing District Plans within the national planning framework which reflect the aspirations of the people of the District. Many of the divisional plans for the special Rural Development Programme were to come from this source. Nellis comments that 'the plans produced by this method were not outstanding. Some of the initial ideas produced by the district staff (many of whom had gone through previous planning exercises which had ultimately come to nothing and were thus weary and cynical) were little more than "shopping lists" of desirable numbers of new staff, houses, vehicles and speciality programmes, such as nursery schools for community development. It rapidly became clear that for plans to be produced on the district level or below would require a major educational effort and a considerable period of time' (Nellis, 1972, p. 175).

The special Rural Development Programme set out to test and design plans for integrated rural development and has met with only a limited degree of success. Strangely the planners seem to have ignored the concept of physical development planning during their testing process, yet this concept is an important part of Kenya's rural development strategy as outlined in the Development Plan. Physical development planning in Kenya entails 'a synthesis of development resources and of the planning proposals of all the various ministries and agencies concerned with physical development' (Republic of Kenya, 1969, p. 82). Planning revolves round the concept of growth centres of which there are four main types: urban centres, rural centres, market centres and local centres in descending order of importance and size. Urban centres are medium-sized towns in the Kenya context which have been designed to serve as the main commercial centres for an entire district. Generally they also contain the district administrative headquarters. The Development Plan estimates that they are likely to have a population of more than 10 000 by the year 2000 and that some, such as Malindi, will probably have a population of over 50 000. There are thirty-six designated urban centres and their distribution is shown on Figure 9.1.

Rural centres have been designated for development to provide at least 40 000 people in rural areas with administrative, social and commercial services. It is expected that with the progressive economic development of the rural areas they will grow into small towns of between 2000 and 5000 within twenty or thirty years. These rural centres are particularly important

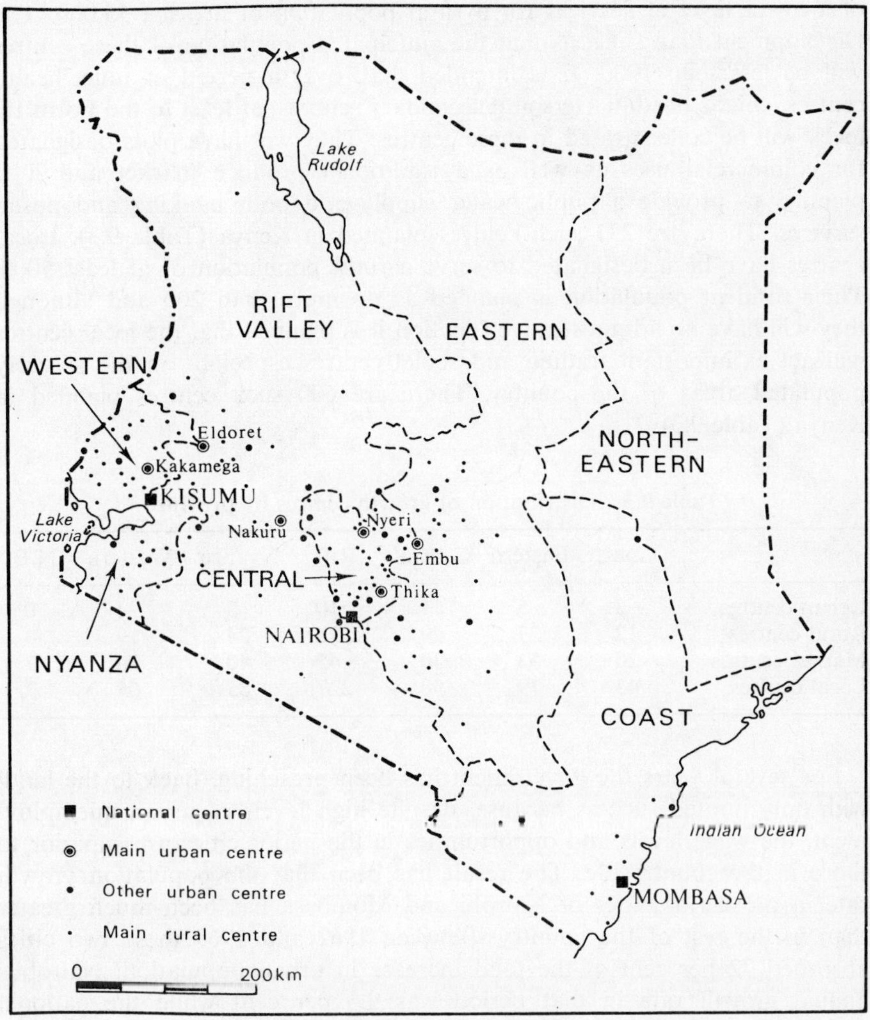

Figure 9.1 Planned growth centres in Kenya

in that they are geared to improving both the quality and quantity of services and amenities in the rural areas. It is planned to concentrate administration and other services in these centres and to service them with public water supply, electricity and all-weather roads. These rural centres are also felt to be the most suited for the development of rural cottage industries. There are 140 such designated centres and the distribution of the main centres is shown in Figure 9.1. A detailed breakdown by province appears in Table 9.3. Market centres have been designated for development of a lower level of services for a rural population of at least 15 000. The Development Plan suggests that the anticipated population of these centres will be less than 1000. It is intended that, over a period of time, health centres, chief's headquarters and secondary schools, at least to the Form III level, will be concentrated in these centres. They will have plots designated for commercial uses as well as a traditional produce market and it is planned to provide a public water supply and both banking and postal services. There are 231 such centres planned in Kenya (Table 9.3). Local centres have been designated to serve a rural population of at least 5000. Their resident population is planned as no more than 200 and although they will have no administrative function it is planned that the local centres will act as important trading and social centres especially in the sparsely populated areas of the country. There are 600 such centres planned in Kenya (Table 9.3).

Table 9.3 Distribution of growth centres by province

	Coast	Eastern	Central	Rift	Nyanza	Western	N.E.
Urban centres	3	5	11	10	3	4	0
Rural centres	12	27	29	26	24	19	3
Market centres	20	53	30	45	40	36	7
Local centres	43	79	98	227	85	65	3

For several years the government has been preaching 'back to the land' with only limited success because, despite high levels of urban unemployment, the wage levels and opportunities in the major cities are superior to those in the countryside. The result has been that the population growth rate in the major cities of Nairobi and Mombasa has been much greater than in the rest of the country. Between 1962 and 1969 these two cities absorbed 72 per cent of the total increase in urban population. Nairobi's annual growth rate in that period was 9·6 per cent while the national population was increasing at 3·3 per cent. The migration from rural areas to urban areas is of course a world-wide phenomenon but in Kenya the problem is intensified by the very wide gulf between the two main cities and the rest of the country. Urban life, except along the coast, was introduced by Europeans and in Kenya the transition from rural to urban is sudden and dramatic. Most towns did not grow naturally—they were artificially created. Over 70 per cent of Kenya's urban population lives in Nairobi and Mombasa and there are only eleven towns with a population of over 6000. The third largest town in Kenya, Nakuru, is less than one-seventh the size of Nairobi.

The concept of growth centres represents an attempt to decentralise and to slow down the movement from countryside to town. The Kenyan Government realises that 'the concentration of all economic, social and political life in the two main cities carries dangers of an economic and cultural gulf being created between them and the rest of the economy' (Republic of Kenya, 1969, p. 85). Seven strategically located towns—Kisumu, Nyeri, Kakamega, Eldoret, Nakuru, Embu and Thika (Figure 9.1)—have been designated as major urban centres and a policy of limited decentralisation of government and commercial services is proposed at the regional level. It is also planned to provide incentives for industrial development.

In the rural areas government services will be concentrated in the designated growth centres. The rationale behind this is that 'in a very real sense rural development will imply rising urbanisation, not in the major cities . . . but through growth of commercial activity in a large number of small trading centres. The plan designates rural growth centres as foci of trade, social services and communications which serve surrounding farm areas and which can significantly alter the pattern of migration and provide more even development of the nation as a whole' (Republic of Kenya, 1969, p. 168). All new infrastructural investment will be concentrated in these centres and 'proposals involving development of service facilities within non-designated centres will therefore be discouraged' (Republic of Kenya, 1969, p. 89).

The growth centre concept and rural development

The designated growth centres could be a vital cog in rural developmental planning, but in order for this to be achieved modifications to the current application of the concept will have to be made. One of the major problems is that the general theory of growth centres is very poorly developed; Berry (1972) refers to it as being 'sadly deficient'. A review of the literature reveals that the label is used not only to describe differing and quite distinct concepts, e.g. central place or growth pole, but also to describe geographical, social, economic and political aspects of developing urban systems. Indeed, as Darwent (1969) notes, unless this concept is defined more rigidly it may prove of little use for analytical, explanatory or planning purposes. Darwent does not suggest such a precise definition but he does provide guidelines within which the theoretical and empirical specifications of the concept of 'growth centres' may be determined. Generally a growth centre is 'a central place structure, which will maximise the interdependency of the peripheries of each polarised region with its respective centre by gathering flows from the periphery and directing them up through the hierarchy, and by bringing the benefits of urban life to the smaller centres' (Darwent, 1969, p. 14). When applied to regional development policy the notion implies 'that investment is best concentrated in growth centres rather than scattered around in some vague quest for "balance" or "equity"' (Darwent, 1969, p. 13). Though related to central place theory the growth centre concept is more closely allied to the idea of a growth pole, which involves economic space defined in terms of forces of attraction and repulsion between one centre and various others. A growth centre in functional terms in rural Kenya has two purposes: firstly, to provide adequate services for the rural areas and, secondly, to stimulate the economic and social development of the area around it. This is to be achieved by creating the hierarchical network of centres described earlier and outlined in Figure 9.1. In theoretical terms the strategy used leans very heavily on central place theory; 'patterned upon central place theory, this schedule was designed to create an hierarchical network of service centres much more complete than that presently existing, with nodes ranging in size from local villages up to Nairobi itself' (Ominde *et al.*, 1971, p. 17). This theoretical framework is adequate to meet the first purpose of providing the most efficient service pattern for the rural areas but is less than adequate for the second purpose of stimulating economic and social development. Kenya has chosen its 'growth centres' from existing towns and villages but in many cases, especially in the rural areas, the centres chosen may be adequate service centres but they are not growth centres. For the purposes of growth a different methodology should have been used which in many cases would have resulted in a different centre being chosen. Van Dusseldorp (1971) stresses the vital need for evaluation of plans of this type and such an evaluation in Kenya may lead to modifications of the plan which will greatly increase its effectiveness.

The methodology used in Kenya to select growth centres in the rural areas has been outlined by Carvalho (U.N. Economic and Social Council, 1969) and was developed first in Central Province. In essence all the existing centres were surveyed. An inventory of all the services judged to have a central place function was prepared. Data were collected for five categories of function: administration and protection; social services; communications and transportation; commerce; and industry and power. Each category was subdivided to give twenty-one specific subservices. Each of these subservices was assigned points on the basis of the quality of the service offered; quantity was not considered. Three points were allotted for a high level service, e.g. a hospital; two points for a medium level service, e.g. a health centre; and one point for a low level service, e.g. a dispensary. The points accumulated in each centre were totalled and it was assumed that the more points a centre had the greater was its centrality and the wider its sphere of influence. Variety of service was also included, consequently a place had to have at least one subservice in four out of the five main categories of function to qualify as a central place. The maximum number of points possible was fifty-one and the minimum was set at eight.

In Central Province out of 250 centres studied 134 qualified as central places. Those qualifying were then ranked according to their total accumulated points and plotted on a graph (Figure 9.2) to 'show the number of service centres at each level and to determine their hierarchical grouping' (U.N.E.S.C., 1969, Appendix, p. 2). Four groupings were distinguished on the graph; 8–17 points, 18–24 points, 25–37 points and 38–51 points (Figure 9.2). 'These groupings were however slightly modified in the final grading of the centres since it was found on inspection that their functional groupings did not correspond exactly with the grouping by level of service as determined from the graph' (U.N.E.S.C., 1969, Appendix, p. 2). These four levels of service centres were classified in descending order of hierarchical

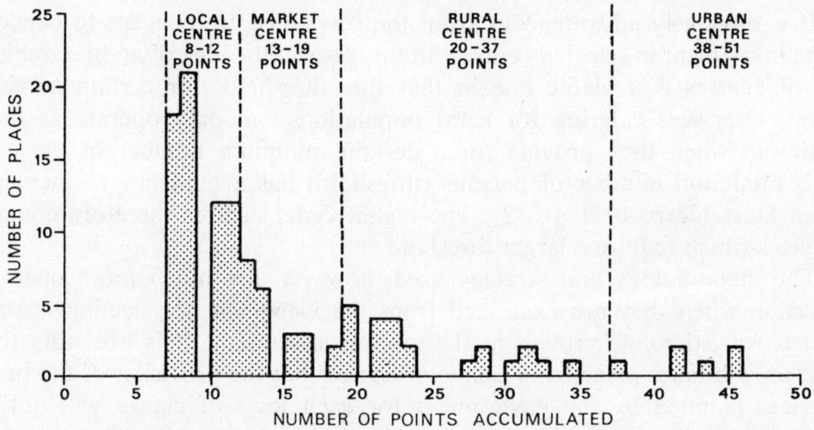

Figure 9.2 Central places in Central Province, Kenya (after UNESC, 1969)

importance as Urban Centres (38–51 points), Rural Centres (20–37 points), Market Centres (13–19 points) and Local Centres (8–12 points). Maps were drawn to show the service area of each centre 'on the basis of topography, road pattern, the postal box location of primary schools and, in a few cases, local survey' (U.N.E.S.C., 1969, Appendix, p. 3), and details of health services and secondary schools were superimposed on a population distribution map.

The final choice of growth centres was then made with the central place data providing the basic material for selection. Other factors were included: the economic potential of the area; transportation accessibility; population distribution and density; topography and water potential. These, however, do not seem to have been considered in as systematic a way as the central place factor which clearly dominated the selection process.

The study in Central Province resulted in a Physical Development Plan for that province being prepared (Republic of Kenya, 1967). Similar plans using the same methodology are being drawn up for the other provinces with those of Nyanza, Western, Eastern and Coast Provinces already published (Republic of Kenya 1970a, c and d; and 1971a). The surveys on which these plans are based revealed an apparent lack of centrality in many areas. Carvalho (U.N.E.S.C., 1969) reports that in Central Province 35 per cent of the dispensaries and 40 per cent of the secondary schools were 'located at random'. In many areas administrative centres were separate from trading centres which in turn were separate from mission centres. Ligale comments that 'the diseconomy of this random scattering of physical development became very evident when the government attempted to plan all weather road access, piped water supply or electricity distributor grids to serve these fractured communities' (Ominde *et al.*, 1971, p. 17).

It is obviously advantageous from the service centre concept to concentrate investment in selected centres. In this respect the idea of an hierarchical set of centres is a viable one in that the 'threshold' for certain services varies—'services catering for rural populations can only operate at their optimum when they provide for a certain minimum number of persons. This minimum number of persons (threshold) has a tendency to increase' (Van Dusseldorp, 1971, p. 12). The higher order centres therefore contain services which require a larger threshold.

The methodology and strategy used, however, are much more open to question when they are examined from the viewpoint of selecting centres which will stimulate growth in the area around them. It is probably true that growth takes place from points or centres but the provision of the basic services planned by the government for each level of centre will not in itself bring about growth. For rural development to occur these centres must be used as injection points for innovation both in the economic and social senses. Roads, water and electricity probably provide the infrastructure 'necessary for rural areas to integrate themselves into the modern economy' (Ominde *et al.*, 1971, p. 17), but more is required. Van Dusseldorp (1971) argues that the creation of such centres is advantageous in itself. The concentration of services in selected places facilitates interaction between these centres. Consequently, if all government services in a given area are concentrated in one centre, when a farmer visits that centre he is more likely to use more than one service. If the agricultural structure is to be improved and production increased, a complex of radical measures must be taken—'only when it forms part of an integrated development plan for an area can a satisfactory plan for the spatial structure of socio-economic services be drawn up' (Van Dusseldorp, 1971, p. 19).

If the growth centres are considered to be core areas for agricultural and rural development then some of the problems of the lack of administrative integration in the rural areas mentioned earlier might be alleviated. In each District, in one of the growth centres, a District Development Centre is to be established which could coordinate all rural development in the area. For the first time an effective means of coordinating the various agencies working in the rural areas will be created. The growth centres are at present peripheral to rural development strategy, being viewed more as service centres than as growth centres. In the special Rural Development Programme they appear to have been completely ignored, and although the Development Plan suggests that the growth centres are focal points this is not at present the case. Berry (1972) argues that growth cannot be separated from the diffusion of innovation. He argues that growth centre theory can be enriched by placing it within the context of innovative diffusion. Berry states that 'diffusion theory provides a sound conceptual base for the growth centre idea. Growth occurs as a consequence of the filtering of innovations downwards through the urban hierarchy, and the spread of use of the innovations among consumers residing within the urban fields of the adopting centres' (Berry, 1972, p. 359). The effect of a given innovation is described as a declining function of time and is also subject to a threshold limitation, a minimum size of region beyond which diffusion will not proceed. Consequently the lowest levels of welfare are in areas peripheral to small urban centres in the outlying hinterland regions. Berry examines the question of how growth centres can be used to induce development and suggests a number of conclusions relating to threshold limitations, diffusion times and accessibility. Berry's view is that any policy that reduces threshold limitations will induce development further down the hierarchy and out into the peripheral regions; and he argues that the extent to which families make use of new innovations in urban centres that have adopted them is a function of their access to these centres. Any decrease in effective distance will therefore tend to increase the use of innovations.

In Kenya, where the aim is to transform the rural areas, Berry's ideas have some interesting implications. If innovation filters down through a hierarchy with both time and distance decay effects, why should an urban hierarchy of the four-fold nature proposed be created at all at this point in time? Could innovations not be introduced at the lowest level of urban centre and should these centres not be given the greatest degree of initial attention? Berry's conclusions on threshold limitations, accessibility and diffusion times lend strength to the argument that, if the stimulation of growth is the primary aim, then perhaps an increased number of rural centres rather than the hierarchy of centres proposed might be a more effective strategy. A hierarchy of centres may emerge later.

If the centres chosen are to be chosen primarily as growth centres rather than service centres then what methodology should be used to select them, given the fact that hundreds of small villages already exist? An understanding of the settlement history in the rural areas is required. The first central places in Kenya were probably the periodic markets. These had no permanent buildings but certainly offered centrality functions (Taylor, 1968); often the marketing system was a well-developed spatial system with a hierarchy of markets existing (Taylor, 1969). When the colonial government took charge, another system of central places was created for administrative purposes. These were the District, Divisional and Locational headquarters which were selected for administrative convenience or because of the personal preference of the earliest administrator. In some instances existing nucleation points such as the periodic markets were deliberately avoided. These colonial centres tended to add service functions as government bodies grew in complexity. A few shops often emerged, but these were to serve the needs of the civil servants rather than those of the people of the area. The earlier spatial system of the traditional markets continued to exist and in some instances grew rather than declined. It is perhaps not surprising that colonial government marketing systems for local agricultural produce based in their administration centres were not very successful: they were in direct competition with an already well-developed system. The mission stations of the various religious groups were often established at the same time and again these often avoided both the colonial centre and the traditional nucleations. In some areas trading centres were gazetted separately from administrative centres. These were dominated by Asian traders and tended to be 'extractive' in character. Again, traditional nucleations were avoided. Thus there emerged in Kenya not one spatial system but several, each serving a different function. There was little integration between them even in cases where all occupied the same site, as sometimes occurred. In such instances the town or village would invariably have two or three parts: an administrative *boma*, a trade centre or market place and a mission station, and each would be physically as well as functionally separate from the others.

A new spatial system is beginning to emerge in some areas of Kenya. In Central Province during the 1960s increasing agricultural progress and wealth brought with it a demand for services and goods which did not exist before. The result was the fairly rapid growth of what had been small villages into much larger centres. The first manifestation of this was an increase in the number of *dukas* or general stores. This was accompanied by a move towards specialisation with general stores giving way to butcher's shops, tailor's shops, bakeries etc. These emerged at sites which were convenient to the people and perhaps not surprisingly these often coincided with the site of the traditional market (Taylor, 1968) with the shops growing up round the market perimeter.

In choosing growth centre locations this pattern of development could be utilised. In the methodology currently used all of the services developed in a centre are treated as if they were functionally integrated. No consideration is given to the time element, yet it is probably vital to an understanding of the centre being assessed. Of the five main categories of function the commercial one is the most vital in choosing a growth centre. An examination of the traditional marketing pattern together with an analysis of where and when shops are emerging would provide vital indicators. Included in this analysis should be both the total numbers of shops and the degree of specialisation. This would give a good indication of those centres which the people found most convenient to visit.

Central Province is perhaps the most advanced area in Kenya and one might question whether a pattern developed there might be appropriate to other areas. An examination of Coast Province (Taylor, 1970) reveals a

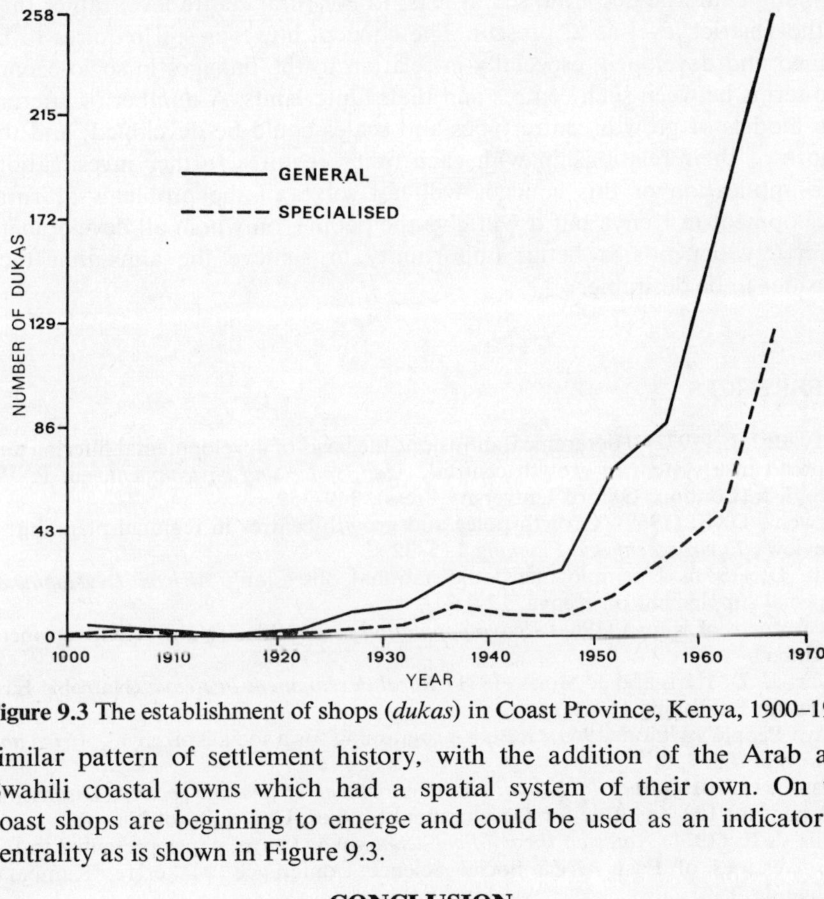

Figure 9.3 The establishment of shops (*dukas*) in Coast Province, Kenya, 1900–1970

similar pattern of settlement history, with the addition of the Arab and Swahili coastal towns which had a spatial system of their own. On the coast shops are beginning to emerge and could be used as an indicator of centrality as is shown in Figure 9.3.

CONCLUSION

The growth centre concept includes both the idea of growth and the notion of service centres. The concept of growth should be the primary one in selecting such centres. This may involve the removal of existing government services in administrative centres formed during the colonial period. Fortunately in Kenya the level of capital investment in such centres is not usually high. Any growth centre can have the service element added; not all service centres can become growth centres. The growth centre concept is an exciting one; it provides an opportunity for integration of innovative inputs by all agencies working in the rural areas. The spatial aspects of the concept emphasise the distributive element which some feel is lacking in Kenya (Leys, 1970) but which should be a central feature of Kenya's rural development strategy. Growth centres can be established all over the country on a small enough scale to allow effective local involvement. Local planning could be decentralised at least to the rural centre level rather than at the District level as at present. The concept however still requires to be refined and developed, especially in relation to the linkages in socio-economic terms between such centres and their hinterlands. A number of alternative models of growth centre types and scales could be developed; and the nature of their relationship with each other requires further investigation. The application of this concept will not solve all the problems of rural development in Kenya but it will give the people, on whom all development ultimately depends, a better opportunity to achieve the aims that they consider to be desirable.

REFERENCES

Berry, B. J. L. (1972) 'Hierarchical diffusion; the basis of developmental filtering and spread in a system of growth centres', *Man, space and environment*, ed. P. W. English (London: Oxford University Press), 340–359.

Darwent, D. F. (1969) 'Growth poles and growth centres in regional planning: a review', *Environment and Planning* **1**, 5–32.

Ghai, D. (1971) 'Unemployment: the national obsession', *African Development*, special supplement on Kenya, 12–15.

Government of Kenya (1964) *Kenya population census 1962* (Nairobi: Government Printer).

Heyer, J., D. Ireri, and J. Moris (1971) *Rural development in Kenya* (Nairobi: East African Publishing House).

Kenya People's Union (1969) 'KPU's programme for a socialist Kenya', *Africa and the World* **5**, 3–8.

Leys, C. (1970) *Politics in Kenya: the development of peasant society*, Discussion Paper No. 102, Institute for Development Studies, University of Nairobi.

Nellis, J. R. (1971) *Notes on the political economy of Kenya*, paper presented to the Universities of East Africa Social Science Conference, Makerere, Kampala, Uganda.

Nellis, J. R. (1972) 'The administration of rural development in Kenya: plan formulation and implementation in the Special Rural Development Programme', *Issues in African Development* (Ottawa: Canadian Association of African Studies) 168–182.

Ominde, S. H., A. N. Ligale and A. B. Cahusac (1971) *Urbanisation and environment in Kenya*, paper prepared for the United Nations Conference on the Human Environment under the direction of the Working Committee on the Human Environment in Kenya (Nairobi: mimeo).

Republic of Kenya (1965) *African socialism and its application to planning in Kenya*, Sessional Paper No. 10 (Nairobi: Government Printer).

Republic of Kenya (1967) *Central Province Regional Physical Development Plan*, (Nairobi: Town Planning Department, Ministry of Lands and Settlement).

Republic of Kenya (1969) *Development Plan 1970–74* (Nairobi: Government Printer).

Republic of Kenya (1970a) *Eastern Province Regional Physical Development Plan* (Nairobi: Town Planning Department, Ministry of Lands and Settlement).

Republic of Kenya (1970b) *Kenya Population Census, 1969* (Nairobi: Statistics Division, Ministry of Finance and Economic Planning).

Republic of Kenya (1970c) *Nyanza Province Regional Physical Development Plan* (Nairobi: Town Planning Department, Ministry of Lands and Settlement).

Republic of Kenya (1970d) Western Province Regional Physical Development Plan (Nairobi: Town Planning Department, Ministry of Lands and Settlement).

Republic of Kenya (1971a) *Coast Province Regional Physical Development Plan* (Nairobi: Town Planning Department, Ministry of Lands and Settlement).

Republic of Kenya (1971b) *Economic Survey 1970* (Nairobi: Government Printer).

Sheffield, J. R. (ed.) (1967) *Education, employment and rural development* (Nairobi: East African Publishing House).

Swynnerton, R. J. M. (1955) *A plan to intensify the development of African agriculture in Kenya* (Nairobi: Government Printer).

Taylor, D. R. F. (1968) 'New central places in East Africa', *African Urban Notes* **111**, 15–29.

Taylor, D. R. F. (1969) 'Agricultural change in Kikuyuland', *Environment and land use in Africa*' ed. M. F. Thomas and G. W. Whittington (London: Methuen), 463–493.

Taylor, D. R. F. (1970) *Development of central places in Coast Province, Kenya* (Ottawa: Carleton University).

Taylor, D. R. F. (1971) 'Changing food habits in Kikuyuland' *Canadian Journal of African Studies* **4**, 333–349.

United Nations Economic and Social Council (1969) *Regional physical planning in Kenya; a case study* (Addis Ababa: E/CN.14/HOV/35) **18.**

Van Dusseldorp, D. B. W. N. (1971) *Planning of service centres in rural areas of developing countries* (Wagenigen, Netherlands: International Institute for Land Reclamation and Improvement).

ACKNOWLEDGEMENT

The author gratefully acknowledges the assistance given by the Canada Council to his research.

Chapter 10

Rural Migrations and the Problem of Agricultural Labour in Western Tropical Africa

Reuben K. Udo

Introduction

The problem of inadequate labour for developing the rural areas of West African countries featured prominently in a seminar on modern migration in western Africa held in April 1972 at Dakar, Senegal. Countries such as Ghana and Ivory Coast depend to a considerable extent upon migrant labour from other countries for the development of their agricultural and mineral resources, and although the authorities of these countries are very concerned about the heavy dependence of their economies on foreign labour, it is quite clear that migrant labour from neighbouring densely settled countries is indispensable for the continued growth of the rural economy of these two states. During the colonial period when labour migrations were relatively unrestricted, certain labour-deficit but developing districts continued to suffer from acute shortage of labour, in spite of increasing immigration into such areas. Stricter immigration controls since the middle 1960s and the more recent restrictions on the employment of non-nationals have, as may be expected, worsened the labour situation in some countries. The loss of able manpower from the areas where the migrants originate has also created some problems of development. Aspects of these problems discussed in this chapter include the pattern and magnitude of labour migrations and the practical implications of these movements for food production for the increasing non-farm population as well as for export agriculture which, for most countries in West Africa, provides the main source of foreign exchange earnings.

Dr. Udo is Professor of Geography at the University of Ibadan, Nigeria.

In spite of the greater attention which rural–urban migration in West Africa has attracted in recent years, the great majority of migrants originating from rural areas still move to other rural areas. In Ghana, for example, between 1900 and 1950, nine out of every ten migrants who moved from a rural area went to another rural area; and according to the 1960 post-census enumeration survey for that country, 70 per cent of all migrants settled in rural areas (Addo, 1972). The situation in other countries is very similar, although marked changes are likely in view of the rapid growth of urban centres since 1960, during which year many West African countries obtained independence from colonial rule.

Background to the labour situation

Under the essentially subsistence economy of the pre-colonial period, agricultural labour in western tropical Africa was provided by the extended family or by slaves. Village and clan chiefs depended largely on labour provided by younger members of the community who were generally organised into age-groups, while elderly people supplemented their family labour from small groups organised by their sons-in-law. Amongst the generality of the people voluntary work groups were common, consisting of members of an age-group or kinsmen who worked together in turns in each member's farm (Ojo, 1963). In such cases the only form of payment consisted of food and drinks which the owner of the farm provided for the workers. Paid labour was not unknown but was restricted and payment was usually in kind.

Labour migrations were also restricted both in volume and in the distance covered by the migrants. In the first place there was no great demand for such labour under the traditional social order except by widows who had neither adult sons nor sons-in-law. Secondly, the general situation of social insecurity prevented long-distance movements, thereby restricting labour migrations to within the village territory or clan area from which the migrants originated. This situation continued during the early years of the colonial period when the colonial administration found it difficult to obtain labour for essential development projects such as the building and maintenance of roads and the construction of public buildings. Hence the adoption of forced labour in many colonial territories, but particularly in French-administered territories, where village and clan chiefs were required to provide a specified number of adult males to perform certain duties, such as carrying loads or working on specific projects, e.g. the construction of public buildings, roads, railways or ports (Skinner, 1965; Anyanwu, 1971).

By 1920, the labour situation had started to change radically in Ghana, Nigeria, Sierra Leone, Niger and Mali; and the concern of the colonial administration was no longer one of obtaining adequate labour for certain projects but, rather, how to provide work opportunities in some districts for the increasing number of unemployed and underemployed people. The economy had become largely monetised and almost everything including food, services and taxes had to be paid for in cash. Farmers were encouraged to produce export crops and in parts of Ivory Coast each farmer was compelled to cultivate small plantations of cocoa (Thompson and Adloff, 1958). But family labour became inadequate as more and more children went to school and the need for wage labour increased. The development of mining, commercial plantations and large-scale exploitation of timber resources also created opportunities for wage labour. An increasing number of people migrated to work in the developing port towns along the coast and in the newly-established administrative towns of the interior; and farmers had to increase their output to meet the rising food demand of the growing proportion of the non-farm population.

The situation since 1960 has been one in which the more prosperous agricultural district of southern Ivory Coast, southern Ghana, south-western Nigeria and the Senegalese coastlands attract large numbers of migrants from those districts where opportunities for earning a livelihood are very restricted as a result of pressure of population on available resources. Often the migrants have to travel very long distances which may exceed four hundred miles and it is quite clear that the development of road transport has played an important part in the migration process.

The environmental basis of West African labour migrations

In this chapter reference is made to a broad division of West Africa into interior and exterior sectors. These terms are used in place of 'forest' and 'savanna '—terms of identification commonly used in the geographical literature of West Africa—because, even though the exterior sector is largely covered by forest, the western part, including Senegal and Gambia, is savanna country, and also because much of the coastlands between Accra (Ghana) and Lagos (Nigeria) support grassland rather than forest vegetation. The exterior sector therefore consists of the coastlands of Guinea and extends inland for about one hundred miles, while the interior sector consists of the vast area of the Guinea and Sudan savannas which extend northwards to the southern edge of the Sahara desert.

Aspects of the natural environment, cultural practices and the range of opportunities for employment vary considerably between the sectors and are basic both to our understanding of the pattern of, and also to some of the reasons for, the migratory flows discussed in this chapter. The most important factor responsible for these differences is location, relative to the sea as well as to the Sahara desert. By virtue of its location near to the sea, the exterior sector which comes under the full influence of the rainbearing south-west monsoon winds receives adequate rainfall for most of the year. Unlike the drier north, it has adequate moisture for tree-crop production and a wide range of crops and therefore offers, in terms of agricultural production, a much greater variety of opportunities. The exterior sector is the zone of large-scale development of tree-crop production on peasant farms and commercial plantations and is an area of intensive exploitation of timber resources both for the home and the export market. It is also much better endowed with mineral resources, including petroleum (Nigeria) gold (Ghana and Sierra Leone) iron ore (Sierra Leone, Guinea and Liberia) and diamonds (Ghana and Sierra Leone). In addition, most of the state capitals and all the seaports are located in this sector, which has a more developed transportation network and most of the industrial establishment of western tropical Africa. Economically, the exterior sector is far more developed and more accessible than the interior sector and has naturally attracted an increasing number of migrants from the interior.

During the pre-colonial period, however, the drier interior sector was more advanced economically and politically than the exterior sector. At that time, the north had trading and cultural contacts with the Arab world, by way of the Saharan caravan routes, and movements of people, ideas and goods were relatively easy, thanks to the use of animal transport which featured the camel, the horse, the donkey and even cattle—animals that are still largely restricted to the interior sector. It was in this open grassland area that the famous and ancient empires of Mali, Ghana, Songhai, Bornu and more recently Sokoto-Gwandu throve (Forde, 1953). The protection that these empires offered their people led to the concentration of population in the vicinity of centres such as Kano, Sokoto and Ouagadougou and in southern Mali; today these areas constitute major source regions of migrants (Figures 10.1 and 10.2).

General pattern of migratory flows

The mainstreams of migration in western tropical Africa flow into the more-developed coastal areas, most of the migrants originating from the less-developed interior sector (Figure 10.1). And since the better endowed forest belt is located east of longitude 13°W, the main component of the migratory flow is from north to south. The more important north-to-south movements include: Mossi migrations into southern Ivory Coast and southern Ghana (Skinner, 1965); Hausa migrations into south-western Nigeria and south-western Ghana (Prothero, 1958); migrant labour from north-eastern Ghana to the Ashanti cocoa belt (Fortes, 1936); and migrations from southern Mali into southern Ivory Coast. Important east-to-west movements include: Ibo migrations into south-western Nigeria (Udo, 1970); the *navétanes* (strange farmers) who migrate seasonally from Mali

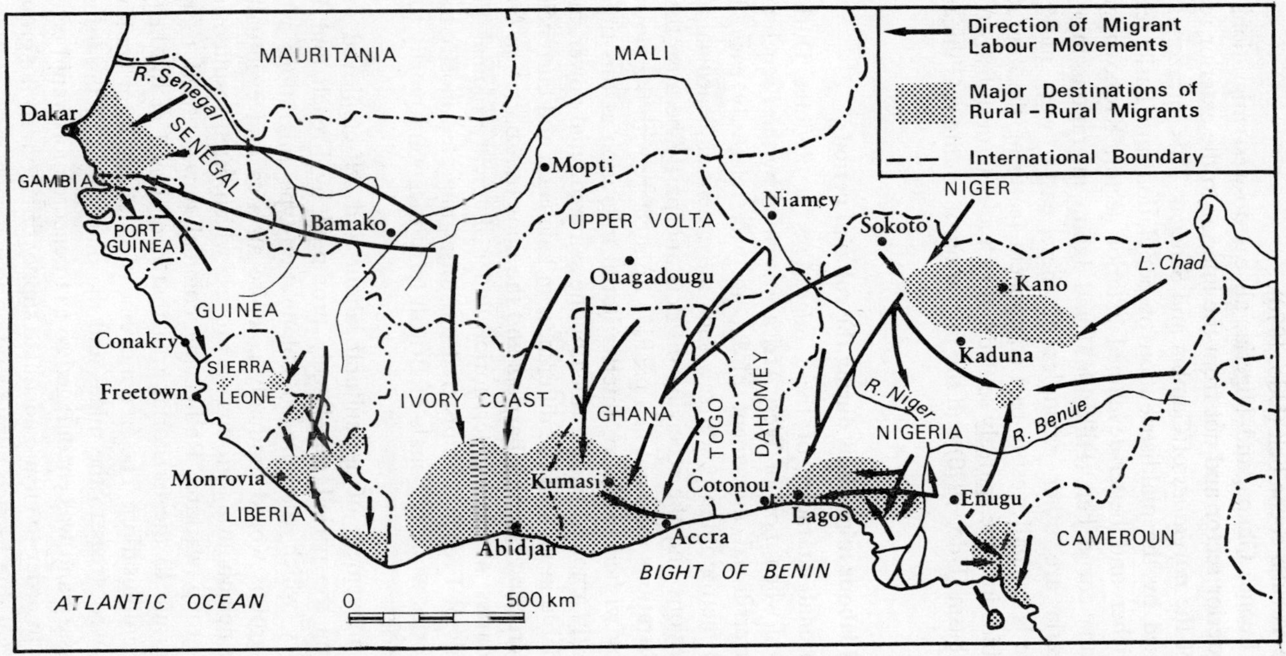

Figure 10.1 West Africa: rural migrations (adapted from R. M. Prothero, 'Migrant labour in West Africa', *Journal of Local Administration Overseas* **1**, no. 3, July 1962, p, 151, reproduced by permission of the Controller of Her Majesty's Stationery Office)

and Guinea into Senegal and Gambia (Jarrett, 1949); and Urhobo migrations into the Yoruba cocoa belt (Udo, 1970).

In Sierra Leone, Ghana and Nigeria, all east-to-west migrations occur within the exterior sector and not from the interior to the exterior sector as in the case of the *navétanes* of Gambia and Senegal. This intrazonal movement is caused by the maldistribution of population and basic resources within the richer and better-developed exterior sector. Krobo migrations into the Ghana cocoa belt (Hill, 1963) and Igbira migrations into south-western Nigeria are other examples of these east-to-west movements. Finally, there is a small west-to-east component featuring the migration of Ibos and Ibibios into the sparsely settled but rich agricultural districts of the Cross River basin (Udo, 1970) and the migration of southern Guineans into Ivory Coast.

International labour migrations during the colonial period

The international character of these migrations is obvious. During the colonial period when most of West Africa was ruled by the French and the British, interterritorial movement of labour was relatively easy, largely because each metropolitan power saw such movement as beneficial to both the source regions and the receiving districts and partly because the territorial administrations were not in a position to interfere with the movements even had they wished to do so. Interterritorial migration at this time was, however, mostly amongst people ruled by the same colonial power, and this is understandable in view of the differences in languages and currencies, the problems of crossing customs barriers and the poor transportation facilities across boundaries separating territories under different colonial powers. Thus nationals of French-administered territories migrated mostly to other French territories while nationals of British territories went mostly to other British territories.

Increasing movement of agricultural labour and self-employed migrant farmers during the period was largely prompted by French and British colonial policy which discouraged colonisation-type migrations whereby land-hungry groups would gradually colonise an area and eventually displace weaker neighbouring groups—a process which often resulted in prolonged inter-group warfare. Rather the colonial powers tried to restrict each ethnic group to the territorial area that it occupied at the beginning of European colonisation. The free movement of labour from land-hungry and less-endowed areas to the mines and the plantations of the forest belt became necessary and was encouraged so as to ensure more rapid economic growth. Thus in order to ensure adequate supply of labour to the groundnut fields of Senegal, the French administration issued each migrant from Mali and Guinea with a card which secured him free transport by rail.

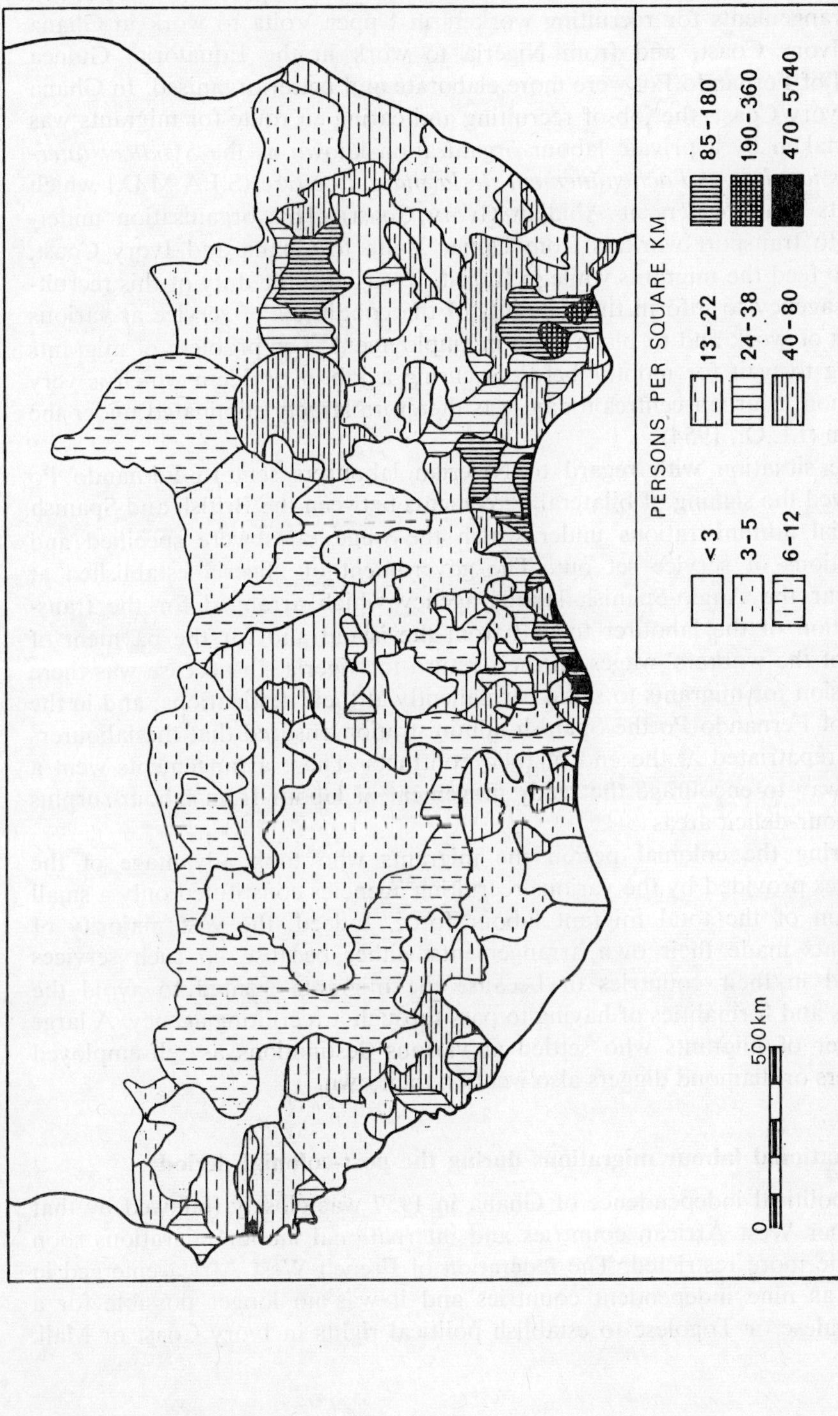

Figure 10.2 West Africa: population density about 1950 (source: W. B. Morgan and J. C. Pugh, *West Africa* (London: Methuen), 1969, p. 8)

Shelter and medical care were also provided at transit points (I.L.O., 1954).

Arrangements for recruiting workers in Upper Volta to work in Ghana and Ivory Coast, and from Nigeria to work in the Equatorial Guinea island of Fernando Po, were more elaborate and better organised. In Ghana and Ivory Coast, the job of recruiting and caring en route for migrants was undertaken by a private labour organization known as the *Syndicat interprofessional pour l'acheminement de la main d'oeuvre* (S.I.A.M.D.) which had its headquarters at Abidjan (Ivory Coast). This organisation undertook to transport workers from Upper Volta to Ghana and Ivory Coast, and to feed the migrants while in transit. It was also the duty of this recruitment agency to inform the migrants of the conditions of service at various places of work and to place them in employment. The problem of migrants having to wait for months before securing a job, a problem which is very common in urban centres today, was therefore largely eliminated under the system (I.L.O., 1954).

The situation with regard to Nigerian labourers sent to Fernando Po involved the signing of bilateral agreements between the British and Spanish colonial administrations under which the wage rates were specified and conditions of service set out. The government of Nigeria established at Calabar the Anglo-Spanish labour agency which arranged for the transportation of the labourer to and from the island and for the payment of part of the workers' wages on their return to Nigeria. In no case was there provision for migrants to settle permanently at their destinations; and in the case of Fernando Po the Spanish administration ensured that the labourers were repatriated at the end of their contracts. These arrangements went a long way to encourage the ready movement of labour from labour-surplus to labour-deficit areas.

During the colonial period the migrants who took advantage of the services provided by the various recruiting agencies constituted only a small fraction of the total migrant labour force. Indeed, the vast majority of migrants made their own arrangements, either because no such services existed in their countries or because the migrants wanted to avoid the delays and formalities of having to pass through a recruiting agency. A large number of migrants who settled at various destinations as self-employed farmers or diamond diggers also went on their own.

International labour migrations during the post-colonial period

The political independence of Ghana in 1957 was closely followed by that of other West African countries and international labour migrations soon became more restricted. The federation of French West Africa emerged in 1960 as nine independent countries and it was no longer possible for a Senegalese or Togolese to establish political rights in Ivory Coast or Mali.

Each country became more concerned with planning its own economy and often came to see the presence of a large number of 'aliens' as a contributory cause of mass unemployment in urban areas. The emergence of a new type of migrant consisting of political refugees or members of an opposition party—who usually fled into neighbouring countries to plot the overthrow of their home government—resulted in strained relations between some countries and in greater immigration controls. There was also a fear of subversion from immigrant groups, a classic example being the case of Mossi migrants from Upper Volta, many of whom were deported for supporting the opposition party in Ghana during the Nkrumah regime.

The climax of these fears, real and imaginary, of economic competition and political subversion by foreigners came with the expulsion in 1967 of some 800 Nigerians from Cameroon (*Africa Diary*, 1967, p. 1013) and many Guineans from Senegal (*Africa Diary*, 1967, p. 3582). In the following year, more than 100 Ghanaians were expelled from Sierra Leone (*West Africa*, 1968, p. 1522) and by January 1970 following the implementation of Ghana's Aliens Law of 1969, more than 300 000 migrants had left Ghana for Nigeria, 250 000 for Upper Volta, 150 000 for Niger Republic and more than 100 000 for the three countries, Togo, Dahomey and Ivory Coast (*West Africa*, 1970, p. 146).

These changes notwithstanding, it is quite clear that the importance of migrant labour in the economy of Ivory Coast, Ghana, Fernando Po and south-west Nigeria cannot be overemphasised. Indeed, whenever restrictions have been placed on migrant labour into Ghana and the Ivory Coast, much cocoa and coffee has remained unharvested because of inadequate labour. To maintain the present level of production and the possible expansion of export agriculture, some labour-deficit countries have, since independence, entered into bilateral labour agreements with the governments of labour-exporting countries who are also anxious to protect the interests of their nationals against ill-treatment and exploitation in the areas to which they migrate.

The first of these bilateral agreements was the 1960 Labour Convention between Upper Volta and Ivory Coast, under which the Labour Department in Upper Volta was charged with responsibility for all matters connected with the recruitment of workers. Employers in Ivory Coast were requested to route their applications for workers through the Ivory Coast Labour Department, which would then transmit their request to Ouagadougou. In addition to various clauses for the protection of the migrants, the government of Ivory Coast agreed to pay the government of Upper Volta the sum of 1500 CFA francs to cover the cost of recruitment in respect of each worker employed in Ivory Coast. This convention has remained largely a dead letter since most Upper Volta workers prefer to travel independently to Ivory Coast to seek employment. Large-scale smuggling by illegal agents of agricultural labourers across the Upper Volta–Ivory Coast border has developed. These agents supply labour direct to many planters in Ivory Coast for a payment of 1500 CFA francs per worker. The Upper Volta government reacted to the situation by accusing Ivory Coast of resorting to illegal recruitment of labour and also of discrimination against its citizens through the Ivorianisation policy (unskilled Upper Volta workers are paid 70 000 CFA francs per annum as compared with 82 000 for Ivorians who are also given preference for skilled and low-skilled jobs). Ivory Coast on the other hand has accused Upper Volta of inability to control the departure of emigrants and the activities of unauthorised recruiters of labour and has blamed Upper Volta for the resulting confused situation (Deniel, in press). For many years, however, Ivory Coast has depended largely on unskilled migrant labour from Upper Volta and it is quite clear that, in view of the dense concentration of people in the Mossi area of Upper Volta and the natural poverty of the area, Upper Volta is likely to remain a net exporter of labour. It is also clear that, as a result of the mass expulsion of Upper Volta citizens from Ghana in 1970 and the tightening of immigration controls into that country, the vast majority of Upper Volta migrants will in future go to Ivory Coast.

Magnitude of international labour migrations

Migration studies in West Africa are handicapped by lack of adequate and reliable data. Reliable and detailed census figures exist for only a few countries while official control of migratory movements has been highly localised and often ineffective, as has been illustrated by the case of labour migrations from Upper Volta to Ivory Coast. The figures given below are therefore largely estimates made at different periods for different countries.

In his study of the migrant labour system in West Africa, Elliot Berg (1965) estimated that about 75 000 *navétanes* are involved in the annual seasonal migrations of farmers into the groundnut growing areas of Senegal and Gambia. Most of these *navétanes* originate from Mali and Guinea and are usually employed on groundnut farms on terms which provide time for the migrants to cultivate farmlands allocated to them in lieu of cash payment of wages for work done on the farms of those who employ them. A large number of foreign migrants from Nigeria, Guinea and Ghana (about 40 000 in 1963) work as self-employed small-scale diamond diggers in Sierra Leone.

The country which depends most on foreign migrant labour is Ivory Coast, where migrants from neighbouring countries constitute about 65 to 70 per cent of the unskilled labour force employed in coffee and cocoa plantations. In 1965, about half the total labour force of the country consisted of foreigners, made up of 670 000 Africans and 30 000 non-Africans. Most of these migrants originated from Mali (220 000), Upper Volta (200 000) and Guinea (150 000). The presence of such a large number of migrants is a source of concern to a government faced with mounting unemployment in urban areas. Most of the migrants are, however, employed on tree-crop plantations and the government is fully aware that without the labour supplied by these migrants, the rapid growth of the country's economy would not have been possible.

Ghana is another country where cocoa farmers have for a long time depended on migrant labour from Nigeria, Upper Volta and Togo. The diamond and gold mines have also been worked by the use of migrant labour from these countries and even in the tertiary sector, migrants from Nigeria were in control of a considerable part of retail trading up to 1969. Before January 1970, by which date more than 800 000 foreign migrants had been expelled from Ghana, the country had the largest number of foreign migrants in West Africa. According to the 1960 census, over 850 000 persons or 12 per cent of the 6·7 million people in the country were of foreign origin, and almost 68 per cent of these 'aliens' were born outside Ghana. As in the case of Ivory Coast, a great majority of these migrants originated from the neighbouring countries of Togo (280 670) and Upper Volta (194 570), and the more distant Nigeria (190 780). Many Nigerians, numbering about 10 000 and originating largely from the Ibo and Ibibio areas of south-eastern Nigeria, also migrate to work in the plantations of Southern Cameroon and Rio Muni, while another 20 000 migrate to work in the cocoa and coffee plantations of the Equatorial Guinea island of Fernando Po.

Internal migrations

In Nigeria, Liberia, Cameroon and Sierra Leone, the vast majority of migrant workers originate from within each country and, considered as a whole, the total number of people involved in internal migrations is far greater than the number involved in international or cross-frontier movements. Recent expulsions of foreign migrants by several countries and the greater restrictions imposed on cross-frontier migration suggest that the volume of internal migrations is likely to increase, as has already been the case in Ghana since 1970. Internal migrations of agricultural workers occur in all the countries of West Africa, but the volume of migration is greatest in Nigeria, Ghana, Liberia and Sierra Leone. Nigeria, by far the most populous country in West Africa (with about 56 million people, more than half the total population of West Africa), has a rather complicated pattern of internal migrations. The main components of the migratory flows include the north-to-south seasonal migrations of thousands of

Hausa workers into the cocoa-belt of south-western Nigeria (Prothero, 1958); the east-to-west permanent migration of self-employed farmers of Igbira, Ibo, Ibibio and Urhobo origins into the cocoa-belt and the Benin rubber belt; and the west-to-east migration of Ibos and Ibibios into the fertile but sparsely settled Cross River basin (Udo, 1970). There is also a compensating south-to-north migration of workers into the Jos tin fields and the towns of northern Nigeria, but migrants in this group usually settle in urban centres, not in rural districts.

Seasonal migrants in Nigeria usually work for wages and do not normally spend more than five months at their destinations, but a large proportion of permanent migrants are self-employed. Migrants in this category include share-croppers who work on cocoa and rubber farms for an agreed percentage of the annual proceeds from the farms, and migrant tenant farmers who pay annual rents varying from two to ten pounds sterling for cultivating a given piece of farmland or for harvesting a demarcated area of oil-palm bush. The total number of people involved in internal migrations in Nigeria was given as 1·4 million in the 1953 census and was probably about 2·5 million in 1970, about 60 per cent of whom were rural migrants.

In Ghana, internal migration of agricultural workers has always been associated with cocoa farming; and as late as 1960, over 70 per cent of migrants aged 15 years and above went to rural areas. International migration in Ghana is directed towards the better-developed south and south-west, the migrants originating from the Frafra district of the north-east and from the Krobo and Ewe areas of the east. The estimated number of internal migrants is about 600 000, a third of whom work as wage-earners in agricultural establishments, including private cocoa farms. Although rural–rural migrations amongst the Frafras of north-east Ghana is reported to be decreasing (Hart, in press), the recent expulsion of foreign migrants from the country has resulted in a general increase in internal migration of farm workers. Daily wages have gone up since 1970 and farm workers now work seven hours a day as compared with ten hours in the past.

There is also a considerable internal movement of labour in Sierra Leone and Liberia. In Sierra Leone, the movement is directed largely towards the iron and diamond mines and in Liberia the workers migrate to the Firestone rubber plantations along the coast as well as to the iron-ore mines of Bomi Hills, Bong Mountains and Mount Nimba district. Internal migration in Upper Volta, which is a major exporter of labour to Ghana and Ivory Coast, is directed towards the sparsely settled and rich farmlands of the west; the migrants originate from the congested central district around Ouagadougou and from the drier northern districts.

Major source regions and destinations of migrants

Most migrants who go from one rural area to another originate from very densely-populated districts which experience considerable pressure on available farmland. Population densities in these source regions are usually above 200 persons per square kilometre although there are also a few source regions with less than 150 persons per square kilometre. In the latter case migration is also caused by shortage of farmland arising from destruction of land by soil erosion and sedimentation or the restriction of cultivable land by the seasonal submergence of vast areas by floodwaters, as in the case of Urhobo and Isoko districts of Midwestern Nigeria. Major source regions are generally characterised by low per capita income, lack of wage-earning opportunities and considerable rural underemployment.

Source regions which suffer from extreme pressure of population on farmland, resulting in soil impoverishment and poor crop yields, include the Mossi district of Upper Volta, north-eastern Ghana, the Sokoto district, parts of Owerri, Onitsha, Nsukka and Enugu districts of the East Central State of Nigeria, and Ikot Ekpene, Abak and Uyo districts of the Southeastern State of Nigeria. In parts of Enugu, Nsukka and Awka districts of Nigeria, the land shortage situation has been aggravated by sheet and gully erosion which have rendered useless extensive areas of farmland (Grove, 1951; Floyd, 1965). In the drier savanna districts of Katsina, Sokoto, northeastern Ghana and the Mossi district of Upper Volta, poor harvests and famine are also caused by the unreliable character of the rainfall in an area where irrigation is not widely adopted, and have been largely responsible for seasonal and permanent out-migration from these districts. In contrast, the districts which attract self-employed migrant farmers are generally sparsely-settled areas with abundant farmland or areas where there are certain types of job such as harvesting oil-palm fruits or palm-wine tapping which the local people, who formerly depended on slave labour, would not undertake. A number of self-employed migrants also settle in the relatively densely-populated cocoa belt of Nigeria as sharecroppers or food-crop farmers who produce local staple food crops for sale to the nearby urban centres as well as to some indigenous cocoa farmers who concentrate on cocoa production to the extent of having to purchase up to 50 per cent of their annual food requirements. Wage-earning migrants, on the other hand, go to the timber concessions or to areas of tree-crop production on commercial or on peasant farms.

Problems of agricultural labour at destinations of migrants

The first problem of agricultural labour in West Africa is that in most areas of commercial cropping for export there is a shortage and, in consequence, labour has often been a limiting factor to the expansion of commercial agriculture. In Liberia, for example, where large concessions of land have been given to the Firestone Rubber Company, the company has always had great problems in recruiting labour. In the early 1950s, the company had a labour force of only 22 000 out of a required total of 300 000 (Jurgens *et al.*, 1966). The result was that not only was expansoin slowed down, but many mature acreages also went unharvested. In recent years, the labour situation has been aggravated by competition from the mines and other foreign concessionaires and the closing down in 1957 of some diamond mines was ordered by government in order to stop a diamond rush which threatened to worsen the labour situation in the plantations and iron-ore mines.

Many employers of labour have had to pay comparatively high wages in addition to offering certain incentives including free medical attention, housing, subsidies for food, and even grants of money to chiefs and town authorities as compensation for labour and services lost while their people are away in the concessions. Such unprecedented forms of inducement have certainly raised production costs but what is more disturbing to employers is the fact that these incentives do not appear to have had the desired effects (Hance, 1958). Labour shortage in Liberia is not restricted to lack of skilled personnel as in most other developing countries, but also includes a lack of adequate unskilled labour. This situation has certainly slowed down development and has made it increasingly difficult for the rural areas to produce enough food crops under the present system of shifting cultivation. At the same time there is, as in many other developing countries, a marked increase in urban unemployment of school leavers.

In Ivory Coast, the shortage of local labour has, as we have seen, been largely contained by large-scale recruitment of foreigners who today constitute more than half the labour force of the country. Other areas which experience shortage of agricultural labour include the cocoa belts of Ghana and Nigeria, and the sparsely-settled Middle Belt of Nigeria. In the Nigerian cocoa belt, which continues to attract an increasing number of migrants, labour shortage has persisted because more and more migrants have ceased to operate as wage-labour on cocoa farms and have become established as self-employed tenant farmers cultivating food crops for sale to the urban population as well as to cocoa farmers. Thus, instead of providing the much-needed labour on the cocoa farms, Igbira migrant food farmers as well as Urhobo and Isoko migrants, whose main occupation is harvesting and processing of oil-palm fruits, now themselves employ wage labour whenever they can afford to do so.

Labour shortage in the sparsely settled Middle Belt remains one of the major problems of developing this area. Farmers have been unable to increase farm size because of lack of farm workers and yet the wages paid to agricultural workers in the region (about 5 shillings per day) are higher than those paid in the far north (about 4 shillings per day) where there is a labour surplus and the only limiting factor in employing wage labour is lack of money on the part of the employer (Luning, 1963). Shortage of farm labour in the Middle Belt appears to be caused by the abundance of farmland, the ease of acquiring adequate farmland and the consequent absence of a landless class which would have been obliged to work for wages.

Agricultural labour in West Africa suffers considerably from the high turnover rate which derives largely from the seasonal character of the labour. In the Firestone rubber estates in Liberia, the turnover in field labour is about 30 per cent per year (Taylor, 1956) and, in the Nigerian cocoa belt, most Hausa migrants settle for only four months before returning home to cultivate their own farms. Few of them work for the same employer for more than one year and an employer who intends to retain the services of an Igbira farm worker has to give him presents in addition to his regular wages.

The human problems arising from the presence of a large number of foreigners in certain areas have sometimes generated considerable bitterness between migrants and the host population and, as in the case of Upper Volta migrants in Ivory Coast, between governments. In Ghana and Ivory Coast where foreign migrants are most numerous, it is in respect of competition for urban employment opportunities that the indigenous population resent the presence of the migrants. The presence of the migrants in rural areas has certainly made possible the present rate of development in these countries, but it is also true that the recruitment of these migrants has contributed to the relatively low wages paid to unskilled Ivorians and Ghanaians, in contrast to Liberia, where chronic labour shortage has resulted in relatively high wages. Where migrants are self-employed, as in many parts of Nigeria, conflicts between migrants and host communities often arise where the two are competing for the same opportunities. Migrant palm exploiters and food farmers, for example, are welcomed in the cocoa belt since the local farmers specialise in cocoa production; but the situation is different in the Cross River basin where the migrants and the indigenous people compete as food crop producers.

Maltreatment of foreign migrants by employers in Ivory Coast and Fernando Po has been a major source of friction between the governments of the host countries and those of the countries where the migrants originate. In the case of Nigeria and Fernando Po, a series of labour agreements has been entered into during the past fifty years, with a view to protecting the interest of the migrants. Employers have been known to abuse labour contracts governing the employment of indigenous migrant labour. In the western Nigeria timber trade for example such abuses were discovered in 1947 by the Federal Labour Department which found that contracts were not fulfilled, that deductions were made from wages and that housing and medical conditions were extremely poor (Roper, 1958). Such abuses encouraged some migrants to withdraw their labour and to establish themselves as self-employed tenant food farmers, while others drifted to the towns to swell the urban unemployed, thereby worsening the labour shortage situation in these rural areas.

Labour problems within the migrants' source regions

Since most migrants are males between the ages of twenty and forty-five, it is reasonable to expect a fall in local food output if the proportion of emigrants is high. In part of rural Nigeria, certain jobs such as house-roofing, the making of yam mounds and pounding of steamed oil palm fruits have suffered as a result of the selective emigration of young men; and women have had to perform tasks formerly done by men. A close study of the labour situation in a few source regions suggests, however, that the fear of labour shortage resulting from emigration appears to be more imaginary than real. For many source regions, it is indeed difficult to sustain the argument that emigration results in local labour shortage, in view of the fact that there is much underemployment in many rural areas. Almost all the major source regions have very scanty natural resources and large concentrations of population with little or no capital. In such areas, the supply of labour usually exceeds the demand, so that, to subsist, many people are obliged to farm the small area of land available to them and are only too ready to emigrate to areas with greater opportunities. For such areas, the loss of some able-bodied men should lead to a fuller utilisation of available manpower and not necessarily to labour shortage.

It is tempting to attribute the neglect of village paths and village squares formerly maintained by young men to loss of manpower through emigration. We must however admit that social conditions have changed and that men who now pay rates for the maintenance of such places would not willingly sacrifice their time to perform communal jobs without payment. The growing economic independence of young men and the consequent decline in the authority of local chiefs appear to be a more rational explanation of the situation, at least in some localities. The decline or complete disappearance of traditional voluntary group labour or communal help in farms has been attributed to loss of manpower through emigration (Skinner, 1960; Ojo, 1963); but this is not an unhealthy development and it appears to fit in with the increasing monetisation of the economy. The displacement of yams, generally considered to be the crop of the man, by cassava (the woman's crop), which is sometimes attributed to the loss of male population through emigration, is caused by market conditions as well as by the fact that cassava does better than yams on the increasingly impoverished soils of the densely-settled source regions. The fact that male migrant farmers to the Calabar district specialise in cassava production (Udo, 1964) may be cited in support of this view. If emigration were to occasion a net loss in the productive capacity of a given source region, this should result in a drop in food production and possibly food shortage in the area. This is, however, not the case in the Sokoto district where the timing of seasonal migrations does not interfere with local demand for farm labour (Prothero, 1958). Amongst the Esu of Cameroun Republic, the Ardeners have reported that although up to 40 per cent of the adult population was absent during their field investigation, there was no noticeable drop in food production in the area (Ardener *et al.*, 1960). If the number of emigrants is large, however, it is reasonable to concede that, given the existing farming techniques, agricultural output will drop, particularly if the migrants have to be away during the local farming season.

CONCLUSION

Large-scale migration of agricultural labour which involves about three million people every year is one of the main consequences of the maldistribution of population in relation to natural resources in western tropical Africa. During the colonial period, the development of agricultural exports and mining depended very much on migrant labour, and today a continued supply of migrant labour has become indispensable to the continued growth of the economy of such labour-deficit countries as Ivory Coast, Liberia and Ghana. The recent restrictions on free movements of labour between the newly independent states of West Africa are therefore economically unsound, even though they may be politically expedient. The situation is, however, not as serious as may appear since these restrictions have never been well received by employers of agricultural labour who have either persuaded the governments to relax the orders in respect of certain categories of migrant labour or have contrived to flout such orders. Internal movement of labour within individual countries continues, and the labour-deficit countries may have to remove all restrictions on the free movement of agricultural labour to ensure the continued growth of agricultural exports which form the major source of foreign exchange earnings for financing capital development.

In central and southern Africa where the system of migrant labour has been established for a longer period, the system has been condemned as uneconomic both to the employer of labour, who bears the adverse consequences of high labour turnover, and to the economy of the village of origin of the migrants, which is alleged to suffer from loss of manpower. These problems exist in West Africa but do not appear to cause so much concern. Presently, migrations are beneficial both to the individual migrants and to the overall economy of western tropical Africa, although the labour-exporting countries appear to lose in providing a basic education for citizens who migrate to work in other countries. Present trends indicate that the migrant labour system will continue as long as the present economic situation remains unchanged.

REFERENCES

Addo, N. O. (1972) 'Differential rural migration in Ghana: spatial, demographic and economic dimensions', *Modern migration in Western Africa*, ed. S. Amin (*forthcoming*).

Africa Diary, 1967.

Anyanwu, C. N. (1971) *Port Harcourt, 1912–1955; A study in the rise and development of a Nigerian municipality* (Unpublished Ph.D. thesis, University of Ibadan), 126–128.

Ardener, E., S. Ardener, and W. A. Warmington, (1960) *Plantation and village in the Cameroons* (Oxford University Press), 210–229.

Berg, E. J. (1965) 'The economics of the migrant labour system', *Urbanisation and migration in West Africa*, ed. H. Kuper (University of California Press), 160–181.

Deniel, R. (in press) 'Divergence of interests between labour exporting and host countries and measures taken by their governments', *Modern migration in Western Africa*, ed. S. Amin.

Floyd, B. N. (1965) 'Soil erosion and deterioration in Eastern Nigeria', *Nigerian Geographical Journal* **8,** 33–43.

Forde, D. (1953) 'The cultural map of West Africa: successive adaptation to tropical forests and grasslands', *Transactions of the New York Academy of Sciences,* Series 2, **15,** 206–219.

Fortes, M. (1936) 'Culture contact as a dynamic process', *Africa* **9,** 24–55.

Grove, A. T. (1951) 'Land use and soil conservation in parts of Onitsha and Owerri Provinces', *Geological Survey of Nigeria,* Bulletin No. 21.

Hance, W. A. (1958) *African economic development* (Oxford) 237.

Hart, J. K. (in press) 'Migration and the opportunity structure: a Ghanaian case study', *Modern migration in Western Africa*, ed. S. Amin.

Hill, P. (1963) *The migrant cocoa farmer of southern Ghana* (Cambridge).

I.L.O. (1954) Report V, *International Labour Conference*, 37th Session, 28.

Jarrett, H. R. (1949) 'The strange farmers of the Gambia', *Geographical Review* **39,** 649–657.

Jurgens, H. W., K. A. Tracey and P. K. Mitchell (1966) 'Internal migration in Liberia', *Journal of the Sierra Leone Geographical Association* **10,** 39–59.

Luning, H. A. (1963) *An agro-economic survey in Katsina Province* (Kaduna: Government Printer), 75.

Ojo, G. J. A. (1963) 'The changing patterns of traditional group farming in Ekiti, north-eastern Yoruba country', *Nigerian Geographical Journal* **6,** 31–38.

Prothero, R. M. (1958) *Migrant labour from Sokoto Province, Northern Nigeria* (Kaduna).

Roper, J. L. (1958) *Labour problems in West Africa* (Harmondsworth: Penguin).

Skinner E. P. (1960) 'Labour migration and its relationship to socio-cultural change in Mossi society', *Africa* (Journal of the International African Institute) **30,** 375–401.

Skinner, E. P. (1965) 'Labour migration among the Mossi of the Upper Volta', *Urbanisation and migration in West Africa*, ed. H. Kuper (University of California Press), 60–84.

Taylor, W. C. (1956) *The Firestone operations in Liberia* (Washington).

Thompson, V. and R. Adloff (1958) *French West Africa* (London), 318.

Udo, R. K. (1964) 'The migrant tenant farmer of Eastern Nigeria', *Africa* **34,** 335.

Udo, R. K. (1970) 'Migrations in Nigeria', *Nigeria Magazine* **103,** 616.

West Africa (1968), 'Sierra Leone: the Ghanaian fishermen affair', *West Africa* no. 2690, 21 December, 1522.

West Africa (1970) 'Ghana: Ibos and the Aliens Order', *West Africa* no. 2748, 31 January, 146.

Chapter 11

Migration and Urbanisatiou in Peru

Stella Lowder

Introduction

An outstanding feature of Third World countries is the magnitude of their internal disparities in the degree of economic development, disparities which are mirrored in their social structures. The contrasts between rural and urban are especially notable in this context. Rural communities, with the possible exception of cash-crop estates or mining camps, generally exhibit great poverty, poor health and ignorance as a result of varying degrees of geographical, social and political isolation. Although urban settlements are far from internally homogeneous, the economic and cultural attainments are concentrated in them to a greater extent than is warranted by the size of their populations. While urbanward migration movements are world-wide phenomena, the Third World streams attract special attention because they exhibit a number of novel characteristics: frequently the urban population is already greatly concentrated in a single, dominant primate city and few of the major cities possess an industrial base comparable to those of Western experience. Furthermore, large influxes of population to these cities will obviously erode their position as a preserve of an élite minority.

The Latin-American case differs from that of the rest of the Third World. The majority of the countries have been independent for over a century and many have a per capita income several times as large as that of their African or Asian counterparts. However, Latin-American primate cities appear to be even more dominant and many Latin-American countries have a social structure in which a major sector of the population is still relegated to quasi-serfdom. When and where social controls have weakened, especially if communications have been extended, migration appears as a network of steady streams mostly converging on major cities. Seasonal and cyclical movements, apart from those associated with natural disasters

---

Dr. Lowder is Research Fellow at the Centre of Latin American Studies, University of Liverpool.

such as droughts or earthquakes, appear to account for a relatively small proportion of the movement. So it is not surprising that research work has concentrated on the movements to particular cities. However, considering the proliferation of these studies and the extent of the period over which they have been carried out, the findings would seem comparatively scanty; we still do not know much about such basic themes as the effect of migration on specific population structures, the volume of migrants in particular migrant streams, or the long term effects of migration on the individuals involved and the 'catchment' and reception areas. This is partly because interest in migration, in Latin America in general, stems from concern with the process of socio-economic development only as reflected in urbanisation; the circumstances which have led to a focus on migration and migrants have had little to do with the nature of these subjects *per se*.

The earliest accounts of urbanisation in the Third World in general and Latin America in particular appeared about twenty-five years ago. These considered each unit of the urban hierarchy separately and analysed its function at a particular point in time; a major city was treated as a microsystem mirroring the socio-economic life of its country and as a centre disseminating culture and innovations. The implications of very rapid growth for part of the city's residents were not ignored; however, authors could still speculate that on the whole urbanisation was not just desirable, but perhaps desirable on an even greater scale than hitherto witnessed (Davis and Casis, 1946).

Towards the end of the 1950s, however, the emphasis in these urban studies shifted from the analysis of functions in isolation to the analysis of processes of urban growth and the impact of urban development on rural areas. This was the period when the United Nations' concern for the Third World became very marked; a deluge of publications stressed the need for planning and the role of the city in promoting, or discouraging, regional and national economic development in such areas (United Nations, 1957). These publications included descriptions of squatter settlements and inner city conditions (Hauser, 1961), and conjectures as to how to ameliorate such situations. Usually the phenomena were seen through Western eyes: migration was viewed as a simple mechanism for evening out inequalities of economic opportunity by forcing out the least successful; urban centres were being invaded by legions of illiterate rural peasants as a consequence; and all squatter settlements were peopled by dropouts and failures, either unwilling or unable to better themselves (Lambert, 1963). In addition the United States became more aware of Latin-American living conditions and suddenly began to associate such blatant poverty with the conditions which preceded the Cuban revolution. Certainly accounts provided by journalists and statesmen after fleeting visits gave the impression that several major Latin-American cities were on the verge of dramatic uprisings (Benton, 1961; Szulc, 1963). Unfortunately, these accounts appeared at a time when there was little local work to set them in context. The social sciences were a late introduction to Latin America and early exponents naturally tended to look abroad for guidance as to the appropriate fields for investigation (Solari, 1969), thus accentuating earlier emphases. Furthermore, a disproportionate amount of time was devoted to ideological themes as opposed to empirical work (Labbens, 1969; Hodara, 1969). Local energy was spent on devising theoretical structures around topics, rather than in studying the nature of their components in order to perceive their interrelationships. Thus the local themes continued to be urbanisation in relation to socio-economic development; migration was viewed only as a problem-generating, or problem-solving, phenomenon related to them. In the first case, the details of the process itself were glossed over, or only sought if relevant to future population estimates or to the prediction of future material conditions in the reception areas (Matos, 1966). Further work produced speculations as to the 'problems of integration' and about the state of affairs these movements were supposedly creating in the source areas (Abate *et al.*, 1962). Developmental themes also focused on the urban unit, especially the primate capitals, which were viewed as 'a negative factor in the development of other regions and of the nation as a whole' (Germani, 1967, p. 175).

It is very difficult to comprehend a phenomenon such as migration by studying only its indirect effects in urban areas, especially if the effects so examined are limited to the physical characteristics of migrant and reception areas detached from their social and areal context. Moreover, if the energy of migration movements is to be channelled into those sectors offering most development potential, what is vital is the comprehension of those contexts that generate the phenomenon. A few studies have made some attempt to explore the social origins of migrants (Mangin, 1970), but spatial concepts, such as those of intervening opportunities, 'distance' decay functions or gravity models, are notably absent from local literature; few native geographers have worked on migration themes. However, spatial variables at a microlevel are also relevant to the migration process—the sort of variables that are incorporated in such concepts as the sociologists' 'life-style', the psychologists' 'life space' and the geographers' 'human environment'. They may include the nature of the local establishment an individual patronises, the amount of commerce a settlement engages in, the number of transport links passing through it and the number of and nature of neighbouring settlements which an individual may have visited or resided in; it is this sort of miscellaneous information that influences an individual's assessment of his condition. A potential migrant must be aware of his surroundings, either through the comparison of disparate conditions, or through an inculcated value system which enables him to distinguish favourable characteristics from the rest. Only then can he consider whether to move, perhaps only temporarily, or to stay. Admittedly, this process is irrelevant to a proportion of any migration stream; soldiers are conscripted, state teachers are allocated, children are taken or sent, wives accompany husbands and a few individuals wander off on the spur of the moment. However, even if many of these persons are not active decision-makers, the fact remains that they usually travel at the instigation of those who are and who are known to them. The latter, by definition, are likely to possess the personal qualities of greatest potential for contributing to economic development and consequently, studies designed to illuminate the source environments and value systems of such persons would be of considerable value in an underdeveloped setting. This chapter serves to illustrate the points made above with reference to a specific underdeveloped country.

THE PERUVIAN CASE

Peru would seem to exhibit all the characteristics likely to generate massive internal migration movements, common to most countries of the Third World in which vast areas possess environments uncongenial to man's modern activities. Despite an average population density of only 4·6 persons per km^2, 11 million Peruvians are concentrated in a relatively small number of major urban centres, a few intermontane valleys in the Andes and the irrigated valleys of the coastal strip. If the rank–size rule is applied to Peruvian towns, the typical pattern of a country suffering from extreme primacy emerges: the Lima metropolis currently has a population in excess of three million, while Arequipa, the second city, is barely one sixth of this. Less than half a dozen other cities exceed 100 000 and the bulk of the people live in settlements of 10 000 or less. Moreover, this pattern is sharpening, rather than blurring; the urban growth rate for the intercensal period (1940–1961) was at a geometric scale, while the sierra (mountain region) stagnated or lost population—a difficult achievement in a country with an annual birth rate verging on 3 per cent. Obviously internal migration has been responsible for many of these disparities; however, knowledge about this phenomenon is scarce. The censuses indicate simple totals for relatively large civil divisions (provinces and departments); these must be treated with caution owing to the vast number of boundary changes in the intercensal period, and to the relative instability of Peruvian boundaries in general. However, if the national growth rate for the intercensal period is adopted as a standard for assessing the performance of individual provinces, it is found that the coastal provinces (excluding the capital) grew at an average rate of 5·2 per cent per annum and the capital at 7·9 per cent per annum, while the sierra ones lost some 1·2 million people (Graumann, 1966).

If this data is then examined in conjunction with two other possible migration indicators from the census, the proportion of the population censused in the province of birth and the provincial sex ratio, the resulting correlations are all highly significant statistically.

Attempts have been made to associate this sort of demographic data with various indices of socio-economic development. Bértoli and Portocarrero correlate the net migration rates pertaining to the largest civil divisions (Departments) with their degree of modernisation; the latter condition was determined by their respective literacy rate and their income and occupation levels (Bértoli and Portocarrero, 1968.) However, the limitations of this sort of data are obvious: these are simple, static descriptions of demographic variables attached to areal units of varied size and character, while migration is a complex, mobile phenomenon, more easily understood in terms of small groups of individuals in a specific socio-economic or cultural context. Most of what we know about the detail of the process has been provided by a number of small-scale case studies.

Three comprehensive works based on studies of this sort are directly concerned with aspects of migration at a national level. The earliest of these (Dobyns and Vázquez, 1963) is a collection of case studies covering parts of the capital, parts of three major towns, a sugar plantation and industrial complex, and a number of communities and hamlets in the sierra. The authors were anthropologists, students from the social services school and sociologists; frequently no publication existed for the settlements they studied. Although the treatment varied with the author, the sort of information provided is a factual description of a small sample of persons who migrated either from or to a particular locality, a brief account of the place, and the interpretation of these facts in the light of general migration theory.

Two detailed papers review the findings of the large number of scattered articles referring to migration. Alers and Appelbaum's (1968) analysis consists of a hundred propositions supported with the evidence of fifty-nine articles. These are arranged under six headings: the causes, the effects, and the demographic, social and economic concomitants of migration. The paper also discusses the various anomalies conveniently ignored by some of these studies, such as the definition of a migrant, and possible ambiguities which might arise if studies are compared without taking these into account. The second paper (Martinez, 1968), a much more general narrative, prefaces the discussion with the statement that migration in Peru can only be understood if one appreciates the national context; in his view, the capital has always benefited from the decisions of biased governments, which is the essence of its attraction for migrants. Any change in this attitude on the part of a government is an expression of their fear of accelerating migration to the metropolis. Those aspects of the topic which should be included in a comprehensive account are outlined: 'What is the context of these movements? What direction do they take? What is their volume in terms of age, sex, income, and what is their cultural context? What are the expulsive and attractive factors? Which mechanisms facilitate such displacements? Do migrants sever all ties with their homeland? How do they adapt to their new milieu? What are the implications and consequences of these movements?'

To some extent Martinez demonstrates how this can be done in another study (Martinez, 1969), which is the only detailed account of a single migration stream. Being a social anthropologist, his main theme is the change in socio-economic structures between source and reception areas. As the colonisation area was unknown, being unmapped and accessible only by trails, the work consisted of descriptive observations, based on eight months' fieldwork, supplemented by a census of the permanent population. In addition, a number of shorter articles treat of a single aspect of migration. Those referring to migration in rural areas are usually carried out by anthropologists, whose major concern is with the life of the indigenous communities. Their interest in the topic stems from the effects and influences of the process on the structure of the future population, and on the viability of maintaining these communities as a form of settlement. Sociologists, planners and members of the social services study themes related to the settlement patterns evolving in the reception areas and to the social links both within these new communities and between them and other sectors of the cities (Cerdan, 1965; Bradfield, 1965). Studies of the metropolis predominate in this class (JNV, 1966). A more technical and highly specialised field is the adaptation process of the individual. Studies in this category range from clinical observations of extreme cases of psychological maladjustment (Fried, 1959) to broader accounts of problems faced by migrants on arrival in a new area (Mangin and Cohen, 1966).

The present situation

During the past decade a considerable number of sample surveys have been carried out which have covered the majority of major towns. Most of these were the first attempts in their respective areas; many collected migration data alongside a wide range of socio-economic data and the studies were frequently carried out for planning authorities, such as the Junta Nacional de la Vivienda. This data is piecemeal and merely indicates useful fields of enquiry rather than being a comprehensive study in itself.

An example of this situation is provided by the town of Chimbote. Chimbote is the most extreme case of rapid population growth in Peru; in 1940 it had only 4000 persons. It subsisted by fishing and a certain amount of port work for, although undeveloped, Ferroles Bay provides the best natural harbour of the coast. The immediate area inland is desert, the Lacramarca bed being almost permanently dry; but the southern margin of the bay was occupied by extensive salt marshes; malaria was rife. By 1961 the town had a population of 59 000, attracted by two major developments: the work of the Corporación Peruana del Santa on a substantial hydroelectric scheme at Huallanca, undertaken with a view to establishing the country's first steel mill (SOGESA) in Chimbote; and the development of the fishmeal industry. Chimbote had possessed several tuna canneries since the Second World War; however, it was the mid-1950s when the fishmeal industry became firmly established, after a number of companies had been dislodged from Callao. By 1961 there were over 20 factories, each employing several hundred unskilled or semi-skilled men in the fishing fleets and on land. Meanwhile, Chimbote as a settlement had been transformed. It had overflowed from the original 60 gridiron blocks along both sides of the Pan-American highway onto the islands in the swamps. Most of this sprawl consisted of primitive shacks made of bulrush mats and the slightly more substantial buildings were of adobe. The public services were still those of a village a tenth the size and water was hawked from door to door by donkey and tank trucks.

The earliest study was carried out in 1956 for the Agrarian Reform and Housing Commission (Matos, 1963). It consisted of a 'general census' of two major *barriadas* (shanty towns) which represented 21·4 per cent of the total population. Over 60 per cent of these 6431 persons were immigrants and 60 per cent of them came from the sierra. Two departments, La Libertad and Ancash, claimed over 80 per cent of these but there was at least one individual from every department in the country. Only 4 per cent of the household heads had been born in Chimbote. The immigrants' motivations were also recorded (Table 11.1).

The classification of economic reasons included 'lack of work, badly paid work, sackings, left the mines, change of work or business'. Social motives

Table 11.1 Migrants' reasons for coming to Chimbote (after Matos)

	%
economic reasons	72·60
social reasons	20·24
health reasons	2·60
educational reasons	1·14
housing reasons	0·49
military service	0·33
other	2·60
	100.00

included 'attracted to the city, death of parents, change of ambience, abandonment (by the spouse), widowhood, marriage, for curiosity, ran away from home, bad treatment and other family conflicts'.

Solis's study (Solis, 1963) took place in 1960 and consisted of migration details of 2625 heads of households employed by the Corporación Peruana del Santa and SOGESA, along with those household heads resident in the two *barriadas* most associated with those places of employment. His work revealed similar percentages coming from the sierra and the two departments mentioned by Matos. Finally a sociologist (Tolentino, 1961) analysed the effects of the organisation of local industry on the lives of those so employed. The fishing industry was seasonal and those concerned were paid a proportion of the catch; thus a considerable part of Chimbote's population was without means of support for several months of the year. Owing in part to this and to the complete inactivity of the local authorities, Chimbote rapidly gained the worst reputation in the country for slum conditions and was an excellent example of 'what should never be allowed to occur' (Matos, 1963).

Unfortunately, these studies do not indicate to what extent their findings are applicable to the population as a whole; not all *barriadas* need remain slums. Work in depth has stressed their variety (Delgado, 1968); however, few studies stress a similar variety in their inhabitants. Furthermore, the data presented is usually based on a single question. Matos's study is based on three—the place of birth, the year of arrival with the intention of staying, and the motive for migrating.

Despite the broad classification of motives, a major problem of such studies is that the majority of respondents are likely to be illiterate, inarticulate and unused to rationalising their actions. If requested to do so on the spur of the moment, such a person will always be able to give straightforward factual information: age, marital status or place of origin. However, it does not follow that he will be able to give a valid answer as to the reason for migration, a decision possibly taken as many as five or ten years previously. This sort of respondent is more likely to interpret past actions in terms of present conditions, as revealed by checking the internal consistency of questionnaires. In a Lima study, such a case was the girl who came 'to study' when in fact she was two years old at the time of arrival; another was the lad who came 'to work' at the age of ten (Lowder, 1970). Very little work has been done on the stability of responses from semiliterate societies. One study, involving the re-questioning at a six-month interval of a very oppressed rural Peruvian population, came to the conclusion that consistency of individual responses was not much better than random, particularly for closed questions (Alers, 1970), although group results were highly reliable. A Brazilian study dealing with rural persons with little formal education agreed that the stability of evaluative responses was low, but that demographic and other factual material was relatively stable or, at least, comparable to such material provided in resurveys in the United States (van Es and Wilkening, 1970).

Some of the inferences drawn from the more reliable factual data also leave room for much improvement. Study after study has been satisfied to classify place of origin simply at departmental level or, occasionally, at provincial. This information has been presented as significant *per se*. However, departments are huge areas and could not be considered internally homogeneous. They include not only a wide range of the settlement hierarchy but also totally different physical, cultural and economic regions. To say a migrant comes from a department like La Libertad, for example, tells us nothing about his background. La Libertad includes Trujillo, one of the major cities of the country with a population of about 100 000, as well as many settlements of less than 1000. It included highly organised workers from sugar estates on the coast as well as vast numbers of subsistence farmers living in isolated mountain valleys barely accessible by trail. The most eastern parts of the department lie on the far side of the Andes in the 'ceja de la montaña', the eyebrow of the jungle. A migrant could be a sophisticated, town-bred businessman or an illiterate peasant speaking a variant of Quechua. Siginificantly, when a few later studies have classified origins according to the size of the actual settlement, migrants to the capital appear to have originated in urban places and, particularly, in large cities (DNEC, 1969). Even the majority of migrants to smaller towns do not appear to have come straight from rural pursuits (Lowder, 1973).

Studies such as these tell us little about migrants' own perceptions and values and thus we can see motivations only in the context of commonsense rationalisation. A few studies which have gone in great depth into the life history of a few individuals (Patch, 1967), have revealed the complexities awaiting those who scratch beneath the surface. However, the danger of these character sketches is that it is only too easy to 'typecast' all migrants of roughly the same age from the same region. So far many studies have tended to make deductions from their data without stressing and making allowances for the narrow front on which the data are based. For example, because some studies of migrants in some sections of Lima have demonstrated that they suffered from personality disorders, it has been commonly assumed that these are the cost born by those migrating. A broader study has suggested that those particular individuals would probably have suffered personality problems wherever they had lived (Rotondo *et al.*, 1963) and that it is essential to distinguish those characteristics pertaining to different ethnic backgrounds from those generated by the strains of the migratory process.

One way to reduce the number of data problems would be for more studies to clarify their principal objective. It should be recognised that the same set of questions cannot be expected to illustrate all aspects of the process. If the chief aim of the work is to assess the material conditions of the migrants' living standards after arrival, then a questionnaire dealing with just that, put to migrants wherever they are resident in the settlement and to a control sample of natives, would seem appropriate. Questions as to motivations and background could be left out on the grounds that they could not be gone into in sufficient depth in this context to be useful, given the usual constraints of time and a tight budget. If the magnitude and trends in the migration flow are the objective, a sample survey referring exclusively to demographic details and covering whole settlements would be more useful than a more detailed survey of only a segment of it. A survey designed to demonstrate the demographic effect of the arrival of a large number of single persons of both sexes in their late teens on the subsequent population structure of a single settlement, investigated at successive points in time, has yet to be undertaken. Finally, if a study intends to investigate the motives for migration, the survey would have to include much more data concerning the individuals' socio-economic context. There is scope for much improvement in future studies, in terms of adequate coverage of the study population and more systematic study of the specific field, and the next steps are not obvious. Details of a specific case can illustrate the reasoning behind one method of extending research in this field.

The Chimbote project

It was proposed to carry out a major investigation into the migration phenomena in the town of Chimbote in 1969; at that time the population was estimated at 120 000.[1] Published sources indicated that the town had grown very rapidly through migration, that the majority of migrants were *Serranos* from the departments of Ancash and La Libertad and that the majority of the population, irrespective of migrant status, was living in deplorable conditions. We could only conjecture about more complex topics associated with the phenomenon, such as the problems of the adaptation of those involved to their new circumstances or the socio-economic conditions which led to the decision to migrate. Furthermore, although errors accruing from post-rationalisation or inability to remember could be reduced by questioning only recent, mature arrivals, even the most intelligent and articulate person could have difficulty in explaining their behaviour on occasion. Undoubtedly the more simple person has a completely different sense of time and set of values than the more sophisticated person, a fact which will not be revealed in a short questionnaire. Nevertheless, if the migration process is to be understood adequately, some attempt to glimpse life and its opportunities as perceived by the persons under study would seem essential. Considering the paucity of data in this field, the shorter questionnaire applied to a large sample seemed more suitable than the more elaborate design tested on fewer individuals; at a later stage the position might be reversed. It was also felt that measurements of character attributes, such as sense of achievement or independence, although obviously relevant, should be left to a later stage. Similarly, a study of migrants' values would be hampered by a lack of knowledge of the culture of the source populations; hence the difficulty in validating results when studying such topics as adaptation to a new location. However, it was felt that more could be achieved than had been in the past by the extension of the simple factual question to new areas.

Thus the aims of the project were as follows: (1) to ascertain the basic characteristics of the population of the town; (2) to select a subsample of recently arrived, mature migrants for additional analysis, and, for these individuals, (3) to explore their social and economic context in their source settlement so as to obtain greater insights into the motives which led them to migrate; and, finally, (4) to incorporate certain checks in the questionnaire that would yield some idea of the validity of the question and the stability of the responses. In order to do this, a two-stage, stratified cluster-area-probability sample covering the whole town was taken. The 1800 dwellings selected from this provided basic demographic data for every member in residence in each dwelling. By this means it was found that migrants constituted 51 per cent of the total population and 70 per cent of that aged over fifteen. In fact the age and sex pyramid turned out to be a tree, consisting of a 'native' root and stem with migrant foliage; the percentage of migrants of either sex above the age of thirty-five exceeded 90 and increased with succeeding age groups. A subsample was selected of those who had arrived since 1963 and who were aged fifteen or older; fifteen is the earliest age (by law) at which boys may work without special dispensation but still sufficiently young for them not to be in danger of the lottery for military service. By this means, 502 individuals from provinces other than the capital were selected and asked to submit to the questionnaire proper. A different questionnaire was submitted to the ninety-eight migrants from the capital who satisfied the same conditions on the grounds that the lifestyle of the capital could not be compared with that of anywhere else in the country.

The area of least knowledge was the lifestyle of the individual prior to migrating. This could be interpreted to cover a number of sectors: the characteristics of the source settlement itself, its economy and society, the factors concerning the individual's family relative to the wider society and to the individual himself, and those relevant to the personal experience of the individual. Most of this information could be acquired *via* direct questions; however, rather than using the answers as significant *per se*, it was felt that they should be examined in groups and that various scaling procedures would aid in reducing them to a manageable number while maintaining the greater validity gained through their broader base. The questionnaire for the provincial migrants included a number of scales, designed to explore various aspects of their life prior to migration. Some of these were intended to be used as variables in their own right, such as the Guttman scale of services; others, such as the social and economic indices, were designed to give insights when other variables within the questionnaire were being examined. A few, although providing interesting information, were impossible to evaluate without a control sample of non-migrants from the same settlements for comparison; however, the original use of the 'information index' was to assess whether there was any difference between those migrants who came on their own initiative and those whose relatives were the prime decision-makers. An example of each is described below.

The purpose of the first example was to enable the rather nebulous material concerning the settlement of origin in the questionnaire to be related to a common yardstick to allow comparison. Otherwise, there would be no way of appreciating the effect of respondents' antecedents: a man whose experience has been limited to hamlets may consider that his is relatively 'a big prosperous place', while a much-travelled businessman could easily describe his far larger town as 'rather a sleepy *pueblo*'. The measure utilised consisted of a list of facilities of varying degrees of sophistication; the respondent was asked to indicate those which he believed the settlement to possess. An expression of uncertainty was treated as a denial. Items for a reduced accumulative scale were then chosen; those accruing from over 80 per cent, or from under 20 per cent of the sample being eliminated so as to avoid invalidating the tests for reproducibility. Thus the items selected for the final Guttman scale were chosen on the grounds of their relatively even spacing between these limits (Table 11.2); item 23, one percentage point below the limit was chosen in preference to item 20 because it was possible to check the reported result with fact in this case. The settlements were then assigned to one of the nine rank classes according to the number of items claimed from the reduced scale.

If the scale had been perfect, all members of a class would be identical with regard to which items they possessed or omitted. However, in reality, a smaller number of settlements in most classes included error patterns (Table 11.3); that is, they included an item not identified with their class and excluded another that was. However, there are a number of ways of assessing whether the number of such error patterns invalidates the use of the scale as a variable or whether they can be considered as a tolerable margin of error. These include the calculation of the coefficients of reproducibility and scalability, the observance of a set of conditions limiting the proportion of error patterns occurring in any particular class or for any particular item, and the estimation of the chance reproducibility of a particular pattern given the frequency of occurrence of each item. In this case under discussion, the coefficients (0·97 and 0·88) were well above the usual accepted levels, and Ford's conditions regarding the distribution of error patterns were also met (Ford, 1954). The items' frequencies of occurrence were highly unlikely to have produced this distribution by chance and, in fact, predicted that 64 per cent of the sample would have assorted error patterns, which did not occur. Thus this scale could be used as a

Table 11.2 Original scale of source settlements' services

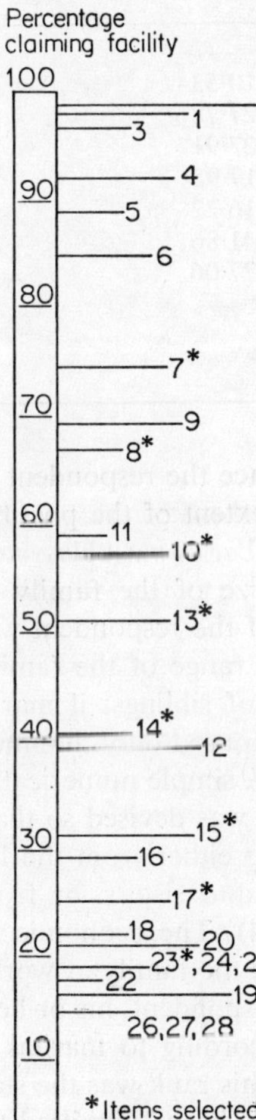

1. Church
2. Elementary school
3. Football pitch
4. *Cantina*
5. Post office
6. Guardia Civil post
7. Resident priest
8. Domestic electricity
9. Mayor
10. Secondary school
11. Medical post
12. Cooperative
13. Cinema
14. Bank
15. Theatre
16. Stadium
17. Radio station
18. Public transport service
19. High court
20. Public library
21. Five or more banks
22. Government hotel
23. Local daily newspaper
24. Cathedral
25. State hospital
26. University
27. Mutual savings bank
28. Supermarket

variable indicating one dimension of the settlements of origin. Moreover, this variable does not seem affected by the subjectivity of other statements concerning the settlements; questionnaires were internally consistent, matched pairs of strangers from the same settlement gave nearly identical answers and the total responses proved to be within one or two per cent of the published data for the few variables it was possible to check in this manner (variables 10, 15 and 23).

Table 11.3 Final Guttman scale of eight items

Rank classes	Percentage perfect patterns	Percentage error patterns	Percentage of total sample
8	100·00	—	18·25
7	89·47	10·53	3·78
6	72·22	27·78	10·71
5	72·09	27·91	8·53
4	82·05	17·95	7·74
3	83·78	16·22	7·34
2	58·14	41·86	8·53
1	72·94	27·06	16·87
0	100·00	—	18·25
			100·00

The second example was an attempt to place the respondent in his social and economic context so as to indicate the extent of the possible pressures bearing on the individual in the home. A number of variables were examined to represent the family background: the size of the family (taking into account the number of brothers and sisters if the respondent was single, or the number of children if married); the age range of the family (if single, the position of the respondent in the order of siblings; if married the age of the oldest child); and whether the home housed only the nuclear family, or whether it included additional members. A simple numerical score (hereafter referred to as a 'social pressure' index) was devised so that the higher scores indicated the higher pressure resulting either from the larger family of dependent children or younger brothers and sisters, or from responsibility for more distant relatives (Table 11.4). The economic context was represented by the number of persons within the family at work (the maximum of three possibilities representing the respondent, his or her father and either a brother or sister or the spouse, according to marital status), and the prestige rank of the work in each case. This rank was the standard scale adopted by most Peruvian surveys which has been submitted to numerous tests. Its six points could be briefly described as: (1) manual, unskilled work; (2) semi-skilled; (3) highly-skilled artisans, foremen; (4) routine white-collar, technicians, petty bourgeoisie; (5) skilled white-collar, professions such as schoolteaching and engineering, businessmen; (6) the higher professions such as medicine, law, university professors and diplomats,

Table 11.4 Marital status and the social pressure index

	Social pressure index							Total	
	1/2/3	4	5	6	7	8	9/10	%	N
S	8·30	11·42	20·42	28·03	19·72	7·96	4·15	100·00	289
M	6·22	3·83	11·96	36·36	27·27	10·53	3·83	100·00	209
									498

Table 11.5 Number in family working and prestige of that work

Number working	Prestige rank 1	1·5/2	2·5/3	3·5/4	4·5/6	Total %	N
1	53·57	6·25	20·54	12·50	7·14	100·00	112
2	34·50	27·91	22·48	10·46	4·65	100·00	258
3	22·85	20·00	30·48	19·05	7·62	100·00	105
							475
(Prestige rank unascertained for one or more members)							23
							498

Table 11.6 Economic index and social pressure index

Economic index	Social pressure index 1/2/3	4	5	6	7	8/9/10	Total %	N
1–2	5·38	10·00	20·77	30·00	23·85	10·00	100·00	130
3–4	8·13	4·88	15·45	30·89	30·08	10·57	100·00	123
5–6	6·41	12·82	20·51	32·05	16·67	11·54	100·00	78
7–8	8·89	8·89	15·55	31·11	17·78	17·78	100·00	45
9–17	7·14	2·38	19·05	23·81	33·33	14·29	100·00	42
								418
(Economic index invalidated by omissions)								80
								498

entrepreneurs of national standing. An 'economic index' was derived by adding the prestige rank of all reported working within the family (Tables 11.5 and 11.6).

These basically descriptive variables are perhaps more useful as an indication of the parent population for the whole sample than as a sample classificatory device, particularly when cross-tabulated with other data within the questionnaire. The social pressure index revealed that the majority of single persons had five or more brothers and sisters; most respondents in this group claimed to be in 'the middle' of the sibling age range and thus they would have to have that number to claim a score of six or more. Similarly, the majority of the married respondents, although still in their early twenties, must have had a minimum of three or four completely dependent childen in order to achieve the same score. Thus the average home family consisted of about six to eight persons; yet in the single respondents' case, 35 per cent of these families were supported by the sole efforts of the father, in two thirds of the cases a manual worker. This seems odd, considering that the respondent and often another sibling were of working age and that only an insignificant proportion of the families had anything to do with the land. Married respondents reported more persons working within each family and, interestingly, the more workers a family possessed, the more prestigious their work was likely to be (significant at ·001 level in a chi-square test).

The relationship between the two indices is slight; however, their principal purpose is to represent the complex set of personal context circumstances when examining responses with more evaluative material, for example, a Likert scale elsewhere in the questionnaire which invited the respondent to express an opinion about a number of statements referring to the ambience of the settlement. Furthermore, they provide a tentative insight into the values of the home settlements; despite the great majority of the sample appearing to come from large families with poor reported means of economic support, only 86·7 per cent claimed their economic position there to be 'poor' or 'very poor' and only 8·13 per cent of the total sample were not working and claimed that this was due to the absence of any work at all, any suitable work or any reasonably paid work. The rest were either working or stated there was no need for them to do so.

Finally, a set of questions explored the means by which a migrant might have acquired knowledge about other settlements; directly, by having visited or lived in them himself, or indirectly, through friends and relatives having left home prior to his own departure. These four variables were arranged in sequence, according to the greater number of positive responses, and their relationship plotted (Figure 11.1). Although this reveals the considerable extent to which the majority had contacts elsewhere, the actual order of these variables may be chance; 45 per cent of the responses could be

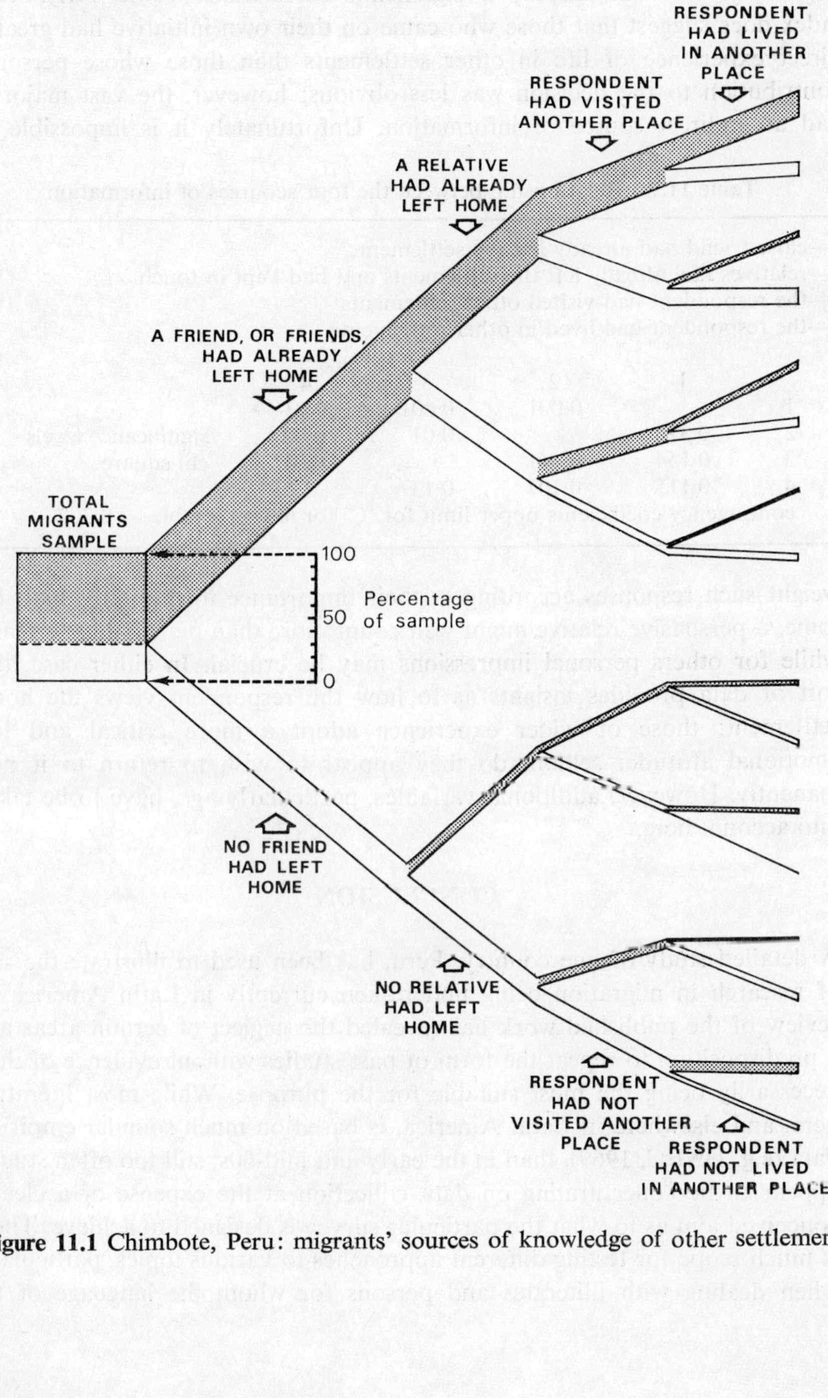

Figure 11.1 Chimbote, Peru: migrants' sources of knowledge of other settlements

predicted by the variables' marginal frequencies alone. However, the contingency coefficients display a significant correlation (Table 11.7). This index does suggest that those who came on their own initiative had greater direct experience of life in other settlements than those whose personal contribution to the decision was less obvious; however, the vast majority had an indirect source of information. Unfortunately it is impossible to

Table 11.7 Relationship between the four scources of information

1—close friend had already left the settlement
2—relatives had already left the settlements and had kept in touch
3—the respondent had visited other settlements
4—the respondent had lived in other settlements

	1	2	3	4	
1		0·001	0·001	0·01	
2	0·316		0·01	0·02	significance levels
3	0·154	0·141		0·01	chi square
4	0·115	0·107	0·137		

contingency coefficients upper limit for 'C' for a 2 × 2 table = 0·707

weight such responses according to their importance to the individual; for some, a persuasive relative might well count more than personal experience, while for others personal impressions may be crucial. In either case, this sort of data provides insights as to how the respondent views the home settlement; those of wider experience adopt a more critical and less emotional attitude; seldom do they appear to wish to return to it permanently. However, additional variables, particularly age, have to be taken into account here.

CONCLUSION

A detailed study of one country, Peru, has been used to illustrate the sort of research in migration being undertaken currently in Latin America. A review of the published work has revealed the neglect of certain areas and a predisposition to repeat the form of past studies without evidence of their necessarily being the most suitable for the purpose. While most literature here, and elsewhere in Latin America, is based on much sounder empirical data (e.g. DNEC, 1969), than in the early and mid-60s, still too often studies appear to be concentrating on data collection at the expense of a clearly conceived aim as to what the particular survey is designed to achieve. There is much scope for testing different approaches to various topics, particularly when dealing with illiterates and persons for whom the language of the questionnaire is foreign and who, as a consequence, may interpret key questions very differently. However, surveys are being called upon to provide data on problems of increasing complexity and the single factual question is no longer appropriate. Gradually more sophisticated descriptive variables must be built up, tested and examined for regional and cultural variations.

The survey described here explored a number of ways of collecting information about some of the less tangible aspects of migration. The examples given are three out of a much larger number designed for exploring each sector: the settlement, the individual in it and the individual's experience. Some assessment of their relative reliability was obtained from checks embedded in the questionnaire and others were provided by external sources. Considerable forethought went into how the study could be placed in context, both in terms of total population of the town and in relation to this town's catchment area. This was particularly necessary as little attention has been given as yet to common geographical variables in such studies; greater attention, for example, to 'distance' decay functions and to both physical and personal communication networks would do much to broaden the understanding of a topic such as migration, apart from displacing the present stereotyped images conjured up by a department's name or a broad physical region. Similar attempts are being made in the current work being carried out elsewhere in Latin America, as is evident from the articles in a recent specialist edition of the *International Migration Review* (1972).

However, it is not merely the depth of such studies that requires attention; we also need to cover a wider range of settlements. We now know much about the capital and something about one or two other major cities in each country; with a few notable exceptions, we know practically nothing about smaller settlements or the movements taking place in rural areas. Few migration streams have been followed from catchment area to area of new settlement and, with the exception of a small number of case histories, we can only speculate in general terms about what puts a person 'at risk' of migration. There is plenty of work for the future.

NOTE

1. The study referred to was a joint venture between the author and CISM, a social survey research institute attached to the Ministry of Work, Lima, and was sponsored by a grant from the Centro de Estudios de Poblacióny Desarrolo, also in Lima. This was made possible by the support given to the project by Dr. B. Samané, at that time director of CISM, and Dr. J. Donayre, director of the Centre, to whom I am very grateful.

REFERENCES

Abate *et al.* (1962) 'La desorganización familiar y social por migración rural–urbana', *Informaciones Sociales* **17,** 16–39.

Alers, J. O. (1970) 'Reliability of survey techniques in highland Peru', *Rural Sociology* **35,** 500–511.

Alers, J. O. and R. P. Appelbaum (1968) 'La migración en el Perú: un inventario de preposiciones', *Centro de Estudios de Población y Desarrollo*, serie oríginal 2.

Arriaga, E. E. (1965) 'Components of city growth in selected Latin American cities', *Milbank Memorial Foundation Fund Quarterly* **46,** 237–252.

Benton, W. (1961) *The voice of Latin America* (New York: Harper).

Bértoli, F. and F. Portocarrero (1968) *La modernización y la migración interna en el Perú* (serie Urbanización, migraciones y cambios; Lima: Instituto de Estudios Peruanos).

Bradfield, S. (1965) 'Some occupational aspects of migration', *Economic Development and Cultural Change* **14,** 61–70.

Cerdan, C. (1965) 'Proceso de migración en la comunidad de Moya', *Cuadernos de Antropología* III, 96–117.

Davis, K. and A. Casis (1946) 'Urbanisation in Latin America', *Milbank Memorial Fund Foundation Quarterly* **24,** 186–207.

Delgado, C. (1968) 'Tres plantéamientos en torno a problemas de urbanización acelerada en áreas metropolitanas: el caso de Lima', *Cuadernos Plandemet*, serie asuntos sociales 1.

DNEC (1969) *Encuesta de immigración Lima Metropolitana* (3 vols.; Lima: Dirección Nacional de Estadística y Censos).

Dobyns, H. F. and M. C. Vázquez (eds.) (1963) *Migración e integración en el Perú* (Lima: Editoriales Andinos).

Ducoff, L. (1965) 'The role of migration in the demographic development of Latin America', *Milbank Memorial Foundation Fund Quarterly* **43,** 197–216.

Elizaga, J. C. (1965) 'Internal migrations in Latin America,' *Milbank Memorial Foundation Fund Quarterly* **43,** 144–158.

Elizaga, J. C. (1970) *Migraciones a las áreas metropolitanas de América Latina* (Santiago de Chile: Centro Latinamericano de Demografía).

Ford, R. N. (1954) 'A rapid scoring procedure for scaling attitude questions', *Sociological studies in scale analysis*, ed. M. W. Riley *et al.* (New Brunswick: Rutgers University Press), 302.

Fried, J. (1959) 'Acculturation and mental health among Indian migrants in Peru', *Culture and mental health*, ed. M. K. Opler (New York: Macmillan), 114–137.

Germani, G. (1967) 'The city as an integrating mechanism', *The urban explosion in Latin America; a continent in process of modernisation*, ed. G. Beyer (Ithaca: Cornell University Press), 175–214.

Graumann, J. (1966) 'Redistribución de la población del Perú', *Boletín de Análisis Demográfico* **4-5,** 22–74.

Hauser, P. M. (ed.) (1961) *Urbanisation in Latin America* (Paris: Unesco).

Hodara, J. (1969) 'La explotación de la sociología', *Mundo Nuevo* **41,** 24–31.

International Migration Review (1972) 'Internal Migration in Latin America', **6,** 121–222.

Junta Nacional de la Vivienda (JNV) (1966) *La migración y el crecimiento urbano* (Lima).

Labbens, J. (1969) 'The role of the sociologist and the growth of sociology in Latin America', *International Social Science Journal* XXI, 428–432.

Lambert, J. (1963) 'Requirements for rapid economic and social development: the view of the historian and sociologist', *Social aspects of economic development in Latin America*, ed. E. de Vries and J. Medina Echavarria (Paris: Unesco), 59–60.

Lowder, S. (1970) 'Lima's population growth and the consequences for Peru', in B. Roberts and S. Lowder *Urban population growth and migration in Latin America; two case studies*, (Monograph series; Liverpool: Centre for Latin American Studies) **2,** 21–33.

Lowder, S. (1973) unpublished Ph.D. thesis.

Mangin, W. P. (1967) 'Latin American squatter settlements: a problem and a solution', *Latin American Research Review* **1,** 101–128.

Mangin, W. P. (1970) *Peasants in cities. Readings in the anthropology of urbanisation* (Boston: Houghton Mifflin).

Mangin, W. P. and J. Cohen (1966) 'Cultural and psychological characteristics of mountain migrants to Lima, Peru', *Sociologus* **14,** 81–88.

Martinez, H. (1968) *Las migraciones internas en el Perú*', (Lima: Centro de Estudios de Población y Desarrollo, serie original 3.

Martinez, H. (1969) *Las migraciones altiplánicas y la colonización del Tambopata* (Lima: Centro de Estudios de Población y Desarrollo).

Matos, J. (1963) 'Consideraciones generales acerca del proceso migratorio en la ciudad de Chimbote', *Migraciónee integración en el Perú*, ed. H. F. Dobyns and M. C. Vázquez (Lima: Editoriales Andinos), 72–76.

Matos, J. (1966) *Estudio de las barriadas limeñas* (1955) (Lima: Departamento de Antropología, Universidad Nacional Mayor de San Marcos).

Morse, R. M. (1965) 'Recent research on Latin American urbanisation: a selective survey with commentary, *Latin-American Research Review* **1,** 35–74.

Patch, R. W. (1967) 'La Parada, Lima's market: a study of class and assimilation', *American Universities Field Staff* (West Coast South American series XIV, Peru 1).

Rabinovitz, F. and F. Trueblood, (eds.) (1970) *Latin American Urban Research 1* (Ithaca: Sage Publications).

Rotondo, H. (1963) 'Problemas de salud mental en el área urbana de Mendocita', *Estudios de psiquiatría social en el Peru*, ed. B. Caravedo *et al.* (Lima: Ediciones el Sol), 13-183.

Solari, A. E. (1969) 'Social crisis as an obstacle to the institutionalisation of sociology in Latin America', *International Social Science Journal* XXI, 445–456.

Solis, C. A. (1963) 'Feuntes de migración al puerto industrial de Chimbote', *Migración e integración en el Perú*, ed. H. F. Dobyns and M. C, Vázquez (Lima: Editoriales Andinos), 78–81.

Szulc, T. (1963) *The winds of revolution: Latin America today and tomorrow* (New York: Praeger).

Testa, J. C. (1970) 'Las migraciones internas en el contexto del desarrollo social latino-americano', *Aportes* **15,** 96–109.

Tolentino, E. (1961) 'El impacto de la industria en Chimbote', *Pesca*, July.

Turner, J. C. (1968) 'Uncontrolled urban settlement: problems and policies', *International Social Development Review* **1,** 101–128.

United Nations (1957) *Report on the world social situation* (Paris).

van Es, J. C. and E. A. Wilkening, (1970) 'Response stability in survey research: a cross-cultural comparison', *Rural Sociology* **36,** 191–205.

Weller, R. H., J. J. Maciso and G. R. Martine (1971) 'The relative importance of the components of urban growth in Latin America', *Demography* **8,** 225–232.

Chapter 12

Land Tenure and Agricultural Development in the Central Altiplano, Bolivia

David A. Preston

Introduction

An underlying aim shared by many research workers in the social sciences is the investigation of problems the solution of which may help to improve the quality of life of the people studied. This goal is especially apparent in relation to work in rural areas in developing countries where countryfolk are stragglers in the race for self-improvement and where a wide variety of institutional barriers permit and even encourage the development of under-development. Social science research in developing countries is also relevant to the more developed countries, particularly to the least-developed areas of such countries. Peasants in Brittany, western Ireland, and highland Bolivia have many problems in common, not least the lack of interest of their central governments. By their very urgency, problems in developing countries often demand more immediate investigation than those at home.

The research on which this essay is based was motivated by a desire to know more about the organisation of farming in a region characterised by severe environmental conditions in order to be able to pinpoint specific problems, solutions to which would satisfy certain needs of the local population. Geographers are primarily concerned with understanding the organisation of phenomena in space: in rural areas this involves an investigation of the role of both physical and socio-cultural constraints on farming and thus its spatial organisation. Although individual investigators in the social sciences—particularly social anthropologists, economic historians and rural sociologists—have included in their analyses the role of areal organisation, it is not characteristic of social scientists other than geographers to do this. Although considerable overlap of interest exists between social scientists the geographer's contribution is the consideration of ecological relationships in a spatial dimension. The present study is distinctively

---

Dr. Preston is Lecturer in Geography at the University of Leeds.

geographical in its orientation and it is suggested that the type of analysis involved results in a more satisfactory delineation of problem areas and solutions to certain problems resulting from land tenure systems than would work undertaken by a rural sociologist or agricultural economist.

The focus of this study is upon the role of land tenure in the agricultural development of a particularly barren highland area. It was believed that since the areas occupied by communities in this zone were particularly large, the types of land tenure that would have evolved would reflect differences in the use to which particular zones were put. In addition, rapid social change in Bolivia in the past two decades since the 1952 Revolution has not been matched by changes in land tenure in freeholding communities which were but little affected by the agrarian reform.[1] Although a number of changes have taken place in the freeholding communities, they are generally inadequately documented and such accounts as do exist refer to the more populous and accessible areas. In this essay an attempt will be made to remedy this by describing the existing tenure situation in the Central Altiplano and by identifying some of the perceptible changes in it. In addition the problems that result from the tenure situation and which affect agricultural development are examined.

The Central Altiplano, that part of the high plateau lying south-west of the city of Oruro (Figure 12.1), suffers from aridity and isolation and attracted only limited attention during the colonial period. Silver mines in the vicinity of Carangas, on the eastern fringe of the Western Cordillera, attracted colonists in the sixteenth century and the deposits were mined until late in the nineteenth century. According to Bacarreza, mercury was deposited at Carangas which was a transit point between the port of Arica and the silver mines of Potosí where mercury was in much demand (Bacarreza, 1931). Even so this mining activity seems to have left little mark on the surrounding agricultural communities, so poor is the land. The income to an *encomendero* (a person holding a right to labour tribute) in this area would have been so paltry that exploitation would have been scarcely worthwhile. The hacienda system thus never became established west of the Río Desaguadero. The land-tenure system has evolved directly from that which existed under the Inca empire without the major changes that accompanied the establishment of large estates. The degree of communality of tenure that existed has gradually declined until today some communities have little land which is not now owned by individuals. The extent to which communal rights to cropland did exist in the early part of the present century is very uncertain, but the most striking finding of recent field work in the Central Altiplano has been the almost complete disappearance of communal cropland. The region does, however, illustrate how land-tenure forms vary in relation to the use to which land is put and to its value in the eyes of the community.

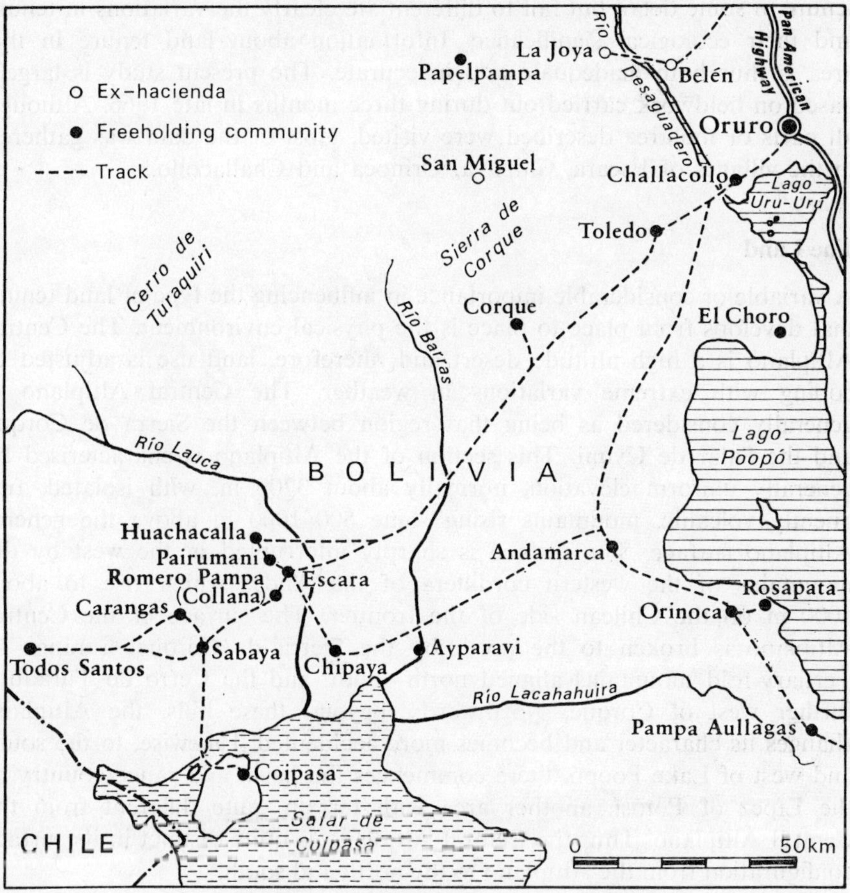

Figure 12.1 The Central Altiplano, Bolivia

Reports by Forbes based on experience in Bolivia in the latter half of the nineteenth century, by Posnansky in the early years of the present century, by Bacarreza in 1910 and by Métraux during several months' field work during 1930–31 supply most of the published information on the communities of the Central Altiplano; only Métraux, who was primarily concerned with the Chipaya, provides specific information about land tenure (Forbes, 1870; Posnansky, 1925; Bacarreza, 1931; Métraux, 1934). More recently, since 1960, the community of San Miguel, north of Corque, has been described; a group of Peruvian social scientists have studied Llica, to the south-west of the Salar de Coipasa; and some mention has been made of the agricultural conditions of Sabaya (McEwan, 1969; Rodriguez *et al.*, 1963; Burela and Burela, 1970). The two former studies treat land tenure in some detail but fail to differentiate clearly the variations in tenure and their ecological significance. Information about land tenure in this area is thus both inadequate and inaccurate. The present study is largely based on fieldwork carried out during three months in late 1966. Although all parts of the area described were visited, most of the data was gathered in the villages of Escara, Chipaya, Orinoca and Challacollo.[2]

The Land

A variable of considerable importance in influencing the type of land tenure that develops from place to place is the physical environment. The Central Altiplano is a high altitude desert and, therefore, land use is adjusted to coping with extreme variations in weather. The Central Altiplano is generally considered as being that region between the Sierra de Corque and the Salar de Uyuni. This section of the Altiplano is characterised by generally uniform elevation, normally about 3700 m, with isolated, frequently volcanic, mountains rising some 500–1000 m above the general Altiplano surface. The surface is sharply interrupted in the west by the mountains of the western cordillera of the Andes which rise to about 6000 m on the Chilean side of the frontier. The surface of the Central Altiplano is broken to the north by the Sierra de Corque, a range of Tertiary fold mountains aligned north–south, and the Cerro de Turaquiri further west of Corque. Northwards, beyond these hills, the Altiplano changes its character and becomes more accidented. Likewise, to the south and west of Lake Poopó, there commences the basin and range country of the Lípez of Potosí, another area with terrain quite different from the Central Altiplano. Thus the Central Altiplano is quite distinct in its surface configuration from the Altiplano to the north and south.

The surface of much of the region is composed of either salt flats or sandy, lime-rich, gently undulating terrain. Mature soils have developed in very few places and cultivation is frequently at the mercy of shifting sands, periodic inundations or saline efflorescences. Many of these characteristics are related to the fact that a large part of the Central Altiplano was covered by Lake Minchin during part of the Quaternary period and evidence of this is the fretting of many hillslopes by former shorelines of this lake, which eventually dried out leaving only Lake Poopó and Lake Coipasa, both of which are almost completely dry during the dry season.

The vegetation of much of the level area of the Central Altiplano is dominated by thola (*Lepidophyllum* spp.), a low evergreen shrub, lampaya (*Lampaya medicinalis*), and various bunchgrasses (*Stipa* spp., *Calumagrostis* spp. and *Festuca* spp.). Although the ecology of the Central Altiplano has never been studied in depth it appears that the bunchgrasses are largely confined to the sandier areas while thola and lampaya occur in less sandy areas. In the shallow natural drainage channels that occur frequently over the area to carry storm waters, and beside the perennial rivers, a dense sward of very short grass predominates. The most common vegetation type, covering as much as 60 per cent of the surface area, is thola which grows to a height of one metre. Thola plants are seldom very close together and frequently have 50–150 cm of bare ground between them, depending on rainfall and soil quality. In the flat pampa area bordering on lake flats, such as between Chipaya and Escara, and south of Andamarca, the cushion-like yareta (*Azorella yareta*) is the only vestige of vegetation to be seen over wide areas.

Most of the plants that predominate in the area under study are eaten by grazing livestock (llamas, sheep and, occasionally, donkeys) but the short grass pasture is most preferred and seems capable of supporting intensive grazing. Ecological work in similar areas of Peru has suggested that the grazing carrying capacity of this terrain is approximately one llama (or three sheep) per 3–6 ha although a Bolivian source suggests one llama per hectare (C.N.R.A., 1953; Tosi, 1960). By comparison Pearson concluded that the density of wild vicuña in south Peru at 4200 m was one animal per 49 ha (Pearson, 1951). This rather scanty evidence may be taken to suggest that the carrying capacity of Central Altiplano scrub pasture is low and existing livestock densities should not be increased.

In many years the crucial factor that determines the success or failure of the crops is rainfall. Over the area studied rainfall amounts vary from less than 100 mm in the west to as much as 400 mm near Oruro. Rainfall as well as being slight is also highly variable, a difference of 200–300 per cent between individual years being not unknown; amounts falling during the early part of the wet season when most crops are sown (between October and December) are particularly unreliable. In the far west, in the region of Coipasa and Escara, rainfall is lowest and irregularity greatest, although there are no rainfall records for the area to prove this. Evaporation is everywhere high and records for Chuquiña, near Toledo, show potential evaporation some 80 per cent in excess of the mean precipitation, giving further evidence of aridity. In addition, the danger of frost for much of the year affords a further hazard for many crops; in years when there is adequate rain for crop growth the growing plants may still be killed, particularly early in the year when they are flowering. In all, the climatic hazards in the area are such as to require either elaborate remedial methods—in particular irrigation—or the storage of harvest in years with adequate rainfall to guard against possible, or even probable, crop failures in dry years. In Escara in late 1966, after a series of dry years, few families had any potatoes left for sowing for the 1967 crop. The periodicity of rainfall is also important in the consideration of both the agricultural potential and the extent of present population pressure. In such an area of semi-desert conditions, the occurrence of groundwater can be the critical factor for the survival of livestock and, in the few areas where any form of irrigation is practised, for the growth of crops. In fact the flow of the two perennial rivers of the area, the Desaguadero in the east and the Lauca in the west, is amply maintained during the dry season, the latter by snow-melt and the former by drainage from Lake Titicaca. In few places is the water table far below the surface, and most of the isolated hills have a few springs near their base; small springs are not uncommon in the large areas of thola scrub. Seldom, however, is there enough water from such springs to support much livestock; and irrigation using spring water, despite the comparative frequency of its occurrence, is a source of dispute in many communities.

Salinity is an important factor affecting crops in areas of pampa throughout the Central Altiplano and it is related directly to the occurrence of groundwater and to rainfall. Many of the areas of lake flats, such as those near Toledo and Challocollo, are temporarily flooded after heavy rains as a result of the clayey nature of the soils. When they dry out the concentration of salts at or near the surface is common and prevents the growth of crops. Similarly, areas of spring seepage near the foot of hills and even in hollows near river courses are frequently affected by saline efflorescence. The most effective method of combating this problem that is used in the area seems to be that adopted by the Chipaya who divert branches of the Río Lauca over areas to be cultivated for several months prior to sowing; the superficial salts are thus washed away and moisture is accumulated to stimulate plant growth.

Human background

In an area with a physical environment relatively hostile to agricultural settlement, cropland is confined to those areas where the most favourable conditions for agriculture are to be found. Population is concentrated in small villages with 200–400 inhabitants. Such villages are for the most part located at the foot of the few hills to be found in the region. Even though the population is sparsely scattered over the area, several settlements are often to be found around the foot of a hill. Thus, for example, around the foot of the hill where Escara is situated there are six other villages even though the nearest is some 40 km away. While such areas with cultivated land contain a majority of the permanent dwellings, large areas of scrub, lake flat pasture or sand dunes contain widely-scattered alternative and temporarily occupied dwellings. A result of this pattern is that, while the cultivated land is generally close to the permanent home, the pasture land is far away, often far enough to necessitate the building of a house there for shelter. Such pasture areas, belonging to individuals, are usually referred to as *estancias*.

Villages are arranged in a regular gridiron pattern of streets, with an old, frequently colonial period church overlooking the central square. Houses are built predominantly of mud bricks and have only a single storey and no windows. Occasionally houses on the central square are more modern, painted, with windows and even with two storeys. In the *estancias*, houses vary from rectangular thatched cottages to conical turf huts (*putucos*) on the lake flats where wood and stone for building are scarce.

The inhabitants of the Central Altiplano are predominantly Aymara-speaking with the exception of the Chipaya people, who are a remnant of a formerly more extensive group that occupied the western part of the Altiplano from Lake Titicaca southwards and which included the now-extinct Urus of Lake Titicaca. They live in two villages on the lake flats north of the Salar de Coipasa. The Aymaras and, to a lesser extent, the Chipaya are highly acculturated; young people speak Spanish with some fluency and dress in commonly accepted *cholo* [3] styles. One of the most interesting cultural traits of the area that relates to agriculture is the complete absence of the plough west of the Río Desaguadero. Everywhere hand hoes and metal-ended digging sticks are used to prepare the land for sowing. The lack of ploughs may in part be attributed to the shortage of pasture for draught animals and to the unsuitability of many of the hillside fields for ploughing. Throughout the region similar cropping systems are practised with potatoes (without exception, native varieties) and quinoa, an indigenous grain, predominating. In a few sheltered places, such as Rosapata, on the western shores of Lake Poopó, frost incidence is sufficiently low for a native dwarf variety of maize to be cultivated.

The extent of cultural change may be readily associated with the degree to which migration has long been a traditional activity during the slack season after the harvests and before planting begins. Migration on foot, carrying salt to the eastern mountain valleys, is very ancient; more recently, seasonal migration has been directed towards areas of more intensive agriculture with high seasonal labour demands; and, lastly, there has been a growing current of temporary and often permanent migration to the towns and mines of the eastern Altiplano.

The migration of men and llamas taking salt to the valleys of Potosí department is very probably of prehistoric origin; salt was one of the first commodities to be traded in Europe during the Neolithic period. Groups of three or four men each with 15–20 llamas set off between June and September on a two or three-month journey of some 300–400 km to the east. Each llama carries 4 blocks of salt, weighing in total about 25 kg, each of which costs about 2 p (4 US cents) and which can be sold for as much as 7 p (15 cents). Other llamas may carry *chuño* (dried potato), *charqui* (dried llama meat) or quinoa, which may be traded in the valleys for temperate products. The purpose of the journey is not so much to make money but to trade highland products for maize, beans and wheat which form an important part of the Altiplano diet. Many people, particularly in the eastern part of the region, have made the journey with such frequency that they also speak Quechua, the predominant language of the eastern valleys, in addition to Aymara and perhaps Spanish. This mobility has long enabled rural people of this area to supplement their meagre food supply and has also provided an occupation at a time of the year when there was little to do in the fields.

Since the Chaco War of the late 1930s, and the more recent rapid expansion of agricultural settlement on the margins of the eastern lowlands, opportunities for casual labour have increased and attracted men from those areas where agricultural incomes are low and where there is a long slack season. Labour is in short supply and wages correspondingly high. Furthermore some crops, in particular sugar cane, need a lot of labour during a short period of the year which often coincides with the very period when there is little to do on the Altiplano. The large sugar cane estates in north-west Argentina, especially around Jujuy and Ledesma, import large numbers of Bolivian workers, hiring railroad wagons and sending trucks to collect them. In the Orinoca area, almost the entire population of some villages goes off to Argentina every year. Finally, the estates in northern Chile, where agriculture is based on irrigation, need additional workers, as do the fish-processing factories of the ports, and many Bolivians are thus attracted to work in Chile. The movement to Chile is particularly widespread from the western parts of the Central Altiplano; in Escara, for example, few families do not have someone who has worked in Chile for a time during the past 20 years and some have relations living there still. It is likely that this migration is also linked with widespread smuggling that exists along the border and which also provides a valuable supplement to incomes.

In the east and south of the region movement is largely to Santa Cruz and northern Argentina respectively, while salt traders come particularly from the central zone, especially from the vicinity of Corque. Migration to the mines and cities is widespread and has increased in importance in the years since the 1952 Revolution. Oruro is the urban centre which has received the bulk of the urbanward migration, not only because it is the only town of size in the area but also because all transportation routes from the Central Altiplano focus upon it. The mines in the Oruro area and further east provide the most important source of industrial employment although probably the largest number of migrants gravitates towards service industries principally in Oruro.

Because the Central Altiplano is a singularly barren environment for agriculture, the population is highly mobile and many young people have lived, for a time at least, away from their community. This range of experience is likely to have been an important factor in stimulating change in land-tenure practices, not only because of the number of former migrants but also because migration in one form or another has for so long been commonplace.

Another factor that is responsible for a number of the characteristics of the Central Altiplano land tenure is the shortage of cultivable land and the increasing population that static land resources have had to support. The number of families in Escara (defined as being Ayllu Camacho of Huachacalla) rose from 34 in 1901 to 67 in 1939 and to 96 in 1966, the population thus tripling in sixty-five years.[4] By contrast the population of Orinoca (defined as Ayllu Collana) has only risen from 162 families in 1910 to 187 families in 1966. It may be that different rates of emigration affect these figures. As might be expected from the nature of the terrain, population densities in the region are low. Data from the 1950 census suggested an overall population density for the Central Altiplano of 2·45 persons per square kilometre. This contrasts with the overall density for the Department of Oruro of 3·9 and a rural density of 2·1 persons per square kilometre. The community of Escara however, the area and population of which is known with some accuracy, had an overall density of 1·14 persons per square kilometre in 1966. Given these data, it is difficult, even taking into account the poor nature of the cultivable land, to state categorically that there is a shortage of land. The Aymara community of Collana (whose centre is Romero Pampa) has successfully occupied about 15 per cent of the total area of neighbouring Chipaya during a period when Chipaya population has increased considerably and this has been possible only as a result of under-utilisation of Chipaya land. It seems that, given such wide fluctuations of rainfall, population density is likely to be low in order to enable production to be sufficient to provide a surplus of produce to offset future dry years. The practice of fallowing for as long as 10–20 years in Escara and Orinoca, and probably much of the Central Altiplano, is probably a result of the need to maximise production during good years, and of the need to allow fertility to build up and to conserve moisture.

Land distribution in freeholding communities

In none of the freeholding communities visited were there families without land. The question as to whether all young families have land is complicated by family arrangements regarding the division of land between sons. Thus, although in many cases informants stated without hesitation that there was nobody in the community without land, this was not incompatible with the emigration of some of the younger sons for lack of land. If a father and two elder sons were working the family property there was little need for the assistance of a third or fourth son, although they too would inherit their share of the land on the death of their father. There is, however, no class of person who has no right to any land at all, in contrast with the situation in the northern Altiplano and the Yungas of La Paz. Two categories of people, *originarios* and *agregados* (original and associate members of a community respectively), exist in all communities but, although this refers to the historical position of the individual families in the community in the form the terms imply, its only significance is usually in the amount of tax paid. In some cases *originarios* do have more land at the present than *agregados* but in other cases the group of people owning most land included both *originarios* and *agregados.*

In those communities that possess lands both on hillsides and relatively level areas, namely Orinoca, Rosapata and Escara, three types of cropland are generally recognised and are known by distinct terms. Houseplots (*patios*) that are cultivated in Rosapata and Escara; walled hill fields (*canchones* or *huyus*) usually found on the middle hillslope, and other fields lower down and frequently unwalled (*sayrus* or *mantas*) are found in each of the three communities. In two of these communities, Escara and Ayllu Collana of Orinoca, certain of the *mantas* on the lower slopes are divided between the families of the community and the same rotation is followed by the whole community. In Escara the land thus divided is in fact only sown when rains are adequate; in times of drought, these lands are not divided up or cultivated for several years and are not fallowed for the twenty-year period common in Orinoca. During a drought, when the area is not cultivated by the community as a whole, anyone is at liberty to run the risk of crop failure and plant as much as they wish in this area. In Orinoca one informant suggested that *mantas* that were fallow could be cultivated at will by individuals who were short of land but this detail was never confirmed.

It is difficult to obtain precise data on land which is redistributed periodically, but in both Escara and Orinoca some land, at the hillfoot in both cases, is held communally and parcelled out to persons paying their land tax (*contribucion territorial*) for the period of cultivation. Each time an area is so subdivided an individual family will not necessarily cultivate the same piece of land.

In practice all families have a number of hillplots: in Escara, of the people interviewed, the average number of hillplots was over five, with no one having less than two, and the largest number being eleven. This was also true of the two neighbouring communities, Pairumani and Romero Pampa. Both the size of hillplots and the proportion of each one that is cultivable varies greatly and in the absence of a topographic survey no exact data can be given of the amount of cultivable land. It can be estimated very approximately that the land cultivated annually by each family would be about half a hectare (see Table 12.1). Fragmentation is considerable but, since in most communities the cropland is in a single zone, the distance which people have to walk to get to their plots does not vary greatly nor is it excessive.

Table 12.1 Cultivated land per head in Escara and Chipaya

Type of land	No. of plots	Size	Area	Rotation	Total cultivated annually
		(m)	(m²)		(m²)
Escara					
Patio	1	25 × 20	500	1 year cultivation 3 years fallow	125
Mantas	1	3 × 100	300	cultivated annually*	300
Canchones	5	50 × 50	12 500	3 years cultivation	4166
					4591
Chipaya 'chias†	4	5 × 170	3400	cultivated annually*	3400
Ayparavi (Eastern Chipaya) 'chias†	6	6 × 117 6 × 136	4554‡	cultivated annually*	4554

Based on a limited number of field interviews in each community.
* A different zone cultivated every three years, but the same size area cultivated annually.
† '*Chias* are strips of land allocated to a family annually.
‡ This figure derived from assuming that each person has three 'chias of each of the two sizes.

Areas of grazing in all communities except Challacollo are very large. In Escara the amounts of grazing occupied by individual families seem to vary considerably and appear to be between 150 and 1200 ha with a mean of about 230 ha on which about thirty llamas are kept. Some individuals with *estancias* near town or who live on their *estancia* keep between fifteen and forty sheep. Although such areas are used almost exclusively for grazing, there exist possibilities in isolated places for cultivation and thus the area of grazing may represent a more valuable asset than would first appear. However, very few individuals make use of this potential, as a result of lack of initiative or, in their terms, of *costumbre* (custom).

Challacollo is a highly anomalous community for the usual division between cultivable and grazing areas is not encountered, largely because crops can often be grown in many places in the surrounding pampa provided that salinity is not too high. Thus in most years every person in Challacollo should have at least some land suitable for planting. Parts of the suitable areas may be rented for the season to those unfortunate enough to have all their land in areas too salty for planting. The area of cropland is thus highly variable since many families have left Challacollo to go to Oruro and the renting of their land is widespread. Although the situation of Challacollo is unusual for these reasons it does serve as a contrast with the communities of the western part of the Central Altiplano where renting is rare and resources of cultivable land not very elastic.

LAND TENURE

The legal tenure situation

Many freeholding communities have ancient documents, referred to as 'titles', dating from the colonial period, in some cases identifying by name the mounds or markers indicating the limits of the community. In no case were personal titles for this period encountered. A number of individuals in various communities have receipts for payments of a land tax in accordance with the Financial Law of 1904 which they regard as titles but which, of course, are not. In Bolivian law, however, customary occupation of the land confers a legal right to it and thus both communities and individuals can be considered to have a legal right to the land that they occupy. This, however, is of little value in preventing disputes over land since the limits of the property of individuals or a community are seldom either recorded or arrived at by a mutual agreement with neighbours. Neither legal positions nor financial judgements are accompanied by a map indicating unequivocally the position of limiting markers. In the case of the ancient dispute between Chipaya and Collana the agrarian judge in Oruro ruled that the Aymara of Collana should withdraw from Chipaya land and that the limits between them 'should be subject of an agreement between the two communities'. Since the boundaries are not identified in the ruling, the problem remains unsolved. Thus, although the legal position of the freeholding communities is well defined, in practice the lack of definition of boundaries inhibits legal security.

There is a considerable variety of tenure forms and these may be best described under the following headings:

(1) *De facto individual occupation.* The individual right to land is not recognised uniformly. In each of the communities visited some land, usually

cropland and houseplot, is recognised as having distinct limits and being the property of one person; other land, frequently pasture land, is accepted as being customarily occupied by the livestock of an individual but without any precise limits being recognised. The limits of personal property in the towns and nearby croplands are well defined and offer few possibilities for misinterpretation and conflict. In the case of lands used for grazing, such as areas of thola scrub in western Oruro, personal rights are centred on a small dwelling, which may have been constructed for shelter, and the grazing of the immediately surrounding areas of scrub is generally recognised as the right of the owner of the dwelling. Further away from the house the personal grazing rights become more obscure and, in general, less clearly defined the further one gets from the house. In cases where an area of grazing is divided among heirs the midpoint between a person's dwelling and those of each of his neighbours is assumed to be the limit of his land and it is divided equally between the heirs. Apart from these personal rights, it seems normal for the right of access of livestock of other owners to wells that occur in an area of private grazing to be respected. In two communities, Romero Pampa and Challacollo, limits to personal grazing areas have been established and markers set up to indicate the boundaries.

(2) *Communal ownership*. In the majority of communities there are some areas where personal rights are not recognised and to which all members of the community have an equal right. Such areas are usually springs, particularly those close to the community, areas of spring seepage where good pasture occurs, also close to the community, and areas of pasture beside perennial streams. The poorest grazing land of all on the hilltops is frequently used communally for grazing male llamas. The rights of individuals to collect firewood or grass for thatching are not very clearly recognised. In most cases it is tacitly assumed that grass and brushwood may be collected anywhere by anyone, but cases were encountered where individuals did object to it being collected from their land. A further form of communal ownership applies to those areas which are used for cultivation by individuals for a specific period. Such lands are distributed periodically among all the people paying tax and all follow the same rotation system, the fields in one area most commonly being fallowed after three years of cropping. So at any time a family may be allocated plots in three areas. Each year when a new area is brought under cultivation plots in it are allocated to the families in the community and another area, after having been cultivated for three years, is then abandoned to fallow. Such lands were the lower hillfoot lands in Escara and Orinoca. Frequently, where much of a community's cropland is held individually, the land held communally (although perhaps only 10 per cent of the total cropland) is potentially among the best and most productive. This is clearly an exception to the previously-stated rule that the best land is that first to be

individually owned. Such land in Escara, however, is apt to suffer from drought and cold and may only be sown when rainfall is adequate. In Chipaya all the land suitable for sowing is held communally during the period in which it is cultivated, although in years when it is not cropped it reverts to pastureland over which individual grazing rights are recognised. (3) *Renting.* In only one of the communities studied, Challacollo, is renting practised on a large scale. Here the reason is quite clearly a high rate of emigration to nearby Oruro and probably, in part, a lack of pastureland. One informant rented three surrounding grazing areas of undetermined size for £18 (US \$70) annually and in several other cases land was rented with its occupant sheep (maybe 40–50 head) to a family that would tend them and receive in payment either all or a proportion of the cheeses that they could make from their milk during the year. In other cases flocks of sheep are rented to landowners who run them with their own flock, receiving the milk for cheese as rental. It is significant that in Challacollo, where a variety of forms of renting exists, there appears to be little sale of land and although maybe half the landowners now live in Oruro little of their land has been sold. This is similar to the situation in other communities further from Oruro where renting is less common. In Escara it was said that renting for cash was known but informants could not indicate anyone who was renting land. Rather it seemed that pastureland was rented for the duration of the pasture more often than cropland was rented. In Escara, sharecropping is on the whole more common than renting for money and is probably the most frequent arrangement made by people who have left the community. They retain ownership of the land but leave it in charge of a close relative or occasionally a neighbour in return for a share of the crop. One probable reason for the lack of renting is that those who have more land than they can work are more likely to hire labourers (in Escara either Chipayas or young men without much land) to work it for them than to rent it.

Sale of land

Actual sale of farmland in communities is very rare. Land that is within the urban radius of the settlement is, however, free to be sold. In Escara and in Orinoca some land that lies within the urban limits is not yet built upon and is used for agriculture. In Orinoca this land also has a water supply from springs on the hillside and from a small irrigation channel which allows quite good crops of beans (*habas*) and barley or wheat to be grown, as well as a few vegetables. Parts of this land are both rented and sold, largely for their agricultural use.

Land owned by people who have left the community to live in Oruro, Chile or elsewhere is farmed in partnership with a relative, deemed to be poor or in need of land, rather than sold. In Orinoca and Rosapata, when a landowner is likely to die without a direct heir, his land is normally consigned by him before he dies either to a close relative or to a neighbour who has very little land. This is a purely verbal agreement and provides opportunity for dishonest appropriation by a third party who may claim that the land, or part of it, was left to him. When the land does pass from one family to another without incident then the new owners may draw up a document to this effect and, in a few cases, have it notarised in the provincial or departmental capital.

TENURE PROBLEMS

The fact that neither communities nor individuals have titles to their lands gives rise to disputes; but these do not seem noticeably to affect agricultural exploitation. In a number of cases, existing conflicts over areas of land do cause severe difficulties for small groups of families and active strife with periodic fights may well impede their full use of the land. In some cases disputes over territory have included bitter fights in which participants have been killed or maimed (as in a battle between Chipaya and Collana in 1965). Conversely a dispute over a particular area of land may lead quite frequently to an intensification of its use to justify ownership. Incursions are often originally the result of an area not being occupied, but after a dispute the area under discussion is often more intensively used and fields are created and houses constructed. This may be observed in Chipaya where an area cultivated by the Aymara of Collana in 1965–66 on nominally Chipaya land is now cultivated by the Chipaya ayllu in whose territory it falls even though it practically doubles the area normally under cultivation by that ayllu. It was observed in the area under dispute between the Lloco section of Ayllu Collana of Orinoca and Andamarca. Here the area concerned is poor pasture but members of both communities have constructed conical turf houses (*putucos*) in the disputed area where few if any houses existed previously.

Tenure problems do, however, remind people of the basic insecurity of their tenure and the loss of land and even deaths resulting from land grabs by neighbouring communities are recurrent events in the history of many Central Altiplano communities. These events alone make security of tenure an important, if often unvoiced, goal of many communities. Three classes of tenure problem occur frequently: lack of definition of boundaries, ill-defined and disputed rights to water, and the problems of property delineation in areas of extensive, unfenced grazing.

(1) *Definition of boundaries*. The most severe problems encountered are

those resulting from an indefinite demarcation of boundaries between communities. Boundary limits are usually indicated by prominent stones, mounds or visible landmarks, the former being easy to remove. In many cases, particularly in extensive areas of pasture, the markers are not commonly known, for such areas may be visited only once or twice in a decade, maybe while looking for a missing sheep. The location of markers is thus easily open to dispute. The existence of extensive areas, largely unoccupied and unvisited, makes it quite feasible that two communities should have legitimate claims, supported by ancient documents, to the same area. Both Orinoca and Andamarca have such a claim to an area some 15–20 km long. The lack of clearly visible, relatively immovable markers accentuates the difficulty of reaching fair legal decisions, and once any legal decision is reached there is little likelihood that the legally determined limits will be marked on the ground unless they are the result of a mutual accord. In some instances, notably that of the conflict between Chipaya and Collana, the failure to enforce legal decisions since 1667 has left a legacy of strife between the two communities. Conflicts resulting from lack of definition of boundaries of individual rather than community lands are not so common and the only case encountered of a dispute between individuals was over an inheritance; the temporary solution was the equal division of the disputed land between the two claimants.

(2) *Water rights*. In theory all families have free access to water, both for household use and for livestock. In practice there are instances where this presents problems. There is no conflict over access to water from the public wells within the town limits, and in Escara there are also a number of wells in the houseplots which are used by the family on whose property a well stands or by a group of neighbouring families. Likewise springs and adjacent pastures that lie near settlements are invariably common property and no conflicts arise concerning them. Understandably, problems seem to arise during dry years when the water table is lowest. Under this circumstance some of the wells in the thola scrub that are used by livestock dry out, and some rivers cease flowing—even the third largest river of the area, the Río Barras, dries completely. During such periods access to water for livestock can be critical and although free access is maintained to streams and the riverside pastures this is not always so in the thola scrub. With the drying-out of some watering places overcrowding occurs around the diminishing number of springs and on the surrounding pasture and this may well lead the person with customary rights over that area to seek to restrict the use of the spring and its adjacent pasture. In one case this resulted in a dispute between Orinoca and Pampa Aullagas. People from the southern part of Orinoca were accustomed, during dry spells, to take their livestock to water in the Río Lacahahuira to the south of Orinoca land in territory belonging to Pampa Aullagas. In 1965 and 1966 people from

Pampa Aullagas had taken advantage of the river water to sow crops alongside the river, thus making it impossible for the livestock to get to the river without damaging the crops. In late 1966 the officials of Orinoca tried to arrange a meeting with Aullagas to attempt to have stock routes left unsown to allow herds to reach the river without harm to crops, but the outcome of any such meeting is not known. The problem of access to springs on cultivated hillsides is somewhat similar, since access for livestock would often be across sown land and there were several instances in Escara of intransigent individuals sowing land around the springs on their property and erecting stone walls around them to keep livestock out. Here no solution had been found and some people had to take their animals off the hill on this account. Legal action had been threatened against individuals depriving others of their customary rights but none had been taken. A further example of conflict over water rights was an isolated case in Rosapata of a person sowing quinoa in the humid area around his spring in the scrub pasture and thus depriving others of their customary use of the spring. No legal action had been taken in this case either.

(3) *Tenure in areas of extensive grazing.* The establishment of individual rights to the use of a specific area is most difficult in areas of extensive grazing such as the thola scrub. In an area where progressively fewer parts of communities remain in communal tenure the last area in which mutually exclusive personal rights are delineated is almost invariably the rough pasture. In Orinoca, the canton with the largest proportion of its land still held communally, the definition of personal grazing rights seemed most indistinct; on the other hand in Challacolla near to Oruro, where many changes had occurred, the boundaries of areas of grazing, admittedly not in the scrub, were clearly defined and generally recognised. A basic difficulty of establishing individual rights in areas of rough pasture is wandering livestock. Although a member of a family, usually a woman, may go into the pasture areas every month or so, there is no continuous supervision of livestock and likewise no system of fencing. Animals are thus free to wander at will. Llamas seem to require very little supervision and return on their own to the corral that is usually adjacent to the *estancia* dwelling each night. Likewise in the early morning one sees, crossing the paths or along river courses, lines of llamas going out to their pasture for the day. On the other hand, sheep and particularly donkeys, need constant watching to prevent straying. For this reason sheep are kept mainly by those people who have pasture close to the settlement where livestock can easily be supervised. In communities where sheep predominate, such as Challacollo or Chipaya, at least one member of a family lives all the time in the *estancia* dwelling. In addition to the problems of straying animals, a further factor that makes the establishment of clearly-defined personal rights to grazing land difficult is the effect of dry years. After a long period

with very little, if any, rainfall, shortage of water becomes acute, pasture deteriorates and, irrespective of personal rights, animals naturally tend to concentrate in those areas where pasture is most plentiful and water available. Thus in places that have good pasture even in dry years, such as areas beside rivers, near perennial springs and along dry river courses, it is impossible to establish individual rights and communal rights prevail. These problems have caused the tenure system in areas of pasture in much of the Central Altoplano to be only semi-individualised; that is to say that certain grazing areas are recognised as belonging to an individual family (or in the case of Rosapata to a group of families) but without the limits to their grazing area being defined. In practice it seems that the clarity of recognition of a family's rights to an area decrease away from their dwelling on it. In some cases, where a grazing area has been divided between heirs, the limits to the areas that they have are established and recognised. In this way it is possible for a family to say that they have their own *estancia* but also that its limits are not fixed. Little question arises of enforcement of individual rights since the essence of semi-individual tenure is its flexibility and the existence of a consensus between landholders. It is not surprising therefore that there were no disputes encountered over this type of land. A variation of this type of tenure occurs in Chipaya grazing lands. Here, although each family has one grazing area, a group of families keep their flocks together, use the pasture of all the group and move them from one grazing area to another as weather and state of pasture demand. Thus although one family may have pasture suitable for the winter period, in summer it may be flooded and thus livestock collectively can always be in an area where the pasture is as good as possible. Chipaya informants thought that this cooperation was highly necessary rather than just desirable.

CONCLUSION

Although, in comparison with other parts of highland Bolivia, the Central Altiplano is a backward area and agricultural innovations are rare, there have nonetheless been changes in the land-tenure system. The most important of these changes that can be clearly identified in the area studied is the gradual move towards an individualisation of land holdings and, as a corollary, a decrease in the area of land to which all have common rights. It was significant that, in the areas nearest to Oruro, subject to most outside influence, common land had all but disappeared. This in part reflects a weakening of community cohesion and an increasing emphasis on the family rather than the community as the prime social unit. A further change of importance is the increase in various forms of renting as a result of land being left by emigrants. Residents thus have access to more cultivated land than before but the practice of sharecropping decreases their benefit from the extra land. The principal problem associated with the tenure system is insecurity. The lack of legally valid titles, the impermanence or even absence of boundary markers and large areas of infrequently visited rough pasture combine to make encroachment across boundaries easy, difficult to detect and impossible to prove. Although in several instances territorial disputes resulted in intensifying land use, on the whole such disputes are a frequent irritant and if they could be eradicated farmers would feel more confident in their use of the land.

To what extent is the Central Altiplano typical of a major ecological zone in Bolivia and are the findings of this study relevant to other areas? Fieldwork of a less intensive nature in other parts of highland Bolivia has tended to confirm that the ill-defined boundaries of freeholding communities have caused strife, particularly where there is a strong demand for land. In the southern Altiplano (the provinces of Nor and Sud Lipez of Potosí), where the physical environment is even harsher, the poverty of the land and the extremely low population densities result in little competition for land and therefore, perhaps, fewer disputes. With an increase in mining activity in this area it is not hard to imagine that a plethora of boundary disputes will occur as farmers struggle to increase output to supply the new mining camps. In this area, where they are at present absent, the tenure problems characteristic of the Central Altiplano may well develop. The basic findings of this study have a universal relevance to all freeholding communities in highland Bolivia since all have a similar tenure status. It is encouraging that the recently created mobile teams of the Agrarian Reform Service (SNRA) are now working, in response to numerous demands from the peasants themselves, principally to demarcate the boundaries of freeholding communities. It is hoped that, in the near future, they will survey the boundaries of individual plots within the communities prior to issuing individual titles.

NOTES

1. Many of these changes are examined in work by Buechler (1969), Marschall (1970), Clark (1968), Erasmus (1969), Heath (1970), Leons (1967), and Preston (1969). More detailed bibliographies are available in these works and in the Bolivian edition of the report of the comprehensive study of the Bolivian agrarian reform (S.N.R.A., 1970).
2. During this period the author was working in Bolivia as Consultant to the Inter-American Committee for Agricultural Development as part of a multi-agency team evaluating the effects of the Bolivian agrarian reform. This notwithstanding, the opinions expressed here are personal and unofficial.

3. *Cholo* is a term of cultural identification denoting a person who has entered the Spanish cultural realm, having largely forsaken indigenous ways of dress, custom and speech.

4. These data come from lists of *contribuyentes* that many of the communities possess. *Contribuyentes* are taxpayers and usually all family heads are included. Precisely at what stage a young man becomes a *contribuyente* depends, however, on a variety of factors; data for different years and different communities are not therefore statistically comparable.

REFERENCES

Bacarreza, Z. (1931) 'Monografía de la provincia de Carangas', *Boletín de la Sociedad Geográfica de La Paz* **61–62,** 73–114.

Buechler, H. C. (1969) 'The northern Altiplano and adjacent temperate and subtropical valleys', in D. B. Heath, C. J. Erasmus and H. C. Buechler, *Land reform and social revolution in Bolivia* (London: Pall Mall Press).

Burela, A. and M. Burela (1970) 'Cultura y muyu-muyu; el efecto de la enfermedad en el proceso de desarrollo', *América Indígena* **30,** 41–60.

Clark, R. J. (1968) 'Land reform and peasant market participation on the northern highlands of Bolivia', *Land Economics* **44,** 153–172.

Consejo Nacional de Reforma Agraria (1953) *Informe de la comisión financiera* (La Paz: Servicio Nacional de Reforma Agraria).

Erasmus, C. J. (1969) 'Southern Bolivia: Departments of Chuquisaca and Tarija', in D. B. Heath, C. J. Erasmus and H. C. Buechler, *Land reform and social revolution in Bolivia* (London: Pall Mall Press).

Forbes, D. (1870) 'On the Aymara Indians of Bolivia and Peru', *Journal of the Ethnological Society* **2,** 193–305.

Heath, D. B. (1970) 'A view from the grass roots: peasant syndicates among the Aymara of the Bolivian Yungas', *Latin American Peasant Movements*, ed. H. Landsberger (Ithaca: Cornell University Press).

Leons, M. B. (1967) 'Land reform in the Bolivian Yungas', *América Indígena* **27,** 689–714.

McEwan, W. J. (1969) *Changing rural Bolivia* (New York: Research Institute for the Study of Man).

Marschall, K. B. (1970) *Revolution and land reform in the Bolivian Yungas of La Paz* (La Paz: Servicio Nacional de Reforma Agraria, Sección de Investigaciones).

Métraux, A. (1934), 'Contribution au folklore andin', *Journal de la Société des Américanistes* **26,** 67–102.

Pearson, O. P. (1951) 'Mammals in the highlands of southern Peru', *Bulletin of the Museum of Comparative Zoology* **106,** 115–174.

Preston, D. A. (1969) 'The revolutionary landscape of highland Bolivia', *Geographical Journal* **135,** 1–16.

Preston, D. A. (1970) 'Freeholding communities and rural development: the case of Bolivia', *Revista Geográfica* **73,** 29–41.

Posnansky, A. (1925) 'Nuevas investigaciones en Carangas', *Compte rendu de la XXI session* Congrès International des Américanistes, Göteburg 1924, 85–102.

Rodriguez, H. *et al.* (1963) *Progresismos y cambios en la comunidad campesina de Llica* (Lima: Universidad Mayor San Marcos, Facultad de Letras).
Servicio Nacional de Reforma Agraria, Sección de Investigaciones (1970) *Reforma Agraria en Bolivia* (La Paz: Servicio Nacional de Reforma Agraria) mimeo.
Tosi, J. A. (1960) *Zonas de vida natural en el Peru* (Lima: Instituto Interamericano de Ciencas Agrícolas).

Chapter 13

Irrigation and Mixed Farming in Malaya

D. E. Short

Introduction

Irrigation policy in Malaya [1] is, and always has been, almost exclusively concerned with rice production on peasant smallholdings. The Drainage and Irrigation Department was formed in 1931 with the specific objectives of increasing rice production and in doing so improving rural standards of living; these objectives have remained fundamentally unchanged for forty years (Malaya, 1931–1971). From 1931 until the 1950s, government placed its chief hopes for improvement in production and standards of living in the rice farm community on the opening up, with water control works, of new riceland. This might be termed a phase of extensive development which, though it increased rice acreage and production, did little to change the underlying causes of rural poverty. On the contrary, it sought to reproduce in the new areas certain 'desirable' characteristics of peasant society, notably stability based on a 'traditional way of life'. From the mid-1950s, and in particular since independence in 1957, rice policy has been concentrated on intensification, partly by established conventional methods such as fertiliser use and variety improvement but in particular by the provision of irrigation facilities for rice double-cropping. This multiple-cropping phase has already involved a vastly increased programme of construction and rehabilitation in both old and new irrigation schemes to permit off-season rice cultivation.

One thing which has remained fundamentally unchanged, not only since 1931 but since the government first took a direct interest in Malayan rice production in the 1890s (Short, 1971), is the tendency to appoint riceland to a special status. At first this tended to be a token attitude but in the later years of the colonial period, particularly when the rubber/rice price ratio fell, it became something of an obsession. In order to mitigate the 'rice problem' and at the same time improve the living conditions of the rural Malay population, without 'upsetting' their traditional way of life,

---

David Short is Fellow in Geography at the Centre for South-East Asian Studies, University of Hull.

considerable efforts were made to preserve, improve and extend ricelands. Legislation was enacted which rendered the conversion of ricefields to other uses illegal. Drainage and irrigation works were initiated and, because of the differing administrative status of the Federated and the non-Federated Malay States, were widely established rather earlier in the more directly controlled states of Perak, Selangor, Pahang and Negeri Sembilan. The early structure of the Drainage and Irrigation Department reflected this state pattern and the result was to distribute a large number of chiefly small but sometimes large-scale schemes throughout established and government-developed ricelands in central Malaya. In more recent years, especially since the Pacific War, the northern 'rice-bowl' states (Kedah, Perlis and Kelantan) have received an increasing share of government and public interest in irrigation development culminating in the huge Muda and Kemubu schemes. Nevertheless there is an established attitude reflected in irrigation policy that ricelands throughout Malaya should be upgraded by irrigation works, currently by the widespread introduction of rice double-cropping facilities wherever technically feasible.

Figure 13.1 shows the distribution of irrigation schemes in Malaya. The size categories are chosen to distinguish: small-scale schemes, often conceived as replacements for traditional structures; medium-size schemes which though providing irrigation works of a more modern type and scale still tend to serve fragmentary areas devoted to rice cultivation; and large-scale schemes in which the objective has been to create or improve an extensive riceland. The 'specialised' zones refer to the northern 'rice-bowl' tracts where rice has been estimated to account for 90 per cent of farmland over wide areas. Schemes outside these zones fall within areas where, on the basis of land-use mapping and fieldwork it is possible to postulate the likelihood of rice cultivation in the context of mixed enterprise farms of the type examined in this essay.

The chief motive behind the rice/irrigation policies has been the 'rice problem': Malaya's persistent inability to produce sufficient rice to satisfy the consumption requirements of its population (Grist, 1955; Malaya, 1953, 1956). Understandably more emphasis has been placed on the 'rural development' aspect since 1957 but with independence 'self-sufficiency' became an emotive and vote-winning catchphrase. As in so many development policies in both developed and developing nations political desirability inhibited a completely rational assessment of the overall impact and significance either for the rice-farm or the whole community. It must be emphasised that this kind of influence in water control development is by no means unique to Malaya (Clark, 1970; Wiener, 1972). The result has been a tremendous investment in irrigation. Other inputs, fertiliser subsidies, seed research, a marketing bureaucracy, farmers' associations, cooperatives and guaranteed prices have been inaugurated and substantially funded by

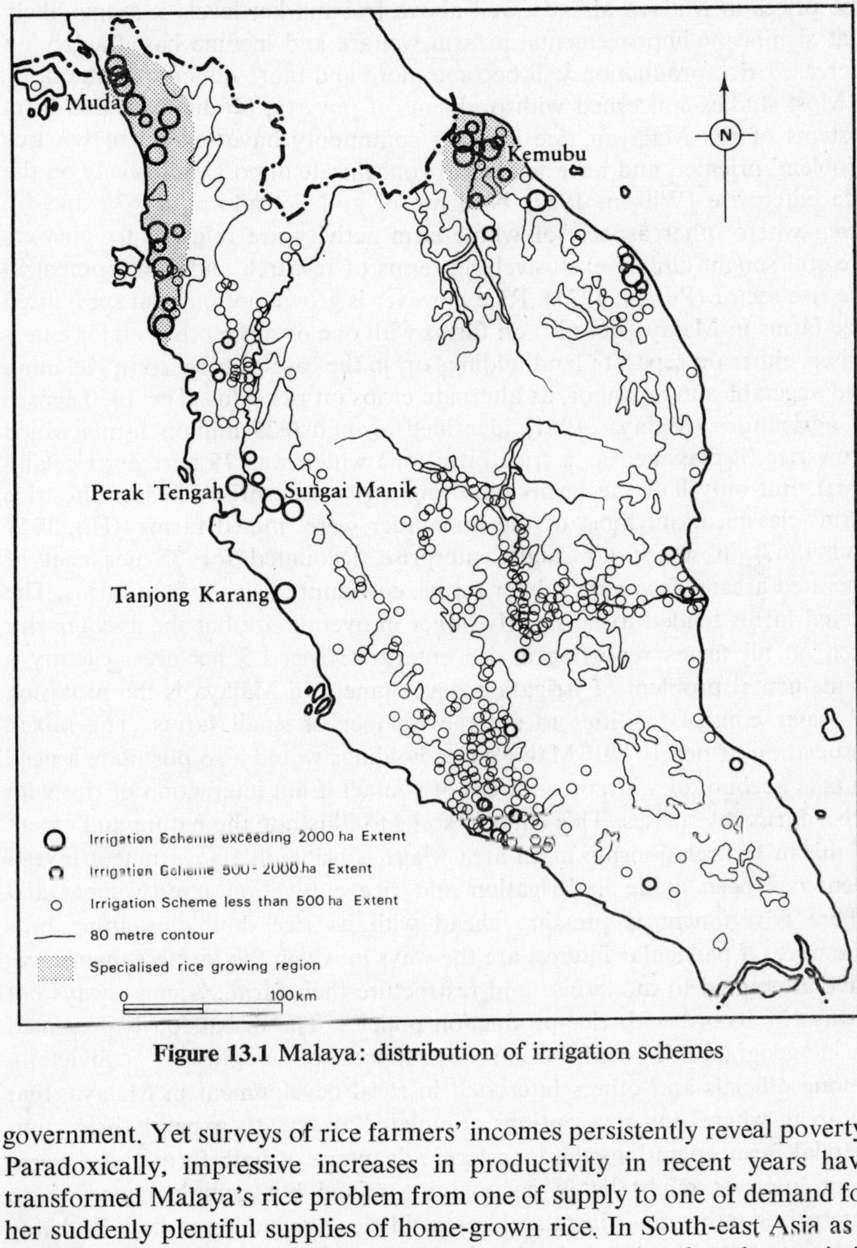

Figure 13.1 Malaya: distribution of irrigation schemes

government. Yet surveys of rice farmers' incomes persistently reveal poverty. Paradoxically, impressive increases in productivity in recent years have transformed Malaya's rice problem from one of supply to one of demand for her suddenly plentiful supplies of home-grown rice. In South-east Asia as a whole the rice trade, one of the few areas of intra-regional exchange, is in difficulties because of widespread trends towards self-sufficiency. Cheaper and more sought-after Thai rice is smuggled across the north Malayan border in operations reminiscent of the prohibition era in the USA. With rice prices in Malaya already well above free market levels it seems likely that significant improvements in farm welfare and income based solely on increased rice production will become more and more difficult to attain.

Most studies concerned with problems of poverty, productivity and farm systems of the Malayan rice farming community have tended to be 'rice problem' oriented and have tended to concentrate almost exclusively on the rice enterprise (Wilson, 1957; Narkswasdi and Selvadurai, 1967a and b). Even where other aspects of whole-farm activity are referred to, answers are still sought almost exclusively in terms of research and development in the rice sector (Purcal, 1971). Rice however is grown not only on specialised rice-farms in Malaya but also on farms with one or more other major enterprises, either on separate landholdings or, in the case of tobacco in Kelantan and vegetables in Selangor, as alternate crops on ricefields. The 1960 census of agriculture (Malaya, 1961) identified around 0·25 million farms which grow rice (it classified as a 'rice farm' one which was 75 per cent riceland area). But only half the farms reporting a rice enterprise fell into the 'rice farm' classification; most of the remainder were 'mixed farms' (Ho, 1961 and 1967) in which no single enterprise accounted for 75 per cent of operated area; a few were rubber farms, coconuut farms or fruit farms. The mixed farms tended to be slightly larger in overall size but the average rice area on all farms reporting a rice enterprise was 1·3 hectares. Clearly a fundamental problem of irrigation development in Malaya is the provision of water control facilities to a large number of small farms. The mixed association of nearly half Malay's rice holdings would also postulate a need to take account of widespread areas of contact if not interaction of rice with other farm enterprises. This chapter seeks to illustrate the nature and extent of this mixed relationship in an area where considerable government investment has been made in irrigation and rice cultivation improvement and where government is pressing ahead with its rice double-cropping programme. Of particular interest are the ways in which the farmers themselves have attempted to modernise and restructure their farm systems—ways not always in accord with rice-production policies. The discussion is presented as a geographer's contribution in support of a growing conviction, among officials and others interested in rural development in Malaya, that in areas where non-rice options stimulate farmers to expand their commercial farm operations the long-term advantage of both farm and government interests might lie in encouraging mixed but complementary farm enterprises rather than in retaining a rigid rice production policy. Probably the central problem of irrigation in Malaya is that it relates to the least profitable crop grown under Malayan conditions. Where options are available it is therefore difficult and possibly unfair to expect peasant farmers to specialise in rice production.

Sungai Manik : the scheme and the survey

In 1967 the author undertook a detailed study of an irrigation scheme in Perak, a state with a long and progressive record in both irrigation and agricultural development, a fact sometimes obscured by its pre-eminence in world tin production. The study was made as part of a wider survey of the role of irrigation in rural development in Malaya. The scheme, Sungai Manik, is situated on the lower Perak river flood plain, in an area offering a fairly wide range of farm enterprise options, and serves 7250 hectares of riceland in a relatively compact and continuous tract. Nearby are extensive areas of land used chiefly for rubber, cassava and mixed fruit trees under smallholder management, with rubber and oil palm on estates. The objectives of this study were: to examine the enterprise structure and income of farms served by irrigation; to attempt measurement of the contribution of the irrigation scheme to farm income and improvement; and to analyse the response of Malay farmers to the opportunities provided by irrigation works. To facilitate these objectives, enquiries were made on farm size, enterprise structure, sources and levels of income as well as more general attitudinal and technical matters. Special emphasis was laid on the spatial aspects of choice: the location of income sources in relation to irrigation; the size and use of land resources as expressions of farmer response. Pioneered from 1932 onwards with government assistance, Sungai Manik is not a 'traditional' riceland but partly a creation of official agencies with their own ideas on ideal patterns of land use and settlement. They did however adopt a traditional peasant-farm framework as the basis of land allocation and operation, as well as recruiting many of the colonists from established ricelands elsewhere. The most important innovations introduced by the scheme were the provision of an 'economic' riceland holding and a regulated water supply to individual farms at low cost to the farmer. The area has, because of government involvement, relatively well-documented and accessible records of planning, development and progress which provide bench-mark data for a researcher interested in processes of change. In Sungai Manik it is possible to examine the way in which peasant farmers have modified and restructured their farms over a period of thirty years as a response to perceived opportunities and restraints. The spatial expressions of these resource-use decisions can be measured by a comparison of information contained in land alienation registers and colonisation records with data collected at farm level. Apart from this there is a compact and closely defined area where irrigation can be seen to play a significant role in farm operation. Rice is the only crop grown under irrigation and no farmer in the survey operated riceland outside the scheme: the boundary between irrigation and other elements in the farm system is relatively clear. Information on Sungai Manik farms given in this chapter was, unless otherwise acknowledged, collected by the author in the 2500 hectare stage I of the scheme during August through November 1967. Although data was gathered on a wider range of topics, consideration is limited here to farm-enterprise structure and income and for this purpose a group of farmers whose rice crop was disastrously affected by flooding in January 1967 has been eliminated from the main sample. This leaves a group of sixty farms and avoids undervaluing the rice enterprise.

Stage I of the Sungai Manik scheme was originally allocated in the form of 1400 officially demarcated lots, chiefly of riceland but including small areas for house sites. Each lot was supposed to form the basis of an individual rice farm, about 2 hectares in extent. It was these surveyed lots portrayed on government cadastral maps which were adopted as the basis for the survey. A 5 per cent random selection of lots was made and then visited for the purpose of interviewing whoever was found to be operating the land—owner, tenant or sharecropper. It had aleady been shown how in the initial colonisation one riceland licence holder in seven sold part of his lot to another settler before receiving permanent title (Wilson, 1957), either because of unwillingness or inability to clear the jungle from the whole area. By 1955 one lot in five was officially subject to subdivision by sale or transfer. It was therefore no surprise to find parts of over ninety farms on the seventy lots visited in 1967. This in itself raises another important problem of irrigation in Malaya: decreasing size of individually operated parcels of land over time, under the pressure of peasant farmer concepts of manageable rice-holding size and traditional forms of land inheritance. This operates against planning goals of surplus production and ideal systems of irrigation layout. The Sungai Manik scheme was designed to provide each farm with a direct irrigation supply at one point on its perimeter and drainage at another: the designers intended that no farmer should have to depend on water access across another's land. This system also simplified the allocation of responsibility for maintenance of field channels, drains and water-control embankments. The breakdown of this pattern results in an increased likelihood of disputes over water distribution, particularly in those areas where micro-relief makes precise water control difficult. This is compounded by failure to accept responsibility for minor water-control structures which could alleviate such problems.

Because of limited time for data collection, a single interview technique was adopted for the survey. The farms visited cultivated only one strictly annual crop, rice; the remainder were either semi-annual (e.g. cassava) or permanent (e.g. rubber). The interviews were undertaken at the beginning of a new rice cultivation cycle so that information on production and income for the previous harvest was relatively easy to recall. For other crop enterprises, the 'memory factor' was reduced as much as possible by the phrasing of questions in terms of systems of operation rather than annual total quantities produced. For example rubber yield enquiries followed the formula: the amount collected per day; the number of tapping days per month; the number of months in the previous twelve tapping was undertaken. Scrap rubber was ignored for its relatively low value and difficulty of measurement. It has been reckoned to contribute 5–15 per cent of net yield by weight but, because of low value, a much smaller proportion of income. On the question of costs, the survey did not account for family labour inputs so that income figures relate to 'family return', defined as net income plus the cost of family labour. As far as possible any data on incomes from other surveys presented for comparison is in these terms. Whereas rice retained for domestic use is included in the family return, valued at the local wholesale prices, similarly-retained fruit, coconut and poultry could not be accounted for because of the haphazard pattern of collection and consumption. This should be borne in mind in considering any final income levels and totals.

SUNGAI MANIK FARM SURVEY

Farm size and enterprise structure

The average size of the interview farms in 1967 was 2·92 hectares. This is relatively large for Malayan smallholdings as a whole but less so in comparison with farms on other irrigated colonisation schemes, for example Tanjong Karang (2·35 hectares) and Perak Tengah (2·94 hectares). In the later stages of Sungai Manik's development the allocation per farm averaged 3·24 hectares (8 acres) made up of 6 acres of riceland and 2 acres of *kampong* (non-irrigated farmland). Rice-farm size averaged 1·65 hectares in the survey. This is closer to the 'normal' size of rice-farms not only on other colonisation schemes but also in major rice growing areas such as the Kedah plain (Wilson, 1958; Kuchiba, 1967). The important difference between Sungai Manik farms and those in the specialised rice growing districts of northern Malaya is the area devoted to crops other than rice: 1·2 hectares or 41 per cent of total farm area. In the Kedah rice bowl the area under 'other crops' has been estimated at 10 per cent of total farm area (Wilson, 1958). At Sungai Manik drained and accessible *kampong* was, at the time of survey, still available to farmers with land rights on the scheme. The reason for this is that early development effort, reflecting government priorities, was concentrated on the irrigated lands; those intended for village construction and dry-land crops had only been opened up since the war and drainage work only completed in 1964. In 1966 a large proportion of this land remained uncultivated: of the newest, the

3078 hectare Redang Ponggor *kampong* area, it was officially estimated that 1985 hectares remained unused. This land could be obtained for an initial survey fee of 5 dollars per acre plus an annually renewable licence fee of 3 dollars. The standard 0·81 hectare (2 acre) allotment could therefore be obtained for an initial outlay of only 13 dollars.[2]

Rice is nevertheless the most important crop on Sungai Manik farms both in terms of total area (99 hectares) and total return. Riceland is the only type common to all 60 farmers, and 13 farmers operated riceland only. Levels of income on individual farms from each major on-farm enterprise are summarised in Table 13.1. According to the survey, yields in Stage I

Table 13.1 On-farm sources: Numbers of households falling into specified monthly income categories (60 farms)

Monthly income M.$	0–14	15–29	30–44	45–59	60–74	75–89	90+
Rice	3	6	17	18	5	4	7
Rubber (26 farms)	1	4	9	7	1	2	2
Cassava (12 farms)	2	4	5	1			
Coconut (6 farms)	2	2		1	1		
Fruit (41 farms)	35	5			1		
Livestock (44 farms)	42	2					

for the 1966–67 season averaged 2072 kg/ha. At the mean selling price received of 25 cents per kilogram, this gave a gross return of 518 dollars per hectare. Total costs excluding family labour but including hired labour and equipment, rent, materials and irrigation charges were 128 dollars per hectare. Thus in a year of high rice prices because of a national flood disaster, the family return to Sungai Manik farmers was 390 dollars per hectare, or just over 647 dollars per farm per annum. Twenty-six of the farmers operated rubber holdings totalling 31 hectares which were productive during the year. A further eight controlled rubber which was either immature or subject to replanting. The rubber land was scattered in areas adjacent to the irrigation area, just under half of it on the government developed *kampong* lands. The remainder was distributed around the edges of the scheme in independently developed mixed *kampong*. The rubber ranged in quality from ageing trees planted in the early years of this century by Indonesian immigrants, to immature high-yielding stock in the recently drained village lands of the scheme. The average size of productive holding was 1·23 hectares yielding 554 kg/ha. Nevertheless it should be pointed out that the yields reported did not reflect the full impact of plantings in the newer *kampong* which had yet to reach overall maturity at the time of the survey. In these areas of high yielding clones, yields approaching 1120 kg/ha were reported and will become commonplace in the future. During the survey year rubber prices (in contrast to those for rice) were declining to their lowest levels for twenty years. Sungai Manik farmers, who invariably sold their product as grade three unsmoked sheet rubber received prices averaging just under one dollar per kilogram. They therefore grossed 548 dollars per hectare, reduced after rent, tapping and maintenance costs to a family return of 454 dollars per hectare or 556 dollars per reporting farm.

Twelve farmers reported the operation of a total of 10 hectares of land planted with cassava during the interview year, all of it on the previously unused newer *kampong* area. As a commercial crop this was relatively new to Sungai Manik farms, though the root has long been regarded as a poor substitute for rice. Chinese entrepreneurs had acted as the catalyst in introducing it as a commercial enterprise on local farms, providing capital in the form of tractors and equipment for hire as well as marketing arrangements. Because of this, and unlike any other enterprise on the farms, cassava cultivation was highly mechanised—with correspondingly high costs. This crop was regarded as mature from 14 to 18 months after planting, so that income during the survey year depended upon the occurrence of a harvest. A factor which complicated yield calculations was the sale of an immature crop still in the ground by farmers unable to wait for maximum returns. Additional variations in income depended upon whether farmers were prepared to invest in simple processing equipment which permitted the more profitable sale of dried cassava flakes. Despite variations and problems of this kind and high cultivation costs an average family return of 410 dollars per hectare was reported. The average income per reporting farm was 320 dollars. Six farms incorporated a total of 6 hectares of coconut as a commercial enterprise, though many more produced coconut as an item of domestic consumption. The commercial coconut holdings were almost entirely located in the coastal districts of Bagan Datok and Sabak Bernam, up to 50 kilometres from the ricelands. They were almost certainly a relic of the inland movement of immigrant Indonesians from their initial settlement areas on the coast. Periodic inspection and collection visits were a necessary part of this operation and in a few cases members of the farm family took up temporary residence on the distant holdings. Payment for harvesting and maintenance costs, plus the practice of selling the nuts still in the husk rather than split and dried, depressed returns. They averaged 395 dollars per hectare or 365 dollars per reporting farm.

Fruit production was possibly the most fragmented and least well organised land use on Sungai Manik farms. It was difficult to analyse its production in areal terms because fruit trees were generally grown in mixed and often small areas; sometimes on odd pieces of land round the house lot or on embankments between ricefields. Only one farmer had a fully commercial fruit enterprise of just under 2 hectares, but 41 farmers reported some kind of income from fruit sales. The identifiable orchard holdings which totalled 12 hectares were mainly in the scheme's *kampong* areas, reflecting the original intention that the farmers should have a fruit and vegetable garden. To this end a special experimental and propagation centre was set up at Degong, on the scheme, to supply farmers with approved budstocks. In many cases, however, fruit trees occupied a minor part of a lot predominantly under rubber. Individual returns were generally low, averaging 100 dollars per annum, and it is perhaps unrealistic, because of rather desultory organisation, to quote the estimated per hectare return of 345 dollars. Threequarters of the total income from fruit was earned from sales of mango (*Mangifera indica*) for which Sungai Maniw is noted. Wholesalers from as far away as the Federal capital were present in the area during sales of this fruit. The character of fruit sales from the *kampong* areas will undoubtedly change as valuable tree varieties such as durian (*Durio zibenthinus*), jak (*Artocarpus heterophyllus*) and rambutan (*Nephelium lappaceum*), planted in what are essentially recently-developed lands, reach maturity. The only other farm enterprise recorded on Sungai Manik farms was livestock-keeping which provided forty-four households with cash income, mainly from poultry and eggs and, to a more limited extent, goats. As in the example of fruit, income was widespread but at low levels. The average return was 52 dollars per annum. The goat farming, employing a stall-fed technique, was on an extremely small scale. The largest 'herd' consisted of 8 animals and involved a turnover of 2 animals during the year for a total of 150 dollars. Goat-keeping was confined to respondents of Javanese ancestry and was apparently a cultural import. Poultry-keeping on the other hand was widespread and revealed little if any evidence of evolved techniques of production. The small rather ragged flocks had their basic forage diet supplemented with rice bran and this plus the use of goat manure on ricefields provided the only examples of a complementary relationship between enterprises on Sungai Manik farms. There is a ready demand for these livestock products which, along with fish, provide the most valuable sources of animal protein widely available and acceptable to the Muslim population.

The sources and levels of income

In addition to the on-farm enterprises listed in the preceding section a number of off-farm and non-farm sources of income were recorded; these are summarised in Table 13.2. It is worth noting the average total income

Table 13.2 Off- and non-farm sources: Numbers of households falling into specified monthly income categories (60 farms)

Monthly income M.$	1–14	15–29	30–44	45–59	60–74	75–89	90+
Hire of farm labour (23 farms)	11	8	4				
Rent on farmland (10 farms)	7	2	1				
Crafts and cottage products (21 farms)	18	2	1				
Fishing and collecting (26 farms)	24	2					
Part-time wage employment (9 farms)	5	1		3			
Full-time wage employment (4 farms)		1	1				2
Trade (7 farms)	2		2	1	1		1

of Sungai Manik farm households: 1475 dollars per annum or 123 dollars per month. Table 13.3 provides some comparative material for farms in some of the more important rice-growing areas in Malaya. Perhaps of more importance in explaining the desperate social, economic and political need for successful rural development policies in Malaya was the per capita annual income in Sungai Manik of 189 dollars, compared with the national average

Table 13.3 Examples of income levels on Malay operated rice growing farms

Area in Malaya	Income (Malaysian dollars p.a.)	Source	Year of survey
Pahang	1656	(Ho, 1967)	1966
Tanjong Karang (Selangor)*	1568	(Narkswasdi, 1967a)	1967
Sungai Manik (Perak)	1475	(*Short*, 1971)	1967
Province Wellesley	1406	(Purcal, 1971)	1962
Province Wellesley	1314	(*Straits Times* 2.1.70)	1970
Province Wellesley*	1271	(Purcal, 1971)	1962
Bachang (Malacca)	1126	(Narkswasdi, 1967b)	1967
Kelantan	647	(*Straits Times* 2.1.70)	1970

* rice double-cropping régime in operation.

(in 1967) of 950. It is the gap between these two figures rather than the 'rice problem' which has proved resistant to policies of alleviation, including irrigation development. Apart from this the most striking aspect of Sungai Manik income is the diversity of its origins on a scheme which was specifically designed to create rice farms and promote rice farming. In view of this the relatively low contribution of the rice enterprise to both area and total income of farms was remarkable, averaging 59 per cent and 44 per cent respectively but with income falling as low as 10 per cent on some farms (Table 13.4). This table also illustrates the extent to which the *kampong* lands on the scheme have assumed an importance far beyond that of the 'garden plot' originally envisaged. It is only because of their important cash crop function that the scheme can claim to provide a large majority of farmers with most of their income.

Table 13.4 Numbers of farms receiving specified percentage of family return from major sources

Percentage of income	1–9	10–19	20–29	30–39	40–49	50–59	60–69	70–79	80–89	90–100
From Padi enterprise	1	5	4	14	12	6	5	4	5	4
From all farm enterprises	1	1	1		2	8	7	8	9	23
From on-scheme sources			1	4	3	8	1	4	4	35

THE IMPLICATIONS FOR GOVERNMENT POLICY

Farmer response and crop diversification

Both government and farmers had sought to improve and secure the position of the farm community in Sungai Manik. Government has opened up and made available extensive areas for the creation of rice farms and has increased rice production and productivity by introducing technical change and innovation, including modern irrigation. The chief flaw in this policy from the point of view of the farmers has been that it presupposes specialisation in rice farming, an arduous activity involving high inputs and relatively low returns, particularly in terms of cash. The farmers on the other hand had sought to diversify their farming activities from an initial rice-dominated orientation towards a more mixed farm structure. Table 13.5 shows the number of farm households having multiple sources of income. The proportion of 15 per cent of total return was selected arbitrarily to exclude those sources of income so minor and so occasionally used as to exaggerate the picture of diversity. Half the households had three or more important sources of income. Sungai Manik farmers had consistenly reduced rice-farm size since the scheme's inception while expanding whole-farm size in terms of a variety of cash-earning enterprises. They had retained a major interest in rice farming but there was a large element of traditional subsistence attachment to this enterprise. Half the return from rice farming was in the form of domestic rice supply. While other farm enterprises showed signs of farm-sponsored modernisation and innovation, rice farming retained many of its persistent traditional associations and techniques (Narkswasdi, 1968). This diversity also offered the farmers security during periods of price fluctuations and natural disaster such as the floods of January 1967. Perhaps more important was the fact that the farmers showed an awareness of the comparative advantage of, and a willingness to respond to, incentives provided by old-established and new cash crop enterprises. As Penny has emphasised, a sound route to improvement in agriculture within a peasant farm framework is via self-motivated change (Penny, 1964). Lack of such motivation has been cited widely as a reason for the failure or only partial realisation of rural development projects (Myrdal, 1968, pp. 977–982). To this extent some of the trends apparent on Sungai Manik farms were soundly based and viable in a modernising society.

Table 13.5 Numbers of families having one or more sources of income accounting for at least 15 per cent of total family return.
(Sources of income based on 13 categories in Tables 13.1 and 13.2)

No. of sources	1	2	3	4	5
All sources (60 farms)	8	22	20	9	1
Farm sources (59 farms)	22	27	6	4	
Off- and non-farm sources (35 farms)	27	8			

For all this there was much room for improvement in their organisation and enterprise structures which were apparently relatively inefficient. For example average return per acre on Sungai Manik farms was 166·36 dollars compared with 196·78 on the double-cropped farms of Tanjong Karang irrigation scheme in Selangor. Similarly farm structure was characterised by an almost complete lack of complementary functions or innovation transfer between enterprises. This phenomenon has been interpreted as evidence of resistance to change in that Malay farmers are reluctant to undertake ventures which would interfere with existing economic arrangements (Swift, 1965). The desire of many farmers to operate a number of enterprises seems to exemplify the contention that the Malay farmer is intent on satisfying immediate consumption goals employing a minimum of long-term organisational and investment efforts for a quick return. This would explain the cultivators who sold unsmoked rubber, unhusked coconuts, and cassava still in the ground. The acquisition of additional land for tree and other low-recurring input crops could also be interpreted as a means to short-term ends rather than long-term investment in improvement and rational extension of holdings. The separateness and even potential conflict of interests between enterprises within the farm structures is perhaps best exemplified by those farms incorporating coconut holdings 50 kilometres away from the ricefields. The implications for the rice enterprise were, however, unequivocal. While responding to perceived incentives elsewhere these farmers had reduced their proportional interest in rice farming. This is a tendency which has disturbed government agencies concerned with increasing rice production in Malaya for at least sixty years since the advent of rubber as a Malay smallholding crop (Short, 1971). It is certainly an indictment of those colonial administrators who projected Malay rice farming as a 'way of life'. Whether in the Sungai Manik case this was more because of increased population pressure and inheritance processes or more the result of a deliberate retreat from rice farming resulting in turn from conflicting demands on time and labour from competing enterprises (Purcal, 1971) cannot be stated with assurance on the basis of the survey. The evidence suggests a combination of both, in that land records show subdivision and sale of holdings from the earliest period of colonisation. Fourteen per cent of the complete sample in 1967 had acquired riceland interests in the area within the previous five years, indicating that it is still a receptor of migrants. At the same time 35 per cent of the farmers cited the scheme itself as their place of origin, indicating considerable absorption by the scheme of internal population increase. With a finite area of irrigated riceland but expanding interest in other land-use types this has inevitably meant a reduction in the area per farm affected by irrigation.

The most recent government response to this tendency has been to introduce double rice-cropping, thereby effectively doubling rice-farm size. In doing so, the departments engaged in agricultural extension work have received a high level of cooperation from the farmers of Sungai Manik. In 1967, 64 per cent of the farmers expressed approval of the impending move into double-cropping, though only half this proportion cited increased cash earnings as their motives for wishing to double-crop. Other significant responses were those which referred to increased rice supplies for domestic consumption (13 per cent) and those which referred to the pressure of family size on available riceland (6·5 per cent). Twenty-one per cent were non-committal, many saying that they would follow their friends and neighbours, leaving only 15 per cent who expressed disapproval of the proposed innovation. Most of the objections were based on quite rational grounds such as problems of water control in part of the scheme which caused difficulties in growing even one rice crop a year. Only two farmers referred specifically to a likely conflict of demands for available labour resources between another rice crop and farm operations elsewhere. A degree of caution has to be exercised in interpreting this response since the farmers, subject to intensive official canvassing on the virtues of rice double-cropping may have felt that a positive response was the 'correct' response. Nevertheless the response would appear to provide further evidence of a high degree of openmindedness towards innovation. It is to be hoped that the double-cropping policy will prove effective in improving household income, though surveys of income in other rice double-cropping areas such as Tanjong Karang and Province Wellesley (Table 13.3) suggest that any improvement emanating from the rice enterprise is likely to be marginal. The Tanjong Karang figure is particularly noteworthy because it relates to a 'showpiece' scheme of similar origins to Sungai Manik with a very similar average rice-farm size. The failure of this policy to improve farm incomes on the other hand could prove very damaging to farmers' faith in government development policies and make them less amenable to innovation emanating from official sources. It would also reinforce the already apparent tendency to diminish their interest in the irrigated lands where options exist.

CONCLUSION: THE NEED FOR REAPPRAISAL

Irrigation in Malaya faces a wide range of problems: financial, engineering, agronomic, social. These have ben examined from a variety of perspectives, particularly since water control became a major element in development policy in the 1930s. In interpreting these problems farmer response has often been alluded to as an area of difficulty: Malay rice farmers have been found to be 'tradition bound', lacking in economic motivation, unwilling to innovate and take risks. Geographers no less than others have contributed to a stereotyped view of rice farms and farming systems in their preoccupation with landscapes and composite images of an apparently aimless and idyllic way of life. Only recently have geographers even begun seriously to question the stereotype of *a* Malay system of rice farming (Jackson, 1969). This chapter has attempted to illustrate by reference to one rice-growing community that a fundamental problem in justifying and successfully implementing irrigation in Malaya has been, and is, a high degree of awareness of the economic limitations of rice farming as a major farm enterprise. The rice farmers studied had responded as individuals to a scheme based on uniformity of farm size, structure and enterprise. In doing so they had achieved widely disparate levels of income through a variety of means. In analysing these the 'whole farm' perspective was essential: the study of the farm as a complex system and without preconceived notions of primary and secondary enterprises. The spatial approach was considered fundamental because it offered a relatively easily quantified, non-controversial and observable dimension which served as a measure of farmers' response to their opportunities. Combined with income studies it helped to locate the frontiers of expansion and incentive on these peasant operated farms. The farms may not have been typical: they were, for example, operated by farmers or the descendants of farmers with proven pioneering ability and with greater than normal opportunities for farm-enterprise development. The relationship between irrigated and non-irrigated lands may similarly be atypical, but only further research on these lines could make this apparent. Nevertheless on farms such as these in Sungai Manik and elsewhere in Malaya a fundamental problem of irrigation must be to retain farmer interest in irrigation and the irrigated lands. Some farmers have already sought to use them for other crops, notably tobacco and vegetables, and government has shown some interest in promoting alternative uses of some ricelands. In 1971 the National Padi and Rice Authority was created, one of its primary tasks being to encourage rice farmers outside the 'rice-bowl' areas to plant alternative off-season crops and seek additional farm enterprises such as cattle and poultry-rearing. Irrespective of this, the overwhelming preoccupation with rice supplies has seen a diminishing area of land on irrigation schemes used for alternative crops in recent years (Malaya, 1968). If the policy of diversification and intensification of use of the ricelands is to succeed, it will have to take account of the competing options at farm level. Most important of all, some commonly held assumptions about rice farms should be subjected to further scrutiny to facilitate a fuller understanding of the complex relationship of factors which promote or retard agricultural development. In this scrutiny, the role of the geographer with his perspective of 'the farm as a spatial system of operations' (Morgan, 1971, p. 27) can be of immense value.

NOTES

1. In order to simplify references to the geographical area under consideration the term Malaya is used throughout this text. This is intended to be synonymous with the current political unit of West Malaysia.
2. Malaysian dollars are the currency units used throughout the text. Approximately 6 Malaysian dollars = 1 pound sterling.

REFERENCES

Clark, C. (1970) *The economics of irrigation* (2nd edition, Oxford: Pergamon Press).

Grist, D. H. (1955) 'The rice problem', *Malayan Journal of Tropical Geography* **5**, 20–25.

Ho, R. (1961) 'Mixed farming and multiple cropping in Malaya', *Journal of Tropical Geography* **16**, 1–17.

Ho, R. (1967) *Farmers of central Malaya* (Canberra: Australian National University Press).

Jackson, J. C. (1969) 'Rice cultivation in Malaya: culture, history, customary practices and recent development', *Conference on South East Asian society and government* (London: School of Oriental and African Studies).

Kuchiba, M. and Y. Tsubouchi (1967) 'Paddy farming and social structure in a Malay village', *The Developing Economies* **5**, 462–485.

Malaya (1931) *Report of the Rice Cultivation Committee* (Kuala Lumpur: Federated Malay States Government Press).

Malaya (1953) *Report of the Rice Production Committee* (Kuala Lumpur: Government Printer).

Malaya (1956) *Final report of the Rice Production Committee* (Kuala Lumpur: Government Printer).

Malaya (1961) *Census of agriculture* 1960*: type, tenure and fragmentation of farms* (Kuala Lumpur: Department of Agriculture).

Malaya (1968) *Annual report, Drainage and Irrigation Department* 1964–66 (Kuala Lumpur: Ministry of Agriculture and Cooperatives).

Malaya (1971) *Second Malaysia plan* (Kuala Lumpur: Malaysian Government Printer).

Morgan, W. B. and R. J. C. Munton (1971) *Agricultural geography* (London: Methuen).

Myrdal, G. (1968) *Asian drama* (Harmondsworth: Penguin).

Narkswasdi, U. (1968) *Rice economy of West Malaysia* (Rome: Food and Agriculture Organisation).

Narkswasdi, U. and S. Selvadurai (1967a) *Economic survey of padi production in West Malaysia: report no. 1. Selangor* (Kuala Lumpur: Ministry of Agriculture).

Narkswasdi, U. and S. Selvadurai (1967b) *Economic survey of padi production in West Malaysia: report no. 2. Bachang, Malacca* (Kuala Lumpur: Ministry of Agriculture).

Penny, D. H. (1964) *The transition from subsistence to commercial family farming in north Sumatra* (Ph.D. thesis; Cornell: University Microfilms).

Purcal, J. T. (1971) *Rice economy* (Kuala Lumpur: University of Malaya Press).

Short, D. E. (1971) *Some aspects of the role of irrigation in rural development in Malaya* (unpublished Ph.D. thesis; University of Hull).

Short, D. E. and J. C. Jackson (1971) 'The origins of irrigation policy in Malaya', *Journal of the Malaysian Branch, Royal Asiatic Society* **44**, 78–103.

Straits Times, 2 January 1970 (Kuala Lumpur: Straits Times Press).

Swift, M. G. (1965) *Malay peasant society in Jelebu* (London: Athlone Press).

Wiener, A. (1972) *The role of water in development* (New York: McGraw-Hill).

Wilson, T. B. (1957) 'Sungai Manik padi farm survey', *Malayan Agricultural Journal* **42**, 92–109.

Wilson, T. B. (1958) *The economics of padi production in north Malaya* (Kuala Lumpur: Department of Agriculture).

Chapter 14

Industrial Location Theory: Its Relevance to an Industrialising Nation

Alan Gilbert

Introduction

Recently several studies have appeared which have subjected industrial location theory to penetrating criticism. While such criticism is not new, this work is interesting in that it has introduced several new concepts into the literature. In particular, it has introduced an approach more closely linked to behavioural and organisation studies than to economics. The advantages of this development, however, must be weighed against the fact that the new concepts have been developed only on the basis of evidence drawn from the industrialised nations and have neglected the experience of the less-developed world. An evaluation of the relevance of these concepts to less-developed countries is overdue, therefore, and this study hopes to fill the gap for at least one poorer nation, Colombia.[1]

RECENT CRITICISM OF INDUSTRIAL LOCATION THEORY

Since economics has long been the main contributor to industrial location theory, it is not surprising that it has borne the brunt of recent criticism. Among other comments, critics have questioned whether entrepreneurs attempt to minimise transport costs and have supported this argument with studies showing that transport costs rarely occupy an important position in the cost structure of modern industrial companies (Törnqvist, 1962). Other critics have argued that economic approaches normally ignored the nature of industrial organisation and thereby failed to understand the framework within which location decisions are made (Wood, 1969; Taylor, 1969). Most damning of all, however, has been the criticism directed at

---

Dr. Gilbert is Lecturer in the Economic Geography of Latin America at University College and the Institute of Latin American Studies, London.

'economic man' with his underpinning assumptions that entrepreneurs seek to maximise profits or minimise costs in an environment of perfect knowledge (Pred, 1967). The tenor of these criticisms has been that 'neo-classical' location theory does not adequately describe *how* companies determine their location. It is no accident, therefore, that a majority of recent studies have attempted to develop new concepts which examine the behaviour of companies more carefully. This does not mean that existing theory has been completely rejected, for several critics have mentioned its value in certain circumstances, but it does mean that the balance of interest has shifted away from economic forces towards the study of entrepreneurial and organisational behaviour.

The most interesting of the new approaches have been based on the argument that personal contacts and information linkages are of major importance in understanding location decisions. These linkages, it is argued, are growing in importance and are exerting an increasing influence upon the location of companies. Accordingly, recent work has examined the social and business links between managers, bankers, financiers, government officials and administrators, and the manner in which information flows between different administrative personnel. Rather than denigrate such factors for their economic inefficiency, as did earlier theory, the new approach has tended to interpret their growing importance as a function of the increasing complexity of society. Within this general area of information and contact studies, the Swedish school has produced several fundamentally new concepts (Hägerstrand, 1952 and 1967; Törnqvist, 1962, 1970; Wärneryd, 1968; Goddard and Anderson, 1970). Törnqvist, for example, has argued that face-to-face contacts are the most effective means of exchanging information, a realisation which is highlighted by his finding that personal contacts occupy a large part of the working time of high administrative personnel. As a result of this development time has become the major constraint upon the efficiency of these personnel and they have become increasingly motivated by the wish to minimise travel time between meetings. The spatial solution to this time constraint has been for managers to locate their offices in close proximity to the offices of other 'contact-intensive' personnel. In an attempt to substantiate these ideas Swedish work has tried to estimate the numbers of 'contact-intensive' personnel, to examine the roles they play within their organisations, and to establish their geographical distribution. The preliminary results suggest that there may be as many as a quarter of a million such personnel in Sweden, of whom 'about half . . . are probably working in the major urban regions' (Törnqvist, 1970, p. 132). Such numbers represent an increase in the total numbers of 'contact-intensive' people from previous years and their concentration suggests that there is a strong inherent spatial tendency towards centralisation in their location decisions. Törnqvist believes that this increase, together with growing urban diseconomies, explains the 'increased functional specialisation within and increased functional differentiation between regions in Sweden. The functions or activities which have become concentrated in just a few expanding core-regions have mainly been those engaged in the handling of information, administration and management. On the other hand, manufacturing and the handling of goods, particularly the routine processes, have been decentralised' (Törnqvist, 1970, pp. 135–136). If these conclusions are correct, therefore, future locational studies should examine these 'contact-intensive' personnel in more detail, noting especially their spatial distributions and effect upon regional employment levels.

Wood's ideas on industrial location support this Swedish work in that he has also stressed the need to examine the nature and scope of information flows. He suggests that earlier work concentrated overmuch on material contacts, but he enlarges slightly on Törnqvist's ideas in suggesting the need for studies of 'material linkages that are individually weak but which form a complex organisational arrangement which must be maintained for successful industrial operation' (Wood, 1969, p. 36). His primary interest, however, is that greater emphasis should be laid upon the way organisations act and how their decisions are made. Little is known, for example, about the ways in which different types of companies make location decisions, who makes the decision within companies or whether there is any particular period when a company's choice of location is of crucial importance.

The similarities between the ideas of Törnqvist and Wood also extend to the work of Pred, particularly on the question of business behaviour and its links with information flows (Pred, 1967). Pred, however, approaches the question from a rather different angle, presenting a scheme by which the spatial behavour of companies may be classified. His basic 'behavioural matrix' interprets locational action as dependent upon two factors, the first being the quantity and quality of information available to each decision-maker, the second the ability of each manager to use this information. A company's success in selecting a viable location will depend upon its position in the behavioural matrix. Those companies with a high level of information open to them and with well-developed ability to use this information are likely to select locations well; on the other hand, those companies lying in the top left-hand corner of the matrix (Figure 14.1) will locate more on a basis of chance and will tend to choose non-viable locations more frequently. In suggesting this matrix, however, Pred is not supporting the idea of an optimal location. Rather he argues strongly that profit maximisation is rarely a guideline for company decisions and should be replaced by the concept of profit 'satisficing' or 'bounded rationality' (Simon, 1957a and b).

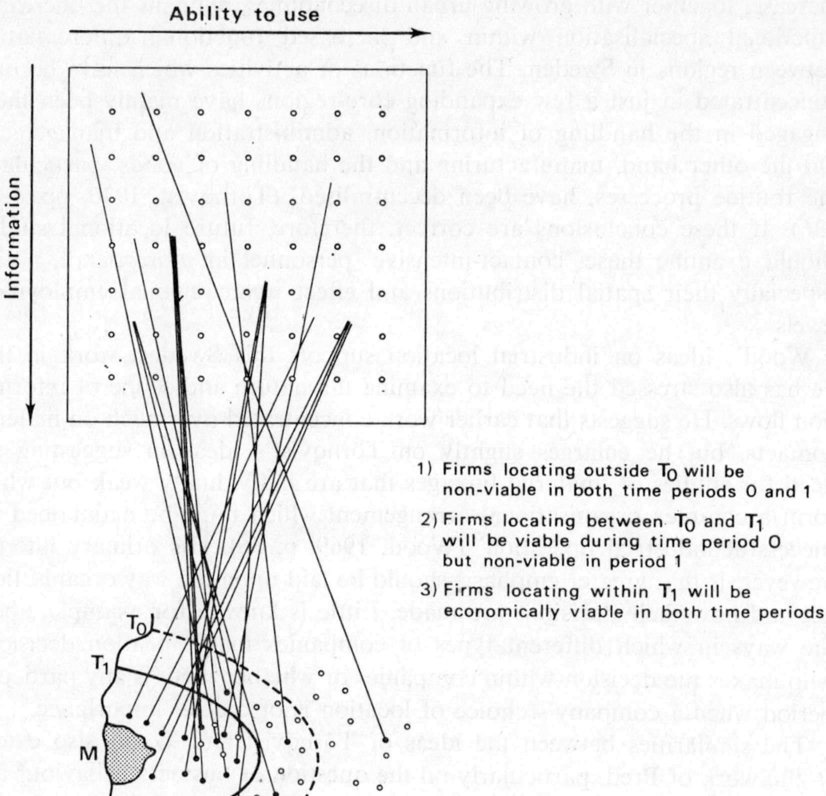

Figure 14.1 Pred's matrix of entrepreneurial behaviour (after Pred, 1967)

Also arising from the demise of 'economic man' has come another 'behavioural' concept about industrial location, 'seed-bed' growth (Taylor, 1969). According to this concept the majority of small companies tend to locate in the vicinity of existing industrial concentrations. The reason for this tendency is that all industrial companies number among their employees potential entrepreneurs knowledgeable about the industrial process. In favourable circumstances these potential entrepreneurs will go into business in closely-related fields. The majority of these entrepreneurs, however, will not have the knowledge or the desire to found a company outside the 'home area'. Accordingly, the majority will set up companies in the area they know best; a tendency supported by the availability of second-hand machinery, potential customers, suppliers and labour forces in that area. The consequence of this process, therefore, is that a large number of small companies are likely to emerge in the vicinity of existing industrial concentrations. The majority of these companies will not have made specific location decisions and will be characterised by a high rate of mortality.

Against this acceptance of the need for a higher 'behavioural' and 'organisational' content, the work of Smith stands in marked isolation (Smith, 1966, 1970, 1971). While other writers have moved away from the assumptions of neo-classical economic theory, Smith has tried to relax the more rigid assumptions of that theory to bring it into closer harmony with real-life complexities. The particular modification which he adds to the work of Greenhut (1956), Weber (1929), Lösch (1954) and others is to introduce the concept of the 'spatial margin'. By this device he attempts to reconcile the economically-optimal location with the range of less profitable, but still viable locations in which companies may operate, a situation depicted in Figure 14.2. In his own words, 'spatial variations in total cost and total revenue create an optimum location at which profits may be maximised, and also spatial margins beyond which profitable operation is not possible; within the margin the firm is free to locate anywhere, providing profit maximisation is not required. This principle provides a means of

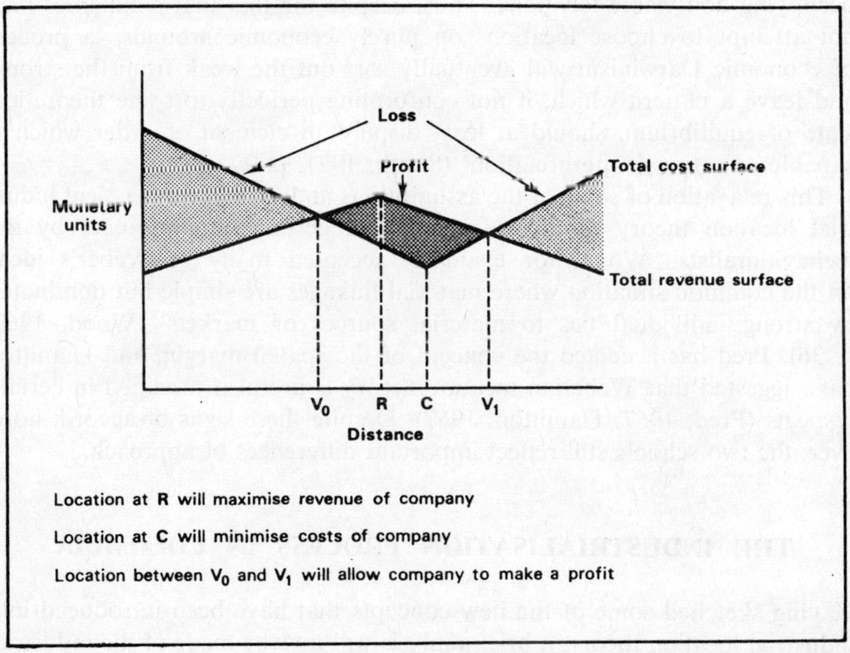

Figure 14.2 Optimum location and the concept of spatial margins to profitability (after D. M. Smith, 1966 and 1971)

incorporating sub-optimal behaviour into location theory, and thus enables the traditional assumption of economic rationality to be relaxed. It represents a very significant refinement of the profit-maximising principle' (Smith, 1971, p. 185). This move towards reconciling economic theories with the newer behavioural concepts is also helped by Smith's acceptance that cost and revenue surfaces for each industry may vary according to the nature, size and organisation of companies. Thus, according to the nature of the business and the skill of the management, the spatial margins to profitability may vary between companies. 'It is unrealistic to think of one cost surface, as the spatial cost situation will vary between firms in relation to differences in entrepreneurial skill. The same is true of the revenue surface. A skilled businessman, using vigorous marketing methods, may be able to raise his total revenue surface above that of his less able competitors. And, like the skilled entrepreneur of the variable-cost model, he may be able to generate enough demand for his product to develop a profitable location which would be beyond the margin of other producers' (Smith, 1971, p. 260). Perhaps more important still, is Smith's recognition that industrial location may be determined by a form of 'economic Darwinism'. While companies may choose their locations on a basis of non-economic criteria, the pressures of free market forces will lead to the disappearance of companies occupying non-viable locations. Thus, despite the fact that companies *may* not attempt to choose locations on purely economic grounds, 'a process of economic Darwinism will eventually sort out the weak from the strong, and leave a pattern which, if not conforming perfectly to some theoretical state of equilibrium, should at least display an element of order which is capable of rational interpretation' (Smith, 1971, p. 273).

This relaxation of some of the assumptions underlying neo-classical industrial location theory has been accepted in certain circumstances by the 'behaviouralists'. Wood, for example, accepted many of Weber's ideas 'in the common situation where material linkages are simple but dominated by strong individual ties to material sources or markets' (Wood, 1969, p. 36). Pred has accepted the concept of the spatial margin, and Hamilton has suggested that Weberian location theory is useful if modified in certain respects (Pred, 1967; Hamilton, 1967). Despite these signs of accord, however, the two schools still reflect important differences of approach.

THE INDUSTRIALISATION PROCESS IN COLOMBIA

Having sketched some of the new concepts that have been introduced into industrial location theory, a brief analysis will now be made of the industrial growth process in Colombia. In examining this process, however, it is not suggested that the country is typical of all less-developed nations. The size

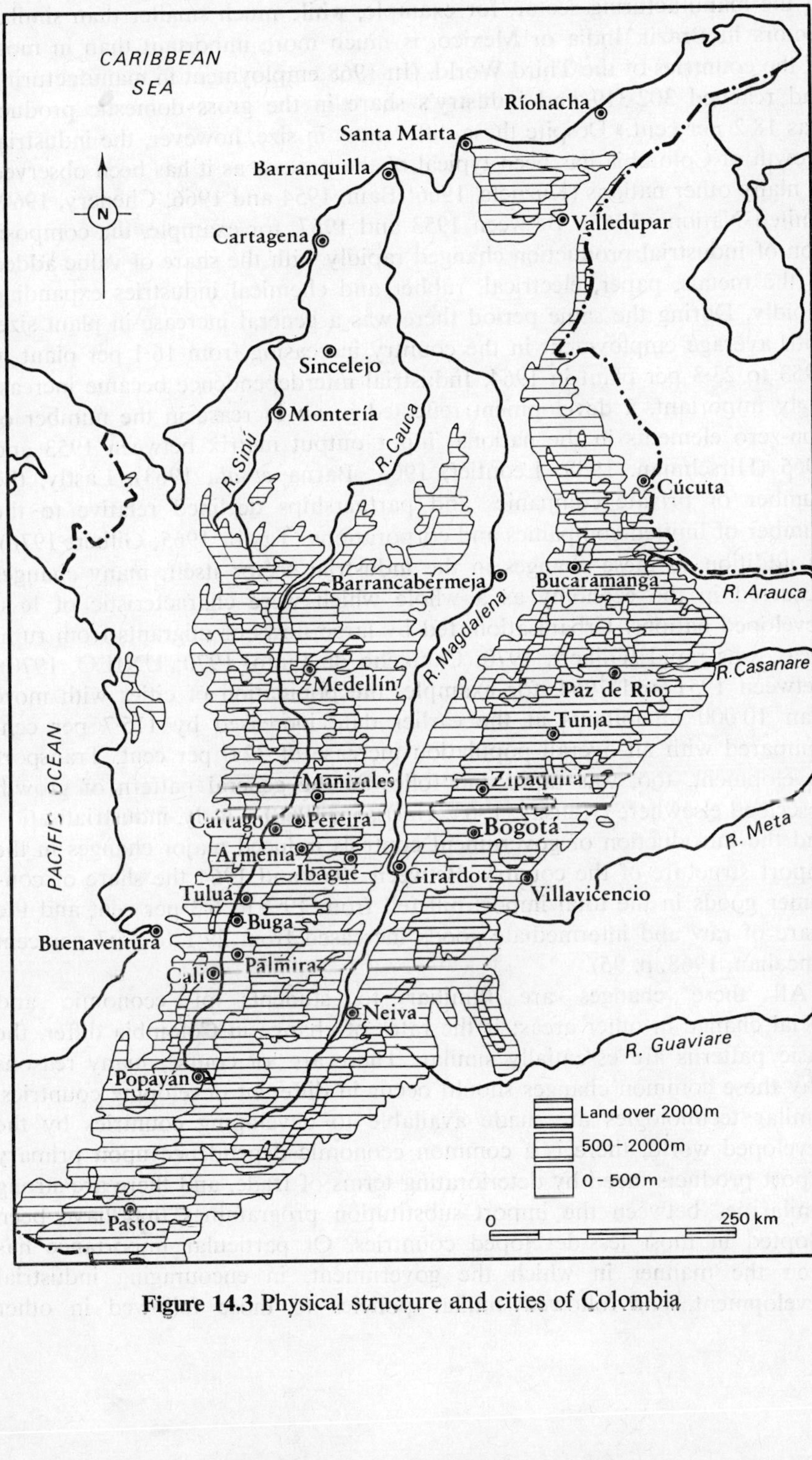

Figure 14.3 Physical structure and cities of Colombia

of the manufacturing sector, for example, while much smaller than similar sectors in Brazil, India or Mexico, is much more important than in most of the countries of the Third World. (In 1968 employment in manufacturing had reached 302 319 and industry's share in the gross domestic product was 18·2 per cent.) Despite these differences in size, however, the industrial growth of Colombia has been typical of the process as it has been observed in many other nations (Kuznets, 1966; Bain, 1954 and 1966; Chenery, 1960; United Nations, 1963). Between 1953 and 1967, for example, the composition of industrial production changed rapidly with the share of value added in the metals, paper, electrical, rubber and chemical industries expanding rapidly. During the same period there was a general increase in plant size, with average employment in the country increasing from 16·1 per plant in 1953 to 23·3 per plant in 1964. Industrial interdependence became increasingly important, a development reflected in an increase in the number of non-zero elements in the national input–output matrix between 1953 and 1965 (Hirschman, 1958; Leontief, 1963; Barna *et al.*, 1963). Lastly, the number of private companies and partnerships declined relative to the number of limited companies and corporations (Taylor 1965; Gilbert 1970). In addition to these changes in the industrial sector itself, many changes occurred in the economy as a whole which were characteristic of less-developed nations. Urbanisation, fed by large flows of migrants from rural areas, was rapid (Gilbert, 1970; Camacho de Pinto, 1970; UNILO, 1970). Between 1951 and 1964, for example, the population of cities with more than 10 000 inhabitants at the earlier date increased by 115·7 per cent compared with an overall population increase of 32·0 per cent. Transport development, too, was rapid and followed the general pattern of growth described elsewhere (Kansky, 1963; Garrison, 1960). Lastly, industrialisation and the introduction of government controls induced major changes in the import structure of the country. Between 1953 and 1964 the share of consumer goods in the total import bill fell from 19·2 to 6·3 per cent and the share of raw and intermediate goods increased from 29·1 to 45·7 per cent (Sheahan, 1968, p. 95).

All these changes are familiar to students of economic and social change in other areas; if the rates of change in Colombia differ, the basic patterns are essentially similar. There are, of course, many reasons why these common changes should occur in different developing countries. Similar technologies are made available to developing countries by the developed world, there is a common economic dependence upon primary export products faced by deteriorating terms of trade, and there are strong similarities between the import–substitution programmes that have been adopted in most less-developed countries. Of particular importance has been the manner in which the government, in encouraging industrial development, has followed similar policies to those followed in other nations. Quantitative restrictions upon industrial imports in the form of import licences, controls upon the treatment of the industrial workforce, controls upon pricing both in the industrial and the transportation sectors, and an increasing supervision over the kinds of industry to be encouraged in the country, represent a common combination in most developing areas (Grunwald, 1970; Furtado, 1970; UNECLA, 1966). So, too, do some of the results of these policies, for the import-substitution strategy in Colombia has created high-cost production and monopolies, has involved the emergence of a privileged urban and industrial workforce and has recently led to a decline in the opportunities for further industrial expansion. In spatial terms, too, there have been similarities between the Colombian experience and those of other developing nations (Gilbert, 1974). The most notable of these tendencies has been a further increase in the already high proportion of industry located in the largest cities. Between 1945 and 1967, the share of total manufacturing employment in Bogotá increased from 16·7 to 25·3 per cent, in Medellín from 21·9 to 22·8 per cent and in Cali from 7·1 to 12·5 per cent (Table 14.1). Taking the industrial employment in the three cities together there was an increase in share from 45·7 to 60·6 per cent between the two dates (DANE, 1947, 1951 and 1964). This tendency was even more marked in growth industries such as paper, metals and chemicals.

The Colombian manufacturing sector, therefore, typifies in many ways industrial growth in other less-developed nations. However, if it is compared with the industrial sector in most developed nations, a number of distinct

Table 14.1 The distribution of manufacturing in Colombia, 1945–1967
Source: Departamento Administrativo Nacional de Estadística (DANE)

City	Share of total employment 1945	1967
Bogotá*	16·7	25·3
Medellín†	21·9	22·8
Cali‡	7·1	12·5
Barranquilla	10·5	8·3
Bucaramanga	3·4	2·3
Pereira	1·7	2·2
Cartagena	2·1	1·8
Manizales	2·4	1·7
Subtotal	65·8	76·9
Total employment in nation	135 400	293 825

* Includes Soacha
† Includes Bello, Envigado, Itagüí and La Estrella
‡ Includes Yumbo

differences emerge. Firstly, there are marked differences in the way in which industry is organised and in which it operates. Compared to most industries in developed nations, for example, the Colombian manufacturing sector is characterised by small average plant size, high production costs, high profits, small numbers of inter-industry linkages and by a high proportion of private companies and partnerships. Secondly, there are marked differences in the environments in which the respective industrial sectors operate. Compared with the situation in Britain, the United States, West Germany or France, for example, there are few urban centres with adequate physical infrastructure, the market for industrial production is limited and the transport system is poorly developed. In addition, the role of government is distinctive in that the Colombian government has attempted to assist, through import-substitu-

Table 14.2 Sectoral characteristics of Colombian industry
Source: Departamento Administrativo Nacional de Estadística (DANE)

Sector	Employment (1967)	Growth of employment (1953–1967) per cent	Growth of plants (1953–1967) per cent	Share of employment (1967) per cent	Coefficient of Geographical Association* (1964)
Food	40 820	10·2	4·2	13·9	0·81
Drinks	15 247	41·9	−11·7	5·2	0·79
Tobacco	3 295	−46·2	−57·2	1·1	0·62
Textiles	46 098	30·0	−21·4	15.7	0·67
Clothing	26 979	11·0	−35.4	9·2	0·91
Wood	5 987	69·6	−14·0	2·0	0·69
Furniture	4 759	38·7	−12·6	1·6	0·80
Paper	6 092	216·1	115·7	2·1	0·67
Printing	11 449	64·9	38·3	3·9	0·82
Leather	4 065	13·3	−8·2	1·4	0·82
Rubber	6 736	144·2	−4·2	2·3	0·69
Chemicals	23 480	151·9	7·0	8·0	0·80
Coal and petroleum	2 058	129·8	30·8	0·7	0·23
Non-metallic metals	33 872	107·4	−0·4	11·6	0·87
Basic metals	4 700	254·4	−48·5	1·6	0·47
Metal products	19 627	219·4	80·3	6·7	0·84
Machinery	5 691	301·3	86·0	2·0	0·85
Electricals	9 329	465·1	102·6	3·2	0·77
Transport	13 373	161·7	40·4	4·6	0·78
Miscellaneous	9 326	252·6	45·4	3·2	0·84
Total	282 983	56·1	−4·1	100·0	—

The smaller the coefficient the higher the degree of spatial concentration

tion policies, the establishment of industries which already exist in the developed nations. Thirdly, the recent growth of Colombian industry has meant that a large share of that growth has not been due to the expansion of existing companies but to the establishment of new ones. In some growth sectors—metals, machinery and electricals for example—the number of plants doubled in the thirteen years after 1953 (Table 14.2).

Different, too, is the way that Colombian industry has become more spatially concentrated, for in many developed nations decentralisation is the more common pattern. In Sweden, for example, Törnqvist has shown that although most administrative functions have become more concentrated, manufacturing has become relatively more dispersed. In Great Britain and the United States various studies have shown a similar tendency (Fuchs, 1962; Cameron and Clark, 1966).

THE NEW CONCEPTS AND THE LOCATION OF INDUSTRY IN COLOMBIA

The question of the relevance of the new concepts in location analysis to the Colombian industrial environment raises several important issues. Can the new concepts explain the varying location patterns exhibited by different industries in Colombia and can they explain why some industries are spatially more concentrated than others? Do they explain why there has been an overall tendency in the country towards further industrial centralisation? Can they explain the manner in which Colombian managers make decisions with respect to location? Can the new concepts assist rational government planning, either by providing guidelines as to where industry should be located or by demonstrating how best to influence the distribution of industry, should this be necessary?

The neo-classical economic approach

In the case of some industries in Colombia the location principles laid down by Weber are still appropriate. The location of the main iron and steel plant at Paz del Rio in the vicinity of iron ore and coal deposits, for example, illustrates perfectly his argument in favour of material orientation. Similarly, the distribution of the main Colombian sugar mills among the canefields of the Cauca Valley justifies his 'perishability-of-materials' principle. There are, in addition, several kinds of industry located in the country's major cities for which Weber prescribed a market location. The widely-dispersed soft-drink and beer industries, for example, support his idea that companies producing goods with a low value-to-weight ratio will locate near to the market (Hamilton, 1967).

The majority of Colombian industries, however, do not fit so conveniently into Weber's framework. In particular, the distribution of light industry, which makes up the majority of the manufacturing sector, cannot easily be accommodated by Weber's uneasy acceptance of agglomeration. Nor can the concentration of this light industry in the major urban centres be satisfactorily linked with the term 'footloose', even though much of it would be classified under such a heading were it located in a developed nation (Estall and Buchanan, 1966, pp. 17–18). But if these concepts are unsatisfactory, do Smith's reformulations of the variable-cost and revenue-maximising approaches represent any marked improvement? According to these reformulations, spatial concentrations are likely to occur in industries whose 'optimal' locations are bounded by steep cost and revenue surfaces, that is, where the spatial margins to profitability are narrow. Are there forces operating in the Colombian environment which prescribe narrow spatial margins to profitability and thereby reduce the number of 'footloose' companies? On the cost side, the distribution of physical infrastructure is heavily biased in favour of the cities so that supplies of electricity, gas, water, transport, telephones, etc. are unavailable or unreliable in areas outside the major urban centres. As a result, industrial location in the smaller towns or rural areas is normally accompanied by higher costs of production. In addition, highly-skilled labour is more readily available in the vicinity of existing industrial plants and, although this labour may be prepared to work in other areas, there is a strong tendency in Colombia for manpower to prefer working in the larger urban areas. On the revenue side, similar kinds of factors are operating to produce a steep spatial gradient. The distribution of income, for example, as in other developing countries, is highly biased towards the urban areas and most rural areas are characterised by low per capita incomes (Williamson, 1965; Kuznets, 1966; Daza Roa, 1967; Marabelli, 1966). These and other similar factors suggest that the cost and revenue surfaces facing Colombian manufacturers are steeper than those facing companies in developed countries. Consequently there are likely to be fewer 'footloose' industries and a stronger tendency for companies to locate in the major cities, a tendency which has also been observed in other developing areas (Alonso, 1968; Crapper, 1968).

These factors explain why many industries locate in the major cities but they are not sufficient to account for the growing concentration of industry since 1945. This can be explained in part by changes occurring in the composition of the manufacturing sector as a whole. If the structure of industry shifts in favour of sectors which have traditionally been oriented towards the largest cities, then increased industrial concentration will occur.

Thus, if city-oriented industries grow more rapidly than traditionally-dispersed industry, the overall level of industrial concentration will increase. To an extent, this has been the situation in Colombia over the past twenty years, where city-based industries such as metals and electricals have expanded more rapidly than foodstuffs and leather. The changing composition of industry certainly has been *one* factor in influencing the distribution of industry. That such marked changes have occurred in the distribution of industry, however, cannot be explained wholly by this phenomenon. They are also closely linked to the fact that the majority of growth industries were those in which large numbers of new plants became established. Although the existing distribution of industry was one factor operating on the location of these new companies, other factors were active which are not accommodated within Smith's theory. Perhaps the most important omission is that Smith's ideas do not cast light on the nature of the decision-making process with regard to industrial location—whether companies make a rigorous study before choosing, whether the decision is part of a long-term investment programme, whether different kinds of company make different kinds of location decision and whether companies even consider the question of location at all? Smith makes no comment on these questions beyond accepting that some companies do not make rigorous location studies. Implicit in much of his work, however, is the assumption that most companies do follow a set of rules in choosing a location and that an economic rationale underlies the overall distribution of industry. If the results of a survey of industry in Colombia are typical (Gilbert, 1970), such assumptions are not wholly valid in many less-developed countries. According to this study less than 10 per cent of companies declared that they had made a definite location study. In addition, there was little sign of a relationship between location and profitability. For, although certain industries were concentrated in cities where their profits were relatively higher than those of identical industries in other cities, there were many other examples of industrial concentrations associated with relatively low profitability. If there is an underlying rationale behind location decisions in Colombia it is more complex than the mere search for profits.

Smith's ideas, however, may help governments to decide how best to influence location decisions. While the theory does not explain how companies actually decide upon a location, it does prescribe limits on where companies can and cannot locate viably. If governments are to influence industrial location effectively they must understand which companies are subject to narrow spatial margins and which are relatively 'footloose'. If they do not appreciate these distinctions they will be faced with the problem of bankrupt companies and will be responsible for creating further unemployment and for wasting scarce capital resources.

The new 'behavioural' and 'organisation' concepts

Smith's reformulation of the older economic models, therefore, has satisfied certain of the criteria for relevance. On the other hand a number of deficiencies are still apparent. How far can the new 'behavioural' and 'organisation' concepts fill the remaining gaps? Some of the concepts, of course, serve little purpose. Pred's behavioural matrix, for example, while useful in finally burying 'economic man', has hardly given birth to a new theoretical giant. On the other hand, the concepts that have focused on personal contacts, on 'seed-bed' growth and on the organisational structure of industry have been very useful indeed.

With respect to the nature of industrial location decisions the new concepts have been particularly helpful. As already shown, the majority of new companies in Colombia do not select their locations after an exhaustive study. One aspect of this is that one of the most characteristic tendencies in the country is for new companies to be founded in the city where the owner(s) previously lived or worked. As an indication of this tendency 88 per cent of 200 companies, which had been founded by Colombians before 1968, were located in cities where the owner had previously lived or worked. Such a tendency is not peculiar to Colombia, however, for a number of studies in developed countries have found a similar tendency for managers to establish industries in familiar areas (Mueller and Morgan, 1962; Taylor, 1969). It may indeed be a universal characteristic of all *small* industry, for Taylor has shown that this kind of 'home-city' or 'seed-bed' growth is common among small firms in the West Midlands of England. Such a hypothesis is certainly appropriate in the Colombian context where as many as 8421 of the 10 689 manufacturing companies in 1967 had less than 20 employees, and only 228 companies more than 200 employees.

This kind of motivation, however, is likely to be less appropriate in the case of larger companies and foreign corporations where concepts that concentrate upon personal contact patterns would seem to have greater utility. Alonso has suggested that personal contacts are even more important to businessmen in developing countries than in Europe or the United States (Alonso, 1968). Such a statement has great validity in Colombia where the structure and size of the social system ensures that many persons in administrative and professional occupations are mutually acquainted. It is also very important in explaining the relationships between government and industry, a factor of crucial importance in explaining the increasing industrial share of Bogotá since 1945.

Since the early 1950s, the role of government in administering the economic system has increased enormously. With growing government involvement in industrial planning, import licensing, labour policies, transport rates, etc., the frequency of contacts between businessmen and government officials has increased. Such a situation was readily acknowledged by the managers of companies interviewed in 1968, many of whom said that such contacts occupied a great deal of their time. Indeed, the spatial implications of these contacts were spelled out by several companies in import-sensitive sectors who stated that a location in Bogotá close to government agencies was in many ways advantageous. There were, in fact, several examples of provincial companies who had decided to relocate their head offices in Bogotá, so as to improve contacts with government agencies and with other companies (Gavirias, Union Carbide, Abbott de Colombia and Celanese Colombiana). Whether companies are correct in believing that successful negotiations with government agencies can be achieved in this way is an open question, but there is some circumstantial evidence in support of this thesis in that the proportion of import licences granted in Bogotá appears to be far higher than its share of industry warrants. In 1953, for example, the department of Cundinamarca, in which Bogotá is located, received 55·8 per cent of all imports by value (1968) while having only 28·2 per cent of industrial employment. Even when allowance is made for the higher concentration of import-sensitive industries in Bogotá, one is led to the conclusion that frequent contact between businessmen and government officials may have encouraged this position.

In one sense, therefore, personal contacts have become more important to industrial companies as a result of increased government participation in the planning process. But such a change has occurred in conjunction with a growing concentration in the whole business, commercial and institutional complex. Just as in Sweden there has been an increase in the numbers and in the relative concentration of administrative, technical and professional personnel over time, so a similar phenomenon is discernible in Colombia. As a result, the numbers of such personnel have increased rapidly in Cali, Medellín and Barranquilla, and especially in Bogotá.

The new behavioural and organisational concepts, therefore, have direct value in explaining the increasing concentration of industry and the nature of the industrial location decision in Colombia. To this extent they will also help the national government to influence this distribution of industry now that it has adopted a policy with this aim in mind. What the new concepts have not done, however, is to tell the government whether a policy of industrial decentralisation should be adopted.

CONCLUSION

Despite the neglect of less-developed nations during the formulation of the new locational concepts, their application to Colombian experience suggests that they may be very useful to such countries. Despite the differences between business behaviour and organisation in developed and less-developed nations, the new concepts have been seen to satisfy several of the criteria of relevance laid down in this chapter. Firstly, the concept of the 'spatial margin' is useful in demonstrating that some industries are inherently less 'footloose' than others. Secondly, 'personal contacts' and 'information linkages' have been seen to play a major role in the tendency for industry to become more concentrated spatially over time. In the Colombian context these ideas have brought out firmly the effect of the government in encouraging industrial concentration; by its presence in the capital city and its centralised system of control it has been a major factor in attracting industry to Bogotá. Thirdly, the 'behavioural' and 'organisational' concepts have thrown more light on the way in which companies make location decisions. In particular, the concept of 'seed-bed' growth is useful in suggesting that few small firms make specific location choices, a conclusion of some importance in developing nations with few large industrial companies. Lastly, the concepts taken together provide governments with a far clearer idea of the extent to which they can influence company location. The concept of the spatial margin, for example, allows them to gauge the locational mobility of companies in different industrial sectors. In addition, the concepts have shown that industrial centralisation follows in part from administrative centralisation, a finding which could eventually be used to foster industrial dispersion in developing nations such as Colombia.

Unfortunately, however, this new understanding is but a small step along the road to satisfying Lösch's much-quoted belief that 'the real duty of the economist is not to explain our sorry reality, but to improve it' (Lösch, 1954). The new concepts have helped to explain our sorry reality but they have done little to tell governments where, and under what conditions, industry should be concentrated or dispersed. That such aid is needed is not doubted by several authors who have stressed that optimal distributions cannot be achieved by companies following the principles they have outlined. Törnqvist, for example, has implied that the spatial processes in Sweden should be modified because they have 'accentuated regional inequalities in income, social structure and education' (Törnqvist, 1970, p. 136). Smith, too, has argued explicitly that industrial location cannot be left wholly to manufacturers if the maximisation of welfare is to be achieved (Smith, 1971, p. 517). But recognition is not resolution; we still seem to be as far away as ever from providing governments with a suitable basis for planning industrial location. The new concepts will certainly help governments in less-developed nations to influence location choices; what they will not do is to tell them where industry should be located.

NOTE

1. Some attempts to produce distributions of industry which satisfy some *economic* optima have been excluded from the discussion; for a good résumé of these alternative concepts see Hamilton (1967).

REFERENCES

Adams, J. (1968) *Review of 'Behaviour and Location'* by A. Pred (Graduate discussion paper no. 20; London School of Economics, Geography Department).

Alonso, W. (1964) 'Classical Location Theory', *Regional Development and Planning*, ed. Friedmann, J. and W. Alonso (Cambridge, Massachusetts: M.I.T. Press).

Alonso, W. (1968) *Industrial location and regional policy in economic development* (Working paper no. 74; Center for Planning and Development Research, Institute of Urban and Regional Development, University of California, Berkeley).

Bain, J. S. (1954) 'Economies of scale, concentration and the condition of entry in twenty manufacturing industries', *American Economic Review* **44**, 15–39.

Bain, J. S. (1966) *International differences in industrial structure: eight nations in the 1950s* (New Haven: Yale University Press).

Barna, T. *et al.* (1963) *Structural interdependence and economic development: Proceedings of International Conference on Input-Output Techniques* (London: Macmillan).

Camacho de Pinto, T. (1970) *Colombia: el proceso de urbanización y sus factores relacionados* (Tunja: Universidad Pedagógica y Tecnológica de Colombia).

Cameron, G. C. and Clark, B. D. (1966) *Industrial movement and the regional problem* (Edinburgh: Oliver and Boyd).

Chenery, H. B. (1960) 'Patterns of industrial growth', *American Economic Review*, **50(4)**, 624–654.

Crapper, D. B. (1968) 'A location study of some British-based industrial firms in East Africa', *East African Geographical Review* **6**, 51–61.

DANE (Departamento Administrativo Nacional de Estadística) (1947) *Primer Censo Industrial de Colombia* (Bogotá).

DANE (1951) *XII Censo Nacional de Población* (Bogotá).

DANE (1964) *XIII Censo Nacional de Población* (Bogotá).

DANE, *Boletin Mensual de Estadística.*

Daza Roa, A. (1967) 'La repartición de los ingresos', *Revista Mensual del Banco de la República*, July, 880–889.

Estall, R. C. and R. O. Buchanan (1966) *Industrial activity and economic geography* (London: Hutchinson).

Fuchs, V. R. (1962) *Changes in the location of manufacturing in the United States since* 1929 (New Haven: Yale University Press).

Furtado, C. (1970) *Economic development of Latin America* (London: Cambridge University Press).

Garrison, W. L. (1960) 'Connectivity of the interstate highway system', *Regional Science Association, Papers and Proceedings*, **6**, 121–137.

Gilbert, A. G. (1970) *Industrial growth in the spatial development of the Colombian economy between* 1951 *and* 1964 (London: Ph.D. Thesis of the University of London).

Gilbert, A. G. (1974) *Latin America: the development process in geographical perspective* (London: Penguin).

Goddard, J. and J. Anderson (1970) *Some current approaches to human geography in Sweden* (Graduate discussion paper no. 33; London School of Economics, Geography Department).

Greenhut, M. L. (1956) *Plant location in theory and practice: the economics of space* (Chapel Hill: University of North Carolina Press).

Grunwald, J. (1970) 'Some reflections on Latin-American industrialisation policy', *Journal of Political Economy* **78**, 826–856.

Hägerstrand, T. (1952) *The propagation of innovation waves* (Lund: Lund Studies in Geography, Series B, No. 4).

Hägerstrand, T. (1967) *Innovation diffusion as a spatial process*, trans. A. Pred (Chicago: University of Chicago Press).

Hamilton, F. E. I. (1967) 'Models of industrial location', *Models in geography*, ed. Chorley, R. J. and P. Haggett (London: Methuen).

Harris, C. D. (1954) 'The market as a factor in the localisation of industry in the United States', *Annals of Association of American Geographers* **44**, 315–348.

Hirschman, A. O. (1958) *The strategy of economic development* (New Haven: Yale University Press).

Hoover, E. M. (1948) *The location of economic activity* (New York: McGraw-Hill).

Isard, W. (1956) *Location and space economy* (Cambridge, Massachusetts: M.I.T. Press).

Isard, W. *et al.* (1960) *Methods of regional analysis* (Cambridge, Massachusetts: M.I.T. Press).

Kansky, K. J. (1963) *Structure of transportation networks: relationship between network geometry and regional characteristics* (Department of Geography Research Paper No. 84; University of Chicago).

Kuznets, S. (1966) *Modern economic growth: rate, structure and spread* (New Haven: Yale University Press).

Leontief, W. (1963) 'The structure of development', *Scientific American*, September, 9.

Lösch, A. (1954) *The economics of location* (New Haven: Yale University Press).

Marabelli, F. (1966) *Tentativa de distribución del producto interno de Colombia por secciones administrativas del país, 1964* (Bogotá), mimeo.

Mueller, E. and J. N. Morgan (1962) 'Location decisions of manufacturers', *American Economic Review* **502**, 204–17.

Nelson, R. R., T. P. Schultz and R. L. Slighton (1971) *Structural change in a developing economy: Colombia's problems and prospects* (Princeton: Princeton University Press).

Pred, A. (1967) *Behaviour and location*, Part One (Lund: Lund Studies in Geography, Series B, No. 27).

Sheahan, J. (1968) 'Imports, investment and growth—Colombia', *Development policy—theory and practice* ed. G. F. Papenek (Cambridge: Harvard University Press).

Simon, H. A. (1957a) *Administrative behaviour* (New York: Macmillan).

Simon, H. A. (1957b) *Models of man: social and rational* (New York: Wiley).

Smith, D. M. (1966) 'A theoretical framework for geographical studies of industrial location', *Economic Geography* **42**, 95–113.

Smith, D. M. (1970) 'On throwing out Weber with the bathwater: a note on industrial location and linkage', *Area* **1**, 15–18.

Smith, D. M. (1971) *Industrial location: an economic geographical analysis* (New York: Wiley).

Taylor, M. C. (ed.) (1965) *Fiscal survey of Colombia* (Baltimore: Johns Hopkins).

Taylor, M. J. (1969) '*Industrial linkage, 'seed-bed' growth and the location of firms*' (Occasional Paper No. 3; University College London, Department of Geography).

Taylor, M. J. (1970) 'Location decisions of small firms', *Area* **2**, 51–54.

Thompson, J. H. (1966) 'Some theoretical considerations for manufacturing geography', *Economic Geography* **42(4)**, 356–365.

Törnqvist, G. (1962) *Transport costs as a location factor for manufacturing industry* (Lund: Lund Studies in Geography, Series B, No. 23).

Törnqvist, G. (1970) *Contact systems and regional development* (Lund: Lund Studies in Geography, Series B, No. 35).

United Nations, Economic Commission for Latin America (1957) *Analyses and projections of economic development: the economic development of Colombia* (New York).

United Nations, Department of Economic and Social Affairs (1963) *A study of industrial growth* (New York).

United Nations, Economic Commission for Latin America (1966) *The process of industrial development in Latin America* (New York).

United Nations, International Labour Office (1970) *Towards full employment* (Geneva).

Wärneryd, O. (1968) *Interdependence in urban systems* (Göteborg: Regionkonsult Aktiebolag).

Weber, A. (1929) *Theory of the location of industries*, trans. C. J. Freidrich (Chicago: University of Chicago Press).

Williamson, J. G. (1965) 'Regional inequality and the process of national development: a description of the patterns', *Economic Development and Cultural Change* **13(4)**, 3–45.

Wood, P. A. (1969) 'industrial linkage and location', *Area* **2**, 32–39.

Chapter 15

Imbalances in Brazil's Industrialisation

J. P. Dickenson

Introduction

Spatial distribution has tended to be a neglected factor in discussions of economic development and its problems (Friedmann and Alonso, 1964; Richardson, 1969). In Latin America, for example, it has been suggested that discussions on industrial development within the Latin American Free Trade Area have assumed that 'each member state is an undifferentiated spatial unit within which either the pattern of distribution of economic activities is of no significance or will take place in an entirely acceptable way' (Odell, 1966, p. 14). Odell argues that these are not acceptable conclusions and do not accord with the experience of the more-developed world, and that if the spatial implications of industrialisation are not considered, investment is likely to concentrate in a very few centres.

There has also been a tendency in the study of development to neglect the significance of geographical size or scale (Demas, 1966; Robinson, 1960). Classic location theory has tended to be based on undifferentiated space (Hansen, 1967); it has also tended to ignore the implications of the size variations of national territories for location and development. An examination of the significance of variations of geographic size for regional planning in Latin America contrasts the problems facing the smaller and larger countries of Latin America (Odell, 1971). It is suggested that, for the smaller countries, patterns of distribution will make little difference to the level of output and its growth rate, but for larger ones, cost variations and spatial competition for investment funds may be of great significance. The problems of size and of the spatial distribution of economic activity in Brazil are particularly acute. With an area of 8·5 million square kilometres the difficulties of securing the development of this vast area and of allocating development within it are profound. As Odell suggests, neither

---

Dr. Dickenson is Lecturer in Geography in the University of Liverpool.

the experience of Western Europe nor that of the USA is of great relevance in tackling these problems.

THE EMERGENCE OF IMBALANCE

In gross terms Brazil has made very considerable economic, and specifically industrial, progress in recent years. In industrial sectors such as automobile manufacturing, steel and cement production and shipbuilding it ranks among the world's twenty leading producers (UN, 1971), substantially above its ranking as measured by per capita income (Streeten, 1971). In 1970 steel output was 5·4 million ingot tons, cement production 8·7 million tons and 411 000 motor vehicles were produced. When these gross figures are measured against a 1970 population of 92·2 millions, however, average figures are low, reflected in a current per capita income level of approximately US $300.

In the post-war period manufacturing industry has experienced strong growth and has been an important element in Brazil's economic advance. During the years 1951 to 1962 industry grew at an average rate of 9 per cent per year and, after a period of slower growth between 1963 and 1966, reattained that level in the period 1967–70. In 1969 industry contributed 34 per cent of GDP as against 21 per cent from agriculture and 45 per cent from the tertiary sector. Its contribution to employment, however, remains modest—14 per cent of the labour force as against 47 per cent in agriculture.

Industry is by no means a recent innovation in Brazil. Craft industries date from the colonial period, cotton textile manufacture from the mid-nineteenth century, a modern iron industry from the 1880s and an integrated steel industry from the 1920s. Textiles, food processing and clothing manufacture were early pioneer industries and continue to be major industrial sectors. There has been a degree of diversification in Brazil's industrial structure since 1920, in that the share of employment in these three pioneers had fallen from 59·5 per cent in that year to 35·2 per cent in 1965, while the share of the 'growth' metallurgical, electrical, engineering, vehicle and chemical industries has risen from 10·4 to 29·8 per cent.

In the twentieth century Brazilian industry has shown, in addition to its expansion and diversification, a strong and continuing tendency to spatial concentration. The main industrial focus which has emerged is that of the south-east, namely the states of Minas Gerais, Guanabara, Rio de Janeiro, and São Paulo. These four states increased their share of the country's industrial labour force [1] from 63·1 per cent in 1920 to over 70 per cent in 1960. In detail this concentration is even more marked in that in 1960 47 per cent of Brazil's industrial employment was located in São Paulo state

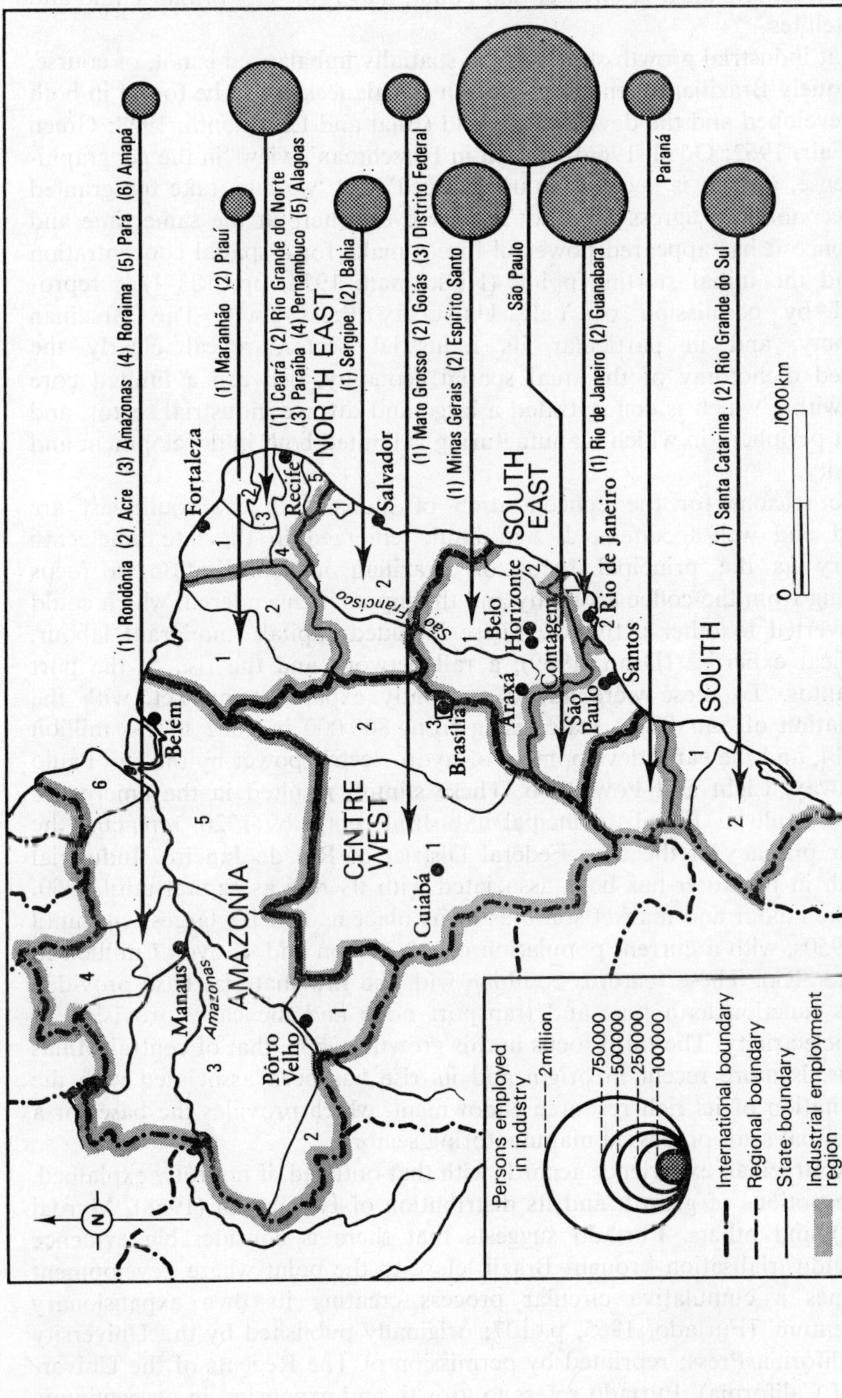

Figure 15.1 Brazil: industrial employment, 1970 (based on VIII Recenseamento Geral 1970, Tabulações Avançadas do Censo Demográfico, Resultados Preliminares, 1971)

and almost one-third in Greater São Paulo, that is the city of São Paulo and its satellites.

That industrial growth should be so spatially imbalanced is not, of course, a uniquely Brazilian phenomenon. Such imbalances are to be found in both the developed and the developing world (Bhat and Learmonth, 1968; Green and Fair, 1962; Odell, 1966). Indeed in Hirschman's view 'in the geographical sense, growth is necessarily unbalanced' and 'we may take for granted that economic progress does not appear everywhere at the same time and that once it has appeared powerful forces make for a spatial concentration around the initial starting point' (Hirschman, 1958, pp. 183–184; reproduced by permission of Yale University Press Inc.). The Brazilian economy, and in particular its industrial sector, reveal clearly the marked dichotomy of the dual spatial economy, between a limited core area within which is concentrated a large and diverse industrial sector, and a vast periphery in which manufacturing is limited both in development and diversity.

The reasons for the concentration of industry in the south-east are varied and well-documented. São Paulo emerged in the late nineteenth century as the principal focus of Brazilian industrialisation, a focus deriving from the coffee economy and the inputs it engendered, which could be diverted to other activities. These included capital, immigrant labour, technical expertise (Dean, 1969), a rail network and the rise of the port of Santos. To these were added a rapidly expanding market, with the population of São Paulo state rising from 800 000 in 1872 to 1·4 million in 1914, and the early development of hydroelectric power by the São Paulo Tramway, Light and Power Co. These stimuli resulted in the emergence of São Paulo as Brazil's principal industrial centre by 1920, replacing the earlier primacy of the then Federal District of Rio de Janeiro. Industrial growth in the latter has been associated with its role as capital until 1960, and the labour and market stimulus of its place as Brazil's largest city until the 1950s, with a current population of 4·3 million and of over 7 million in Greater Rio. These features combine with the raw material base provided by its function as a port and transport node and the early provision of hydroelectricity. The third focus in this growth region, that of central Minas Gerais, is more recent in origin and its rise has been associated with the exploitation of its rich resource endowment, which provides the base for a substantial semi-processed manufacturing sector.

The Brazilian experience accords with that outlined, if not fully explained, in the notions of growth and its distribution of Hirschman (1958), Myrdal (1963) and others. Furtado suggests that there is considerable evidence that industrialisation brought Brazil 'close to the point where development becomes a cumulative circular process creating its own expansionary momentum' (Furtado, 1965, p. 107; originally published by the University of California Press; reprinted by permission of The Regents of the University of California). Furtado refers to growth and expansion in an economic sense; and if Brazilian industry has grown and diversified as a result of 'circular causation' much of the growth has been spatially concentrated. It appears that the 'backwash effects' of Myrdal or the 'polarisation effects' of Hirschman have operated in favour of the south-east and especially São Paulo as against the remainder of the country. Between 1950 and 1960 the region's share of the total national increase in industrial employment was over 50 per cent in all industrial sectors except woodworking and the tobacco industry, and well above its share in the increase in total population.[2] In the 'growth' industries of metallurgy, engineering and the electrical, transport and chemical industries, over 85 per cent of the additional employment was generated in the south-east.

The impact of government policy

This concentration of growth is not surprising in the light of the experience of the developed world and the conclusions of location theory regarding agglomerative tendencies (Isard, 1956; Smith, 1971). In Brazil, however, the concentration has been compounded by the impact of government encouragement of industrialisation. In this, to some extent in a negative way, federal encouragement of industrial development has tended to increase concentration in the south-east. In the absence of clearly defined spatial objectives in national planning, much state-fostered industrialisation has tended to concentrate in this region. Early government participation in the industrial sector, in the establishment of a nationalised steelworks, a lorry factory and an alkali plant, initiated this pattern in that all three concerns were located in Rio de Janeiro state. These tendencies have been intensified by more broadly based economic planning from 1950 onwards, particularly by the Programa de Metas 1956–61, which envisaged a major expansion in Brazil's industrialisation with a growth rate of 12·7 per cent per annum. Eleven industrial sectors were singled out for stimulus and the bulk of the new additional productive capacity which was created was located in the south-east. Thus over 95 per cent of a projected increase in steel capacity of 3·2 million tons between 1956 and 1965 was to be located in the region, as was the bulk of the expansion in non-ferrous metal working, shipbuilding, and alkali and cement production. The entire automobile industry was located in São Paulo. In only two industries, paper and rubber, was there any degree of development elsewhere, a response to particular raw material attractions. The 1963–65 Plano Trienal recognised the disequilibria which the earlier development had created and included some counter-measures (Presidência da República, 1962). However, this plan was never implemented and it was not until the 1964–66 Programa de Ação Economica do Gôverno that specific steps were taken to correct regional imbalances. Even so, there has been a continuing tendency in the industrial sector to divorce national from regional planning (Presidência da República, 1970).

The nature of this imbalance in government action is revealed in the pattern of distribution of funds by the Banco Nacional do Desenvolvimento Econômico which was created in 1952 to administer funds for development. Between 1952 and 1965 the south-east received over 86 per cent of the total domestic and 77 per cent of the foreign capital administered by the Bank, and 96·6 per cent and 94 per cent respectively of the capital invested in manufacturing. Although the Bank acknowledged these imbalances from an early date (B.N.D.E., 1956), it was not until 1965 that active concern was shown for the spatial allocation of funds, as a result of increasing national concern for the less-developed parts of the country. The Bank recognised that attention to the needs of the underdeveloped areas of the country provided a way of maximising the national rate of economic growth (B.N.D.E., 1966). The 1st Congresso da Desenvolvimento Regional at Araxá, Minas Gerais, examined the polarising impact of import substitution industries and the growing regional disequilibrium within the country, and suggested the creation of growth poles (*polos regionais*) as a means of countering this trend (B.N.D.E., 1966; B.D.M.G., 1965). The Bank began to assist the removal of infrastructural obstacles to regional progress and to provide aid in various forms to regional and state agencies. Despite this greater awareness, between 1966 and 1970 63·4 per cent of the Bank's total investment and 60 per cent of its investment in manufacturing went to the south-east (B.N.D.E., 1971).

The role of foreign investment

In addition to the impact of national planning, foreign private investment has played a significant role in Brazil's industrialisation (Furtado, 1970), and particularly in the tendency towards concentration. Foreign investment in Brazil has been of some significance since the nineteenth century but has increased in importance in recent years (Baklanoff, 1966). By the 1960s substantial proportions of some Brazilian industries, particularly in the 'growth' sector, were controlled by foreign capital; 69 per cent of the automobile industry, 57 per cent of auto parts, 37 per cent of chemicals, 22 per cent of plastics and 17 per cent of the steel industry were foreign controlled (Baklanoff, 1966). In 1968 at least 13 of Brazil's largest private companies were associated with foreign concerns (Visão, 1968) including Ford, General Motors and General Electric. They were mainly located in the south-east, especially in São Paulo where, in the early 1960s, about half the private industrial capital was foreign-owned or controlled (Dean, 1969). Such concerns form an integral part of industrial groups based outside Brazil and their corporate decisions may not accord with the interests and aspirations of Brazil (Furtado, 1970). Thus locational decisions taken in the interests of the multinational corporation may not be wholly appropriate to Brazilian development. Servan-Schreiber (1968) has explored the implications of the recognition by North American firms of the potentialities of development within the European Common Market. If such firms were to discern that similar opportunities might be derived from locating in Brazil, in LAFTA or some other economic union in order to exploit the markets of Latin America, then the resulting industrial growth would have, in addition to its economic implications, geographical ones which accord with those explored by Odell (1966). Without locational controls within a Latin-American economic union, industrial growth is likely to concentrate in the areas of Buenos Aires, Mexico City–Veracruz and in south-east Brazil. It may be significant that over 75 per cent of US investment in manufacturing in Latin America is in Argentina, Mexico and Brazil (Furtado, 1970). Thus the initial growth of manufacturing in Brazil, the impact of national planning programmes and the role of foreign capital have all favoured a marked concentration of industry in the south-east. The Programa Estratégico de Desenvolvimento 1968–70 recognised these factors and noted the continuing concentration in the region between 1955 and 1964, when its share of total industrial product rose from 79·5 per cent to 83·2 per cent, and its share of industrial employment from 70·8 per cent to 75·4 (Ministério do Planejamento 1969).

REGIONAL DEVELOPMENT POLICIES

The relative merits of industrial concentration and industrial dispersion is a well-established theme in the literature of development (Friedmann, 1966) and there are strong arguments on economic grounds in favour of concentration. Nonetheless 'the decision of where to locate a new project is as important as the decision to invest in it. The question of social justice in the distribution of the fruits of economic development are as important and as difficult in terms of regions as in terms of social classes' (Friedmann and Alonso, 1964, p. 1; reproduced by permission of The Massachusetts Institute of Technology). It is a curious paradox of Brazilian development that as expansion in the dynamic industrial sector has been encouraged, or at least permitted, to concentrate in the south-east, there has been a policy aimed at improving the economic and social position of the less-developed regions. In consequence, although industrialisation has been regarded as a tool for regional development, the end of securing *national* industrial progress has not necessarily been used as the means to secure *regional* advancement.

Formal regional planning was instituted in 1946 with the earmarking of federal funds for the development of Amazonia and the São Francisco valley, and currently much of the national territory is under the aegis of a regional planning agency. The scale of these regional agencies must be stressed since, with but one exception, their territories are larger than those of any country in Western Europe. South-east Brazil [3] has an area of over 900 000 square kilometres and São Paulo alone is larger than the U.K. Thus, to see the Brazilian situation in terms of a single core-periphery model or a north–south interaction (Hirschman, 1958) is an oversimplification. Furtado noted that the country's size had resulted in the creation of 'a constellation of regional economies' (Furtado, 1965, p. xvii), and these regions persist, distinct in character, degree of development and problems but sharing the common difficulty that their sheer size is an obstacle to their development. The contrasts between the regions have given rise to contrasting approaches by their development agencies. In the relatively more-developed southern states the role of the agency, SUDESUL, has been essentially one of research and publicity to attract additional growth to the region; in Amazonia the role of SUDAM has been more profound, seeking to initiate development in an essentially undeveloped area, while in north-east Brazil the task of SUDENE has been to stimulate growth in a large, heavily-populated region which may be regarded as underdeveloped or depressed. The enormity of the task of developing Amazonia led the predecessor of the present agency, SPVEA, to concentrate growth in a number of more favourable areas selected on economic or political grounds. In recent years such industrial growth as has taken place has been concentrated in the two principal cities, so that 65 per cent of the industrial concerns assisted by SUDAM have located in Belém and Manaus.

SUDENE's task has been more wide-ranging, seeking to rehabilitate and stimulate the economy of the drought polygon of the north-east. In this area industrial development has been an important element in the agency's programme, but this development has shown the same tendency to concentrate in a few centres. Over seventy per cent of the jobs created and of the capital invested between 1960 and 1970 were located in the three states of Bahia, Pernambuco and Ceará, which in 1970 had 50 per cent of the regional population. In fact over half the new jobs were created in the three state capitals, Salvador, Recife and Fortaleza and adjacent *municipios*,[4] with less than 20 per cent of the regional population. In consequence the less-developed areas and states have received limited benefit from programmes which have achieved considerable advances in gross terms. The other regional agencies have had less impact on patterns of industrial distribution, either due to their recent creation, as in the case of the agency for the centre-west region, or because industrialisation is excluded from their objectives, as in the case of the São Francisco valley agency.

A third planning element has been instrumental in encouraging industrial development in Brazil, namely the various state governments, many of which have formulated development plans and agencies of their own. These have tended to a degree to function independently of national and regional planning, but they have had similar consequences in that they have encouraged or facilitated the concentration of industrial growth in limited areas and have fostered especially the concentration of 'growth' industries in the state capital.

TRENDS AND POSSIBILITIES

Brazil's industrial growth since 1945 has resulted in a situation whereby in 1968 only 7·2 per cent of all manufactured goods for the domestic market were imported; only 1 per cent of traditional consumer goods, 9 per cent of intermediate goods and 16 per cent of capital goods were imported in that year (Winpenny, 1971). Industry and the infrastructure for industrialisation have been given high priorities in planning; but, despite the absolute success achieved, industrial growth has not resulted in substantial labour absorption; industry's share of employment remains low and the problems of urban unemployment and underemployment remain marked. In the growth which has taken place, especially in the dynamic sectors, there has been a pronounced tendency towards spatial concentration. Measures for regional and state development, of much less significance when compared to the 'national' growth concentrating in the south-east, have had similar consequences.

The tendency towards spatial concentration is an expected and probably necessary feature of the growth process; yet there is an implicit notion that there will be a trend via 'spread' or 'trickling down' effects towards an equalisation in the economic condition of the various regions (Myrdal, 1963; Hirschman, 1958). In the Brazilian case, however, it has been suggested (Baer, 1965) that the impact of industrialisation has been centripetal, and that in particular government industrialisation policy has worked to the detriment of the north-east. Hirschman argues that if the market forces result in polarisation effects, this will be a temporary feature and will be countered by the implementation of deliberate economic policy (Hirschman, 1958).

In Brazil national and regional policies have shown a tendency to operate in conflict, and even schemes favourable to regional advance have tended to create or perpetuate disparities in what in Brazilian terms are development regions, but which would be, in European terms, national territories. Thus the policies aimed at helping north-east Brazil have sought to ameliorate the conditions of 29 million people, living in an area of 1·5 million square kilometres. In fact much of the assistance went to the relatively more favoured and prosperous regions. The important '34/18' legislation (Winpenny, 1971), which permitted the transfer of 50 per cent of tax dues for investment in development projects, resulted in 53 per cent of the industrial projects approved and 70 per cent of the industrial investment in the period 1963–67 going to the states of Pernambuco and

Bahia, mainly to Recife and Salvador and, to a lesser degree, to Fortaleza (Hirschman, 1967). The device appears to have been a relatively successful one in securing the transfer of investment capital out of the south-east, but the recent extension of the area eligible for '34/18' assistance to include Amazonia may diffuse its potential, while experience in the north-east suggests that, though valuable in stimulating growth, it is an unrefined tool in influencing detailed distribution. It may be that just as imbalances on a national scale are a necessary facet of the development process, they must be accepted also at a regional level, certainly at the scale on which 'regional' development planning is carried on within Brazil. The distinct implementation of national and regional planning to date has resulted in the preservation and creation of economic disparities. Recent government planning has shown a greater awareness of this and a desire to bring about a more diffuse development with 'expansion of the national economic frontier, principally with regard to the Planalto Central, the humid valleys of the north-east and progressively, Amazonia, amplifying the internal market by the incorporation of new areas and, simultaneously by the better use of abundant labour' (Presidência da República, 1970, p. 25). There remains, however, within the policy of regional development a notion of differentiation between the *nucleo basico* of agricultural and industrial development focused on São Paulo–Rio de Janeiro–Belo Horizonte and the surrounding areas of Rio Grande do Sul, Minas Gerais and Espírito Santo, which have grown somewhat less rapidly in recent years, as against the north-east, Amazonia and the centre-west. These are all substantial units and development policy to date has failed to secure any degree of balanced development within them. If there is to be some notion of equalisation, rather than the perpetuation of a number of regions each associated with limited cores—the 'golden triangle' of Rio de Janeiro–São Paulo and Belo Horizonte, Recife–Salvador, Belém—and very large peripheries, then some greater degree of dispersal than that obtaining at present must be aimed at.

Growth poles and growth centres

Within the regional development literature of Brazil, the potentialities of the growth pole (*polo de crescimento*) as a means of achieving such decentralisation is a recurrent theme (B.N.D.E., 1966; B.D.M.G., 1965; SUDENE, 1966). The notion of the growth pole is still somewhat ill-defined and the confusion deriving from Perroux's use of space as an economic rather than a geographic sense (Perroux, 1950) has been discussed in detail in the literature (Darwent, 1969; Hansen, 1967; Gilbert, 1971). If the notion is accepted of the growth pole as a growth-impelling industry and of the growth centre as the point at which such an industry might be located and about which growth might take place spatially, it may be asked what industries and locations are appropriate to Brazilian policies of industrial decentralisation. In fact such growth poles and centres might have differing characters and objectives in the various regions. In the south-east and, to a lesser extent, in the south, new points of growth may be desirable to assist the peripheral areas within these regions and to relieve the growing pressures of the deglomerative forces which are beginning to emerge within the main urban–industrial complexes. For example, a population of 15 million is envisaged for Rio de Janeiro by the year 2000, with the associated problems of congestion, pollution etc. (Comissão do ano 2000, 1970). In the north-east the need is for bases to provide alternative employment in the interior *sertão* and stability in times of drought. In Amazonia the need is for locations from which this territory may be explored and integrated more closely into the national economy. This pattern accords to a reasonable degree with Friedmann's 'development regions' (Friedmann, 1966) in which the south-east exemplifies a core region, the south an upward transitional area, Amazonia and the centre-west resource frontier regions and the north-east either a downward transitional or a special problem region. The experience of the north-east (Baer, 1965; SUDENE, 1961) shows many of the characteristics of downward transition outlined by Friedmann, but given its other characteristics (James, 1953; Robock, 1963) it might be presented as a special problem region.

If industrialisation has a role to play in the development and equalisation of growth within Brazil and its regions, decisions must be made as to the most appropriate locations for new industry (Chapman, 1969), as growth points or location points (Friedmann, 1966) or growth centres (Darwent, 1969). Outside the south-east the short term possibility of developing growth poles in the sense used by Perroux, that is, industries or firms which are large with high interaction with other firms and a high degree of dominance, is limited. The establishment of 'principal poles' such as metallurgical, metal manufacturing and chemical industries (Davin, 1964) may not be feasible outside the south-east at the present level of development, although the chemical and petroleum resources of Bahia could provide for the base for a growth pole/growth centre development, albeit close to Salvador (Presidência da República, 1970).

Industrial estates

Some recent literature (Darwent, 1969; Hansen, 1967; Körner, 1967) has suggested that the growth pole need not be restricted to a single large industry or firm but might consist of numerous small or medium-sized concerns. Recent developments in Brazil may help to sustain this view, where the creation of industrial estates (*cidades industriais*) has been instrumental in creating points of growth. The evolution of the pioneer Cidade Industrial of Contagem in Minas Gerais has resembled closely Myrdal's process of cumulative causation as interpreted by Keeble (Keeble, 1967), although with the initial impetus deriving from infrastructural provision for industry rather than the location of new industry *per se* (Dickenson, 1970a). Within the Cidade Industrial important linkage effects have been created, even to Hirschman's 'classic' multiwall bag factory (Hirschman, 1958), located to take advantage of the incentives of the Cidade and to serve not simply the cement works on the site but also others in the state, as well as other industries using its product. Other backward and forward linkages in the cement industry and in the metallurgical, electrical and food processing industries have developed. Growth at Contagem was perhaps particularly favoured by its location within south-eastern Brazil and its proximity to a major city, Belo Horizonte, but the success of the project has been such that *cidades* have been utilised with some success elsewhere in Minas Gerais and Brazil, to stimulate industrial development in less favoured areas (Dickenson, 1970b). Such developments provide a means to focus industrial growth within the planning regions, reduce development costs by concentrating infrastructural provision at specific points and, as infrastructural provision is an integral part of their establishment, serve to overcome the reluctance of companies to leave the infrastructural levels of the primate city (Gilbert, 1971). The potentialities of the *cidades industriais* need further appraisal, but they do appear to provide a means whereby Brazilian industrial growth can be encouraged and disseminated.

Within the south-east, industrial estates provide the means to attract industry to less, favoured areas and to facilitate the deconcentration of firms from the large cities. In the north-east they would provide a means of achieving economic diversification, of stabilising a migrant population and of countering the influence of the three main cities. The case of the north-east raises the still unresolved question of the appropriate size of urban settlement with which growth centres should be associated. In the state of Ceará, seven development poles have been postulated (CODEC, no date), not all as yet associated with *cidades industriais*, based on towns ranging in size from 10 000 to 842 000 people. Various authors have claimed that population totals ranging from 10 000 to 1 million are appropriate optima for the size of growth centres (Darwent, 1969), and further work is clearly necessary to ascertain the sizes of settlement suitable for designation as growth centres and able to sustain a *cidade industrial*.

Other stimuli to growth

In the interior of Brazil the need is for bases from which its development may be achieved (Friedmann, 1966). Brasília is already regarded as a growth centre for the centre-west (Chapman, 1969), and although manufacturing was not envisaged as an element in the form of the new capital, by 1967 there were over 2000 industrial employees in the Distrito Federal (Fundação IBGE, 1969). The rapid growth of population to 544 000 in 1970 may provide the base for further industrial growth, while Belém and Manaus have already emerged as the principal industrial foci in the north. Further potential growth points may emerge with the penetration of the resource frontier by the federal highways from Belém to Brasília, from Cuiabá to Pôrto Velho and the Transamazonica highway. The potential impact of these highways is indicated by the claims that population along the route of the Belém–Brasília highway has increased by 1·9 million in a decade, over 100 new settlements have been established and substantial pastoralism and some cultivation and industry have developed (Ministry of Transportation, 1970). Much of this growth, however, was spontaneous (Valverde and Vergolino Dias, 1967). Just as it is vital that the scarce resources being used in these highway projects yield maximum benefit to the areas and population they influence, and to the nation, so also must Brazil's industrialisation be utilised in such a way as to provide maximum economic and social benefit to the population of all the areas of the country.

CONCLUSION

The Brazilian case clearly presents a singular problem deriving from the vast size of the country. In its industrialisation to date the play of the market forces and the stimulus of state intervention in the economy have served to favour one region as against the rest of the country. Over time, however, some notion, if not of equalisation of development levels, at least of improvement in the condition of the poorer regions and their inhabitants is necessary. In Brazil, therefore, as elsewhere in the developing world, a regional strategy must be formulated and the place of industrialisation within such a strategy determined. Within this manufacturing sector, industries appropriate to the stimulation of growth within the various regions, the means to facilitate and encourage their emergence and the location in which they will have maximum impact must be identified. The latter consideration is of particular significance given the scale at which Brazilian regional policy has operated to date.

NOTES

1. Labour force has been used as the main measure of industrialisation throughout the essay, notwithstanding its limitations (Alexander, 1958), in order to indicate the pattern of employment as well as of production.
2. In 1970 the south-east had 43 per cent of the national population.

3. There is no development agency responsible for the south-east, but parts of its area fall under the influence of agencies for the São Francisco and Paraíba valleys and of SUDENE.
4. The *municipio* in the basic administrative unit in Brazil, resembling the US county.

REFERENCES

Alexander, J. W. (1958) 'Location of manufacturing: methods of measurement', *Annals of the Association of American Geographers* **48**, 20–26.

Baer, W. (1965) *Industrialisation and economic development in Brazil* (Homewood: Irwin).

Baklanoff, E. N. (1966) 'Foreign private investment and industrialisation in Brazil', *New perspectives of Brazil*, ed. E. N. Baklanoff (Nashville: Vanderbilt University Press), 101–136.

B.D.M.G. (Banco de Desenvolvimento de Minas Gerais) (1965) *Anais do 1° Congresso Brasileiro de Desenvolvimento Regional* (Belo Horizonte: B.D.M.G.).

Bhat, L. S. and A. T. A. Learmonth (1968) 'Recent contributions to the economic geography of India: some current preoccupations', *Economic Geography* **44**, 189–209.

B.N.D.E. (Banco Nacional do Desenvolvimento Econômico) (1956) *Exposição sôbre o Programa de Reaparelhamento Econômico* (Rio de Janeiro: B.N.D.E.).

B.N.D.E. (1966) *XIV Exposição sôbre o Programa de Reaparelhamento Econômico* (Rio de Janeiro: B.N.D.E.).

B.N.D.E. (1971) *Relatório das atividades em 1970* (Rio de Janeiro: B.N.D.E.).

Chapman, M. (1969) 'Geography and the study of development', *Journal of Developing Areas* **3**, 319–338.

Comissão do ano 2000 (1970) *Rio ano 2000* (Rio de Janeiro: Secretaria de Ciência e Technologia).

Companhia do Desenvolvimento Econômico do Ceará (CODEC) (no date) *Estamos esperando por vôce . . .* (Fortaleza: CODEC).

Darwent, D. F. (1969) 'Growth poles and growth centres in regional planning—a review', *Environment and Planning* **1**, 5–32.

Davin, L. (1964) *Economie régionale et croissance* (Paris: Génin).

Dean, W. (1969) *The industrialisation of São Paulo 1880–1945* (Latin American Monographs; Austin: Texas University Press).

Demas, W. G. (1966) *The economics of development in small countries with special reference to the Caribbean* (Montreal: McGill University Press).

Dickenson, J. P. (1970a) *Zona Metalúrgica: a study of the geography of industrial development in Minas Gerais, Brazil* (Unpublished Ph.D. thesis: University of Liverpool).

Dickenson, J. P. (1970b) 'Industrial estates in Brazil', *Geography* **55**, 326–329.

Friedmann, J. (1966) *Regional development policy: a case study of Venezuela* (Cambridge: M.I.T. Press).

Friedmann, J. and W. Alonso (1964) 'Introduction', *Regional development and planning:* ed. J. Friedmann and W. Alonso (Cambridge: M.I.T. Press).

Fundação IBGE (1969) *Brasília* (Rio de Janeiro: Fundação IBGE).

Furtado, C. (1965) *Diagnosis of the Brazilian crisis*, trans. S. Macedo (Berkeley: California University Press).

Furtado, C. (1970) *Obstacles to development in Latin America*, trans. C. Ekker (New York: Anchor).

Gilbert, A. (1971) 'Growth poles—the instant solution to regional problems?', Paper given to *I.G.U. Commission on regional aspects of economic development; Colloquium on regional inequalities of development* (Vitória), mimeo.

Green, L. P. and J. D. Fair (1962) *Development in Africa: a study in regional analysis with special reference to southern Africa* (Johannesburg: Witwatersrand University Press).

Hansen, N. (1967) 'Development pole theory in a regional context', *Kyklos* **20**, 709–725.

Hirschman, A. O. (1958) *The strategy of economic development* (New Haven: Yale University Press).

Hirschman, A. O. (1967) 'Desenvolvimento industrial no Nordeste Brasileiro e o mechanismo de crédito fiscal do Artigo 34/18', *Revista Brasileira de Economia* **21**, 1–32.

Isard, W. (1956) *Location and space economy* (Cambridge: M.I.T. Press).

James, P. E. (1953) 'Patterns of land use in north-east Brazil', *Annals of the Association of American Geographers* **42**, 153–176.

Keeble, D. E. (1967) 'Models of economic development', *Models in Geography*, ed. R. J. Chorley and P. Haggett (London: Methuen), 243–302.

Körner, H. (1967) 'Industrielle Entwicklungspole als Instrumente der Regional politik in Entwicklungsländern,' *Kyklos* **20**, 684–708.

Ministério do Planejamento e Coordenação Geral (1969) *Programa Estratégico de Desenvolvimento 1968–70. Estudo especial A industrialização Brasileira: diagnóstico e perspectivas* (Rio de Janeiro: Ministério do Planejamento).

Ministry of Transportation of Federative Republic of Brazil (1970) *Transamazonian highways* (Montreal: Ministry of Transportation).

Myrdal, G. (1963) *Economic theory and underdeveloped regions* (London: Methuen).

Odell, P. R. (1966) 'La integracion economica Latinoamericana y la situation de sus atividades industrales', *Union Geografica Internacional Conferencia Regional Latinoamericana Temas Geografico-economicos* (Mexico, D.F.: Sociedad Mexicana de Geografica e Estadistica) Tomo II, 14–18.

Odell, P. R. (1971) 'A European view on regional development and planning in Latin America', *International Review of Community Development* **25–26**, 3–22.

Perroux, F. (1950) 'Economic space: theory and applications' *Quarterly Journal of Economics* **64**, 89–104.

Presidência da República (1962) *Plano Trienal de Desenvolvimento Econômico e Social 1963–65* (Rio de Janeiro: Presidência da República).

Presidência da República (1970) *Metas e bases para a ação de Governo* (Rio de Janeiro: Presidência da República).

Richardson, H. W. (1969) *Elements of regional economics* (Harmondsworth: Penguin).

Robinson, E. A. G. (ed.) (1960) *The economic consequences of the size of nations* (London: Macmillan).

Robock, S. H. (1963) *Brazil's developing northeast* (Washington, D.C.: Brookings).

Servan-Schreiber, J-J. (1968) *The American challenge* (London: Hamilton).

Smith, D. (1971) *Industrial location* (New York: Wiley).

Streeten, P. (1971) 'How poor are the poor countries?', *Development in a divided world*, ed. D. Seers and L. Joy (Harmondsworth: Penguin), 67–83.

SUDENE (1961) *The bases of development policy for the northeast of Brazil and scheme of SUDENE's Five Year Plan* (SUDENE), mimeo.

SUDENE (1966) *III Plano Diretor de desenvolvimento econômico e social do Nordeste* 1966–68 (Recife: SUDENE).

UN (1971) *Statistical Yearbook 1970* (New York: Statistical Office of the UN).
Valverde, O. and C. Vergolino Dias (1967) *A rodovia Belém—Brasília* (Rio de Janeiro: Instituto Brasileiro de Geografia).
Visão (1968) 'As maiores empresas do Brasil', *Visão* **33**, no. 5, 164–167.
Winpenny, J. T. (1971) 'Brazil: the industrial sector: its situation and prospects', *BOLSA Review* **5**, 642–650.

Chapter 16

Regional Inequalities in Internal Communications: The Case of Uganda

José A. Smith

Introduction

Regional inequalities in growth are an important aspect of the spatial dimension of development and there are few, if any, developing countries that do not suffer from this problem. These inequalities are more marked in the developing countries than elsewhere (Soja, 1968; Berry, 1969; Friedmann, 1966), and Williamson (1965) has suggested that further economic growth in the developing countries only serves to encourage these regional inequalities so that an economy must achieve a fairly advanced stage before they begin to lessen. Islands of initial economic advancement have emerged within the developing countries and are growing at a far greater rate than, and often at the expense of, the intervening areas. National governments have shown increasing concern for this situation and development plans are devoting greater attention to regional planning as, for example, in Tanzania and Zambia (Berry, 1971; Davis, 1971). The marked nature of regional inequalities in many developing countries has made them both economically and politically unacceptable to people and government.

The apparent importance of these regional inequalities is not adequately reflected in the geographical literature. Quantitative studies of the spatial aspects of regional development are few and usually apply to individual countries rather than to common problems (Soja, 1968; Riddell, 1970; Berry, 1969). Few studies are concerned with the causes of regional inequalities, but they must be explained if geographers are to make a contribution towards predicting spatial trends or towards regional planning. Major problems are involved here because there is a general dearth of

---

José A. Smith is currently lecturing in the geography department of Kingston Polytechnic and was formerly on the staff of the geography department of Makerere University, Kampala.

data at a regional level and a complex problem of a multitude of interdependent factors varying in both space and time. This chapter attempts not to overcome all the problems but to isolate one aspect of development and to apply statistical techniques both to give a detailed description of the spatial inequalities and to suggest some of their underlying causes. The aspect selected is road transport, including an analysis of the network and of the volume of traffic flow; this aspect was chosen partly because data are available and partly because the road transport infrastructure is one of the crucial factors of development; its spatial pattern is closely related to that of development as a whole. Moreover, road transport in the developing countries is rapidly becoming the preferred mode of transport for people and goods (Taaffe *et al.*, 1963). The road network often represents a lead–lag relationship with development as it can be either a prerequisite or a result of growth. Traffic flow, however, is almost entirely a reflection of development and the spatial aspects of flow within a country should represent fairly accurately the spatial pattern of development in general.

Uganda, the case study area, exhibits all the general problems associated with regional inequalities. The more-developed areas of Uganda are concentrated in the Districts bordering Lake Victoria and in Bugisu and include Kampala, the major commercial and administrative centre, and the major industrial centres of Tororo and Jinja. In the other Districts, especially in the west and north, development has lagged far behind that of the lakeshore Districts, the most marked contrast being between the rural areas of the north and the urban centre of Kampala. The Ugandan Government has emphasised regional planning and has inaugurated schemes in the north and west in an attempt to spread development. So far, little has been achieved and a more radical approach is necessary if a more balanced pattern of regional growth is to be established.

THE ROAD NETWORK IN UGANDA

The growth of the network

The main road network of Uganda is well-developed and well-connected for a country at Uganda's stage of development. This may be attributed to several factors amongst which the small size of the country is important. The network was well-developed prior to the arrival of the British and several early explorers remarked on the good wide roads, especially in Buganda. The network was expanded and improved under the British administration with emphasis on opening areas for cash crop production. Uganda is well supplied with raw materials for the construction of gravel roads especially from the lateritic soil, referred to locally as 'murram', and this enabled the administration to provide a good network for the light amount of traffic in operation at the time. This network forms the basic road system of Uganda today as new links have not been constructed since Independence in 1962. The network consists of all the interregional roads and the main feeder roads and comes under the control of the Ministry of Works. In addition there are numerous minor roads and feeder roads which carry small amounts of local traffic and many of which are seasonal in use. These are maintained by the local authorities but only until a road becomes of regional importance when responsibility for it is transferred to the Ministry of Works. Consequently, the main road network has shown a gradually increasing mileage as small feeder roads have attained regional importance; but with these exceptions the network may be regarded as stable.

While the Ugandan Government has not added new links to the main road network, significant changes have occurred in the existing network with the upgrading and bitumenising of existing roads. Progress was very rapid during the period 1968–72 and this work has been undertaken partly to promote development in the peripheral regions and partly to accommodate an increasing volume of traffic. In the period since Independence the traffic population has increased and individual vehicles have increased in weight and speed; on several main roads the murram surface has become totally inadequate. Prior to these recent projects, most of the bitumen roads were those linking Kampala with its adjacent District centres and with eastern Uganda. The most recent developments have been the extension of these links and consequently the lakeshore area and eastern Uganda are very well served by bitumen roads. Some of the latest projects reach the periphery of Uganda, but large areas of western and northern Uganda still lack bitumenised main roads.

Regional inequalities in the network

The distribution of the main road network in Uganda is very uneven (Figure 16.1). Main road density per District has been derived by dividing the mileage of main roads in each District by the area (in square miles) of each District. High densities occur in the Districts bordering Lake Victoria and there is a decrease in density both northwards and westwards. West Mengo District has the highest density reflecting the importance of Kampala as the major node in the network, while the very low densities of the northern Districts of Acholi and Karamoja are a reflection of large areas with little development. This distribution repeats the generally uneven pattern of economic development outlined earlier.

For the whole country, however, main road network density reflects many factors, of which four appear to be of great significance. Distribution of

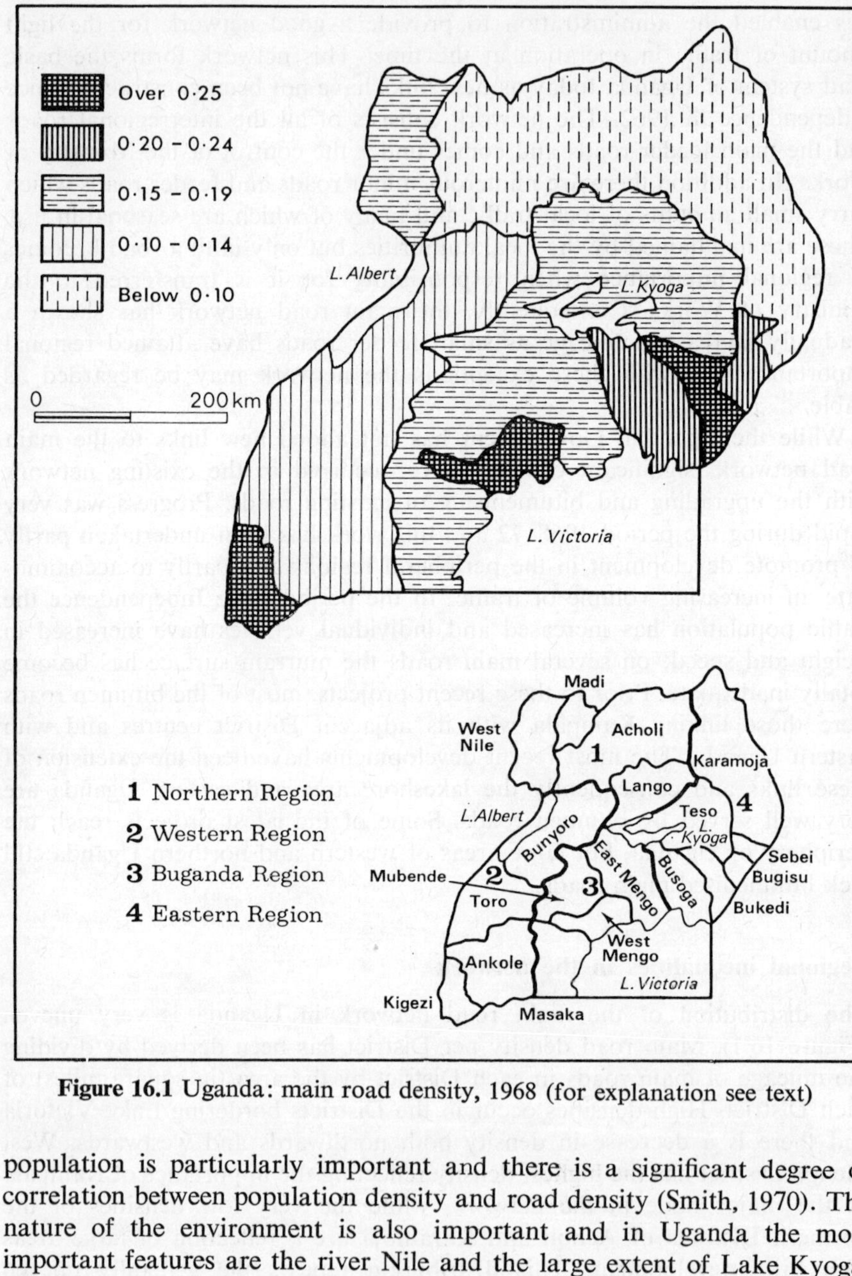

Figure 16.1 Uganda: main road density, 1968 (for explanation see text)

population is particularly important and there is a significant degree of correlation between population density and road density (Smith, 1970). The nature of the environment is also important and in Uganda the most important features are the river Nile and the large extent of Lake Kyoga. These combine both to restrict the number of north–south routes and to deflect the present north–south routes either to the far western or far eastern Districts of the country. The Nile, which flows from Lake Victoria at Jinja across Uganda to the western border and hence northwards, is bridged at only three points. Lake Kyoga presents an even greater problem and attempts to bridge branches of the lake have only just begun. The Tirinyi ferry on the Iganga–Mbale road was replaced by a bridge in 1972 but routes from Kampala to the northern towns (Lira, Gulu, Soroti) will continue to make a long detour through either Bunyoro or Teso District. Consequently, northern Districts suffer from inadequate connection and great distances to the main centre, Kampala, and this has contributed to their lack of development. The dominant position of Kampala is the third major feature of the pattern, resulting in the very high road density of West Mengo District. This reflects the early development of what is now the Kampala area during the nineteenth century which was enforced and encouraged by the British administration. Kampala emerged as the major commercial and administrative centre and this is apparent in the network. Moreover, this high density in West Mengo District consists almost entirely of bitumen roads and if bitumen roads alone were taken into account West Mengo would show even greater contrasts with the north and west of the country. Finally, the area of the administrative units must be taken into account although this is possibly more a problem of cartographic representation. Kigezi District's exceptionally high density is partially explained by the small size of the District [1] whilst the low density of Karamoja is partially linked with the very large size of the District.[2]

TRAFFIC FLOW IN UGANDA

The growth of road transport and the traffic census

A measure of the importance of road transport in Uganda is the rise in the vehicle population in the postwar period. This has increased from 14 000 in 1951 to 50 000 in 1970 and has risen steadily each year, with the exception of 1962 and 1968 (Uganda Government, 1955, 1962, 1971). The fall in 1962 coincided with Independence when many expatriates left the country, taking their vehicles with them. The fall in 1968 is not so easily explained as it occurred through an exceptionally large number of vehicles leaving the road and not being replaced. This, however, was only a temporary check and the vehicle population continued to rise in 1969 and 1970.

The flow of this vehicle population on the main road network forms the main source of data for this study. A traffic census is taken once every six months by the Ministry of Works and consists of seven consecutive days, each day being from 07.00 h to 07.00 h. The census is taken at various points on the main road network (223 in 1968) but these points are by no means evenly spread and the density of census points varies significantly;

for example, West Mengo District despite its small size has the largest number of census points.

There are some disadvantages involved in the use of this traffic census data. Although the census has been taken since 1950, prior to 1966 the census points were very sparse and located predominantly in Buganda; so there is insufficient data to project trends on the basis of past years. In addition, the traffic census gives no indication of the effect of seasonality in traffic flow and there are conflicting views as to the importance of this. In the cash cropping Districts, for example the cotton growing District of Bukedi, traffic certainly increases during the picking season. This is thought to have a considerable impact on local traffic but relatively little impact on the main road traffic. Despite these drawbacks, the census provides a very useful source of data and comparable data is available for other countries.

Regional inequalities in traffic flow

Regional inequalities in traffic flow are even more marked than inequalities in road density (Figure 16.2). The average flow of traffic per mile of road is a generalised description and was derived by assuming that each census

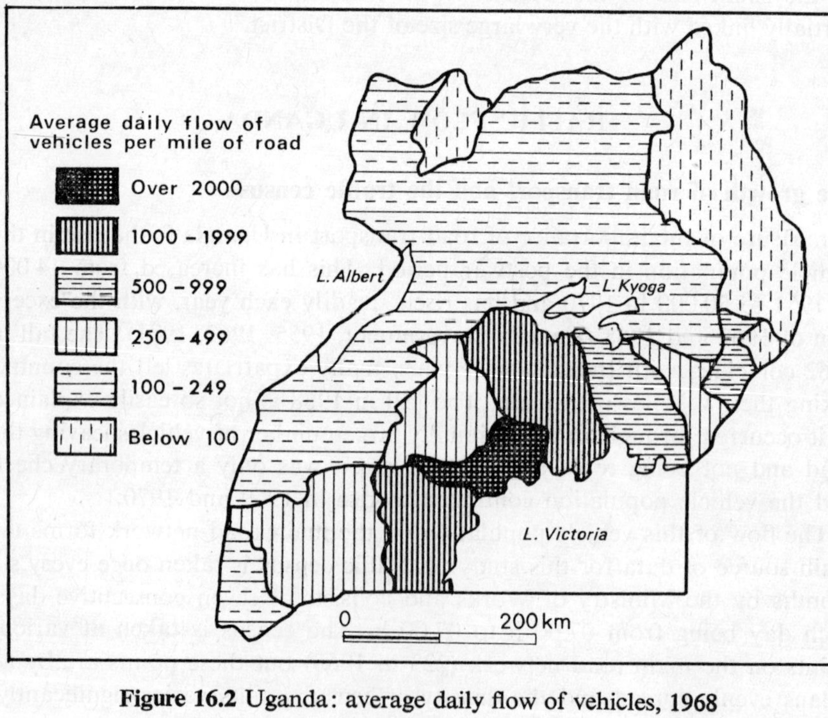

Figure 16.2 Uganda: average daily flow of vehicles, 1968

point was representative of a certain mileage of road. Total traffic flow at each census point was multiplied by the mileage of road it represented and the multiplied traffic flow figures and the distances were summed within the individual Districts. For each District, the total number of vehicles was divided by the total mileage to give the daily average flow of vehicles per mile of road in each District. The contrast between Districts is very marked as there are three Districts (Madi, Karamoja and Sebei) recording flows below 100 whilst West Mengo District has over 2000 vehicles per day. Regional differences are also very marked as there is a strong distinction between the lakeshore Districts and Bugisu on the one hand and the western and northern Districts on the other. To gain a deeper insight into this pattern of traffic flow, and to assist in understanding its possible future growth, the pattern was examined in relation to selected phenomena. This necessitated reducing the main road network to a graph by defining nodes as all towns with a population of over 2000 and the links as the roads between the nodes on which a traffic census was taken. This is a realistic assumption in the case of Uganda as the graph includes all but a few main feeder roads on which the Ministry of Works takes a traffic census. Moreover, the main road network in Uganda is one net rather than a group of sub-nets and may be regarded virtually as a closed system as entry points are few and except for two links with Kenya are little used. Slight modifications in the links were made as it was essential for each link to terminate at a node, but knowledge of the Uganda network confirms that again this is not unrealistic. The resultant graph is shown in Figure 16.3a, which consists of thirty-one nodes and forty links, thus allowing some degree of sophistication but still retaining a study at the national level. On each of these links average daily traffic flow per mile of road was calculated in the same way as for Districts except that multiplied traffic flow and distances were summed to form links and then divided to give average daily flow on each link. For the purpose of comparison with the selected phenomena it was necessary to have the proportion of total flow on each link and this was calculated; all the following flow values are therefore relative, not absolute, terms.[3]

The structure of the network is the first variable used as it has been suggested that there is a relationship between the structure of a network, particularly connectivity, and economic development (Kansky, 1963; Tinkler, 1971). As already stated, the Ugandan network is quite well-developed and connected (Funnell, 1969) and comparison of road density (Figure 16.1) and traffic flow (Figure 16.2) shows regional similarities, but the direct relationship is not known. Relationships are also known to exist between population and many forms of interactions, including traffic flow. It is not known, however, to what extent this is so in Uganda, or to what extent there is a relationship between urban population and traffic flow as

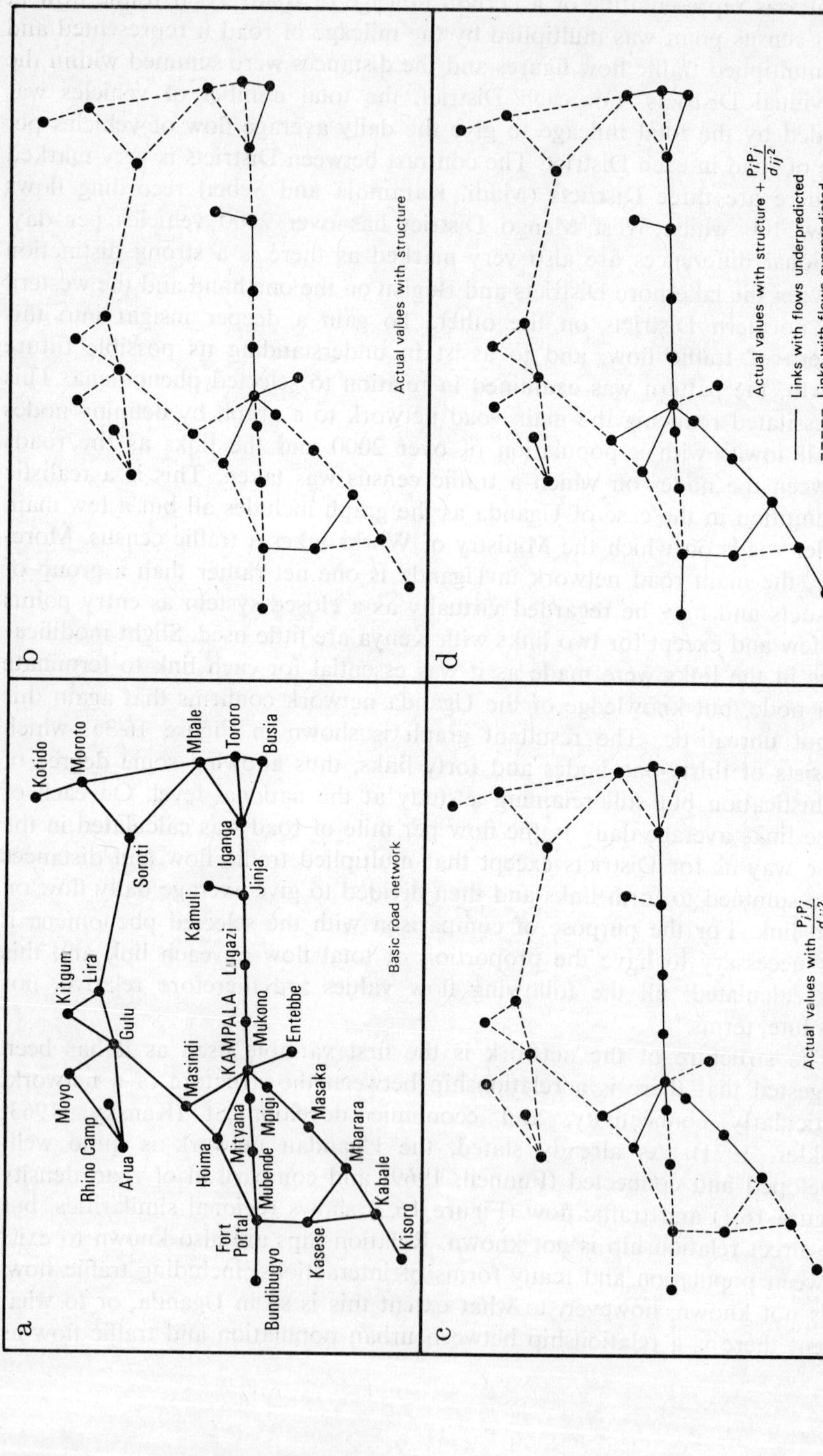

Figure 16.3 Uganda: (a) basic road network; (b), (c), (d) residuals from regression

the census figures suggest that much of the traffic is generated by the urban areas (Smith, 1970). Finally, there may be a relationship with distance which varies greatly on individual links in the Ugandan network; on the main road network links vary from 13 miles to 130 miles. The structure of the Ugandan main road network is used to generate one set of theoretical flows and population and distance are combined in a gravity model to generate a second set of theoretical flows on the forty links of Figure 16.3a. The theoretical flows (independent variables) are then correlated with the actual flows (dependent variable), first individually, then in combination, regression lines are calculated and residuals measured.[4]

Calculation of theoretical flows generated by the structure of the network uses a method suggested by Tinkler (1972). The size of the nodes and the links remain static and only the structure is allowed to introduce inequalities in the flows on the links. The structure is expressed in the form of a (1,0) matrix and random flows are introduced into this structure until a state of equilibrium, measured by the principal eigenvector of the matrix representing the network structure, is reached. The probability vector derived from the principal eigenvector represents the proportion of the total flow in the network to be found at each node when the total flow sums to one. The equilibrium flow at each node is distributed evenly between the routes out of the node and the probability vector value for each node has been divided by the number of routes out of each node to give a value to the flow on each link:

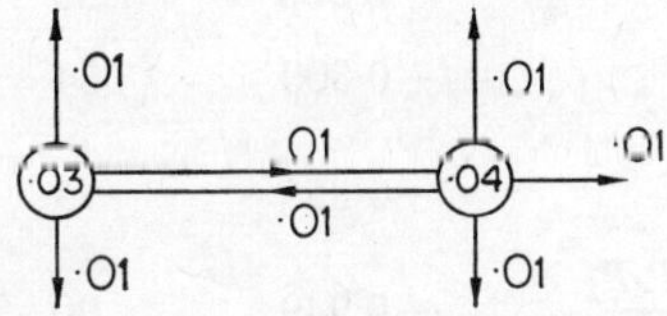

The value on the link between the nodes is ·02.

The basic drawback of this analysis for generating traffic flow is that it does not account for through traffic but only for flow between adjacent nodes. In the absence of studies of traffic generation and destination, Tinkler's model was modified to incorporate extra links whereby every District centre was linked to Kampala. The eigenvector and the probability vector for the modified network were calculated and the flows on the individual links were summed as before. Where a link would receive one or more through links these were summed and, once calculated, the values on the links were renormalised so that they once again summed to one. These form the first set of forty theoretical values to correlate with the actual values.

A gravity model was used to generate a second set of forty flows which were again recalculated to sum to one, giving the proportion of total flow on each link. This allows node and link size to vary, used in the following equation:

$$T_{ij} = \frac{P_i P_j}{d_{ij}} \qquad (1)$$

where T = theoretical traffic flow between two towns i and j
P = population size of i and j
d = distance between the two towns i and j

Following the discussion concerning the relationship of distance to interaction between the nodes, a modification was introduced in which d^2 replaces d as a measure of distance (Isard, 1960):

$$T_{ij} = \frac{P_i P_j}{d_{ij}^2} \qquad (2)$$

Correlation coefficients and regression lines were calculated assuming that the actual traffic flow was the dependent variable and the theoretical flows were the independent variables. The correlations between two sets of variables and the multiple correlations resulted in high positive values for the correlation coefficients (Table 16.1).

Table 16.1 Correlation of actual and theoretical flows

Actual traffic flow with	r	explained percentage	significance percentage
structural model	+0·586	35	0·1
$\frac{P_i P_j}{d_{ij}}$	+0·800	64	0·1
$\frac{P_i P_j}{d_{ij}^2}$	+0·855	73	0·1
structural model + $\frac{P_i P_j}{d_{ij}}$	+0·949	90	0·1
structural model + $\frac{P_i P_j}{d_{ij}^2}$	+0·964	93	0·1

Combining the structural models with the two gravity models results in very high multiple correlation coefficients because separately they give high values of r but there is very little correlation between the two independent variables. The correlation coefficient of structure with $\frac{P_i P_j}{d_{ij}}$ is +0·098 and with $\frac{P_i P_j}{d_{ij}^2}$ is +0·172. While this analysis does not prove the non-existence of other factors, it does suggest from the high values of r that structure, urban population and the spacing of nodes are important features in the spatial distribution of traffic flow.

Assuming that actual values were the dependent variables in each case, least square and multiple regression lines were calculated and the residuals of actual values on the links from the theoretical regression line were measured. Three sets of results, actual values with structure, actual values with $\frac{P_i P_j}{d_{ij}^2}$ and actual values with structure and $\frac{P_i P_j}{d_{ij}^2}$ are discussed here. In all three cases the majority of links are close to the regression line, especially the middle values, but it is the largest residuals and their location which is of interest. The three theoretical regression lines tend to over-estimate the links which have the smallest flow values (negative residuals) and underestimate the links with the greatest flow values (positive residuals). The regression line from the structural model overestimates the ten smallest actual flows, underestimates four of the five largest actual flows; the regression line from $\frac{P_i P_j}{d_{ij}^2}$ overestimates the ten smallest actual flows, under-estimates nine of the ten largest actual flows; the regression line from structure and $\frac{P_i P_j}{d_{ij}^2}$ combined overestimates six of the ten smallest actual flows, underestimates nine of the ten largest actual flows. The spatial aspects of these residuals are illustrated by Figures 16.3b, c and d, where positive and negative residuals have been mapped,[5] and the five largest positive and negative residuals in each figure are indicated by Tables 16.2, 16.3 and 16.4.

Table 16.2 Residuals from structural model regression[6]

5 largest negative residuals		5 largest positive residuals	
Location	Difference	Location	Difference
Iganga–Tororo	−0·037	Kampala–Entebbe	+0·155
Gulu–Masindi	−0·031	Kampala–Mukono	+0·077
Mbale–Soroti	−0·030	Kampala–Mpigi	+0·027
Arua–Moyo	0·027	Jinja–Kamuli	+0·011
Arua–Gulu	−0·026	Kampala–Hoima	+0·010

Residuals in Figure 16.3b (structure) show a somewhat broken pattern but generally overestimate links from Fort Portal in the west, through northern Uganda, to Tororo in the east, whilst four of the six links out of Kampala are underestimated. Table 16.2 shows that the four links out of Kampala occur amongst the five largest positive residuals and two are very large indeed. The largest negative residuals are more scattered. It is of note that these positive residuals occur around Kampala despite the additional weighting given to the capital in the structural model. Positive residuals from the regression line from $\frac{P_i P_j}{d_{ij}^2}$ (Figure 16.3c) are again concentrated around Kampala but in addition they extend along the main roads to the south-west and to Mbale in the east.

Table 16.3 Residuals from $\frac{P_i P_j}{d_{ij}^2}$ regression[7]

5 largest negative residuals		5 largest positive residuals	
Location	Difference	Location	Difference
Kampala–Entebbe	−0·026	Jinja–Lugazi	+0·060
Kotido–Moroto	−0·015	Lugazi–Mukono	+0·058
Kabale–Kasese	−0·014	Iganga–Jinja	+0·036
Iganga–Mbale	−0·014	Kampala–Mukono	+0·034
Arua–Moyo	−0·014	Mbale–Tororo	+0·025

The major difference between the positive residuals in Figures 16.3b and 16.3c is the Kampala–Entebbe link which is negative in Figure 16.3c and the largest negative residual (Table 16.3). The other negative residuals form a continuous arc from Kisoro in south-west Uganda, through the periphery of western Uganda and across northern Uganda to Mbale in the east. This again repeats the pattern of contrast between lakeshore districts and western and northern Uganda. The very large positive residuals of the structural model are lacking (Table 16.3) and both large positive and large negative residuals are scattered in occurrence. The only contiguity occurs on the three links joining Kampala and Jinja which include three of the five largest residuals.

The combined effects of structure and $\frac{P_i P_j}{d_{ij}^2}$ are illustrated in Figure 16.3d and Table 16.4. Again there is a distinct pattern of positive and negative residuals. The positive residuals are concentrated around Kampala, with the exception of the Kampala–Entebbe link and on the main routes to the south-west and east of Uganda. The positive residuals thus occur in the areas that may be regarded as the developed parts of Uganda. The largest positive residuals (Table 16.4) however are no longer concentrated on the links immediately around Kampala, as only one link, Kampala–Hoima, occurs in the five largest positive residuals. The actual values on the links around Kampala are very close to the theoretical regression line as in fact is the Kampala–Entebbe link which although negative, has a very small residual, a direct contrast to the previous results. The negative residuals again form an almost continuous arc through northern Uganda from Fort Portal to Mbale. With the exception of the Iganga–Tororo link, the highest negative residuals occur on the links between District centres in the north—

Table 16.4 Residuals from multiple regression $\left\{ \text{structure} + \frac{P_i P_j}{d_{ij}^2} \right\}^8$

5 largest negative residuals		5 largest positive residuals	
Location	Difference	Location	Difference
Iganga–Tororo	−0·021	Jinja–Lugazi	+0·024
Gulu–Masindi	−0·019	Lugazi–Mukono	+0·017
Mbale–Soroti	−0·018	Masaka–Mbarara	+0·014
Arua–Moyo	−0·018	Hoima–Kampala	+0·014
Arua–Gulu	−0·016	Fort Portal–Kasese	+0·011

Gulu–Masindi, Mbale–Soroti, Arua–Moyo, Arua–Gulu and, in addition, the Moroto–Soroti and Lira–Soroti links show high negative residuals.

Despite the high degree of correlation in the multiple regression analysis, the distinct pattern of residuals in Figure 16.3d suggests that there may be other factors determining the pattern of traffic flow. Road surface may be such a factor since the positive residuals correspond very closely to the distribution of bitumen (of the twenty-one links showing positive residuals only five have no bitmuen, whilst of the nineteen negative residuals only three are completely bitumen). It is difficult to judge, however, whether road surfaces have been improved to accommodate heavier traffic flow or whether better roads attract more traffic. The answer is probably a combination of the two as is illustrated by the links from Kampala to western Uganda. There are two routes to western Uganda, one from Kampala to Fort Portal direct and the other from Fort Portal and Kasese via Mbarara. The longer route via Mbarara was completely bitumenised in 1971 whilst the direct route to Fort Portal is a poor road with 150 miles of murram. Upgrading the final Mbarara–Kasese link was justifiable on the grounds of traffic flow but completion of this bitumen link to the west is thought to be directly competitive to the Kampala–Fort Portal link. Figure 16.3d illustrates that the links to the west via Mbarara have positive residuals whilst the direct links have negative residuals; now the bitumen is complete those residuals are expected to increase.

CONCLUSION

The road network and traffic flow have been used to illustrate the marked regional inequalities in the Ugandan economy. The greatest discrepancies occur in the traffic-flow pattern and these may be related to the structure of the network and the size and spacing of towns. This suggests that regional traffic flow is related to connectivity of the network and to urban population rather than the whole population. In addition there is a marked regional pattern of residuals with positive residuals in the developed parts of Uganda and negative residuals in the northern Districts where the major inter-District roads are overestimated by the theoretical models.

The wider implications of these results may be viewed within a planning context as the reduction of regional inequalities receives high priority in the various planning departments of Uganda and other countries. The method is tested only in the case of Uganda but the type of data utilised here is available in many developing countries and it should be possible to recognise common similarities and differences in spatial patterns and their explanation. Moreover, the theoretical models used in this analysis may be used to predict patterns of flow, which could prove of great value where data are too sparse to calculate trends. The immediate results for Uganda, however, suggest that the dominance of Kampala and the lakeshore Districts is unlikely to be lessened to any great extent. Present policies emphasise the improvement of communications by upgrading roads and by encouraging economic growth in the existing urban centres. This pattern is unlikely to change the regional inequalities in traffic flow to any great extent since it will only slightly modify the underlying causes of these inequalities—the inaccessibility of northern and parts of western Uganda and the regional differences in the degree of urbanisation.

NOTES

1. Kigezi is an area of mountainous upland in which the roads follow the valleys. This and a high population density have encouraged road development as each road can only serve one valley.
2. These factors of network density are very similar to those discussed for Ghana and Nigeria by Taaffe *et al.* (1963).
3. This limitation is imposed by the following theoretical model based on structure which predicts the proportion of total flow on each link not the actual flow. For the purpose of correlation the actual traffic flows and the subsequent gravity model are expressed in the same way. In the latter case it overcomes the problem of using a constant.
4. All calculations were made on standard programs for an Elliot 4120 computer.
5. For a discussion of mapping residuals see Thomas (1960).
6. The regression equation is $y = -0{\cdot}0049 + 0{\cdot}0120x$ and the standard error of the regression is 0·0311.
7. $y = 0{\cdot}0161 + 0{\cdot}0036x$, standard error 0·0190.
8. $y = -0{\cdot}0061 + 0{\cdot}0092x_1 + 0{\cdot}0032x_2$, standard error 0·0102.

REFERENCES

Berry, B. J. L. (1969) 'Relationships between regional development and the urban system: the case of Chile', *Tijdschrift voor Economische en Sociale Geografie* **60**, 283–307.

Berry, L. (1971) *Tanzania in maps* (London University Press).

Davis, D. H. (1971) *Zambia in maps* (London University Press).

Friedmann, J. (1966) *Regional development policy: a case study of Venezuela* (The Massachusetts Institute of Technology).

Funnell, D. C. (1969) *Nodality and perception* (Occasional paper no. 15; Department of Geography, Makerere University, Kampala).

Isard, W. (1960) *Methods of regional analysis: an introduction to regional science* (The Massachusetts Institute of Technology).

Kansky, K. J. (1963) *Structure of transportation networks* (Research paper no. 84; Department of Geography, University of Chicago).

Riddell, J. B. (1970) *The spatial dynamics of modernisation in Sierra Leone: structure, diffusion and response* (Northwestern University Press).

Smith, J. A. (1970) *Road transport in Uganda* (Occasional paper no. 18; Department of Geography, Makerere University, Kampala).

Soja, E. W. (1968) *The geography of modernisation in Kenya* (Syracuse University Press).

Taaffe, E. J., R. L. Morrill and P. R. Gould (1963) 'Transport expansion in underdeveloped countries: a comparative analysis', *The Geographical Review* **53**, 503–529.

Thomas, E. N. (1960) *Maps of residuals from regression: their characteristics and uses* (Publication No. 2; Department of Geography, State University of Iowa).

Tinkler, K. J. (1971) 'A structural model for regional economic inequality', *Graph theory approaches to regional deveolpment in Uganda*, ed. K. J. Tinkler (Occasional paper no. 20; Department of Geography, Makerere University, Kampala).

Tinkler, K. J. (1972) 'Physical interpretation of eigenfunctions of dichotomous matrices', *Transactions of the Institute of British Geographers* **55**, 17–46.

Uganda Government (1955, 1962, 1971) *Statistical Abstract* (Entebbe: Government Printer).

Williamson, J. G. (1965) 'Regional inequality and the process of national development: a description of the patterns', *Economic Development and Cultural Change* **13**, 3–45.

Chapter 17

River Transport in the Less-Developed Countries

B. J. Turton

Introduction

During the nineteenth and early twentieth centuries the commercial exploitation of many of the less-developed regions of Africa and South America was based almost entirely upon the facilities for inland transport offered by river systems. Within Africa the provision of commercial river transport by European traders played an important part in the initial development of mineral and agricultural resources of the tropical rainforest and savanna regions and several of the leading interior trading towns were first established as river ports. Many of the unnavigable sections of the Zaire and Niger systems were subsequently bypassed by railways and a comprehensive series of integrated river–rail routes linking the productive areas of the interior with the coast gradually evolved. River navigation, supplemented where necessary by railways and later by road services, continued to meet a large proportion of all transport requirements in several west and central African territories until the mid-twentieth century, but in recent years the commercial importance of water transport has declined steadily with the rapid expansion of road haulage.

Today the economic significance of many tropical rivers in Africa is based largely upon their potential for hydroelectricity generation and irrigation so the improvements to inland navigation which large-scale river regulation projects can yield are now often regarded as secondary benefits. While the acquisition of a modern road network and an efficient road haulage industry is clearly an important aim of the new independent states of West Africa the waterway is still the cheapest means of conveying the bulky, low-value agricultural and mineral export commodities which form the basis of the economy of most of these developing countries. The

Dr. Turton is Lecturer in Geography, University of Keele.

potential importance of river transport within the modern communications systems of West Africa and other tropical regions is frequently underestimated (Hodder, 1973) and investment in inland navigation is normally accorded a very low priority in the current development budgets of most African nations. Many economic planners now accept, however, that a continuing heavy investment in road systems in less-developed countries, where river transport exists as an alternative, may not always meet the increasing demands for haulage facilities in the most effective way, and there is a growing realisation of the need for a reappraisal of inland waterway systems (United Nations, 1963).

The claims of these systems for a reassessment of their potential has also been reinforced in recent years by the successful development of new methods of cargo handling, such as containerisation, and the introduction of hydrofoils and hovercraft as alternatives to the traditional displacement vessels of river fleets. The independent West African states of Gambia and Nigeria provide contrasting illustrations of the current status of river transport in developing countries. Commercial navigation on the Niger, both within Nigeria and in the states of Niger, Mali and Guinea, has declined drastically in the last decade and now faces an uncertain future, whereas the Gambia river is still the principal transport artery for groundnut exports and the upstream movement of many imports. The future of the Niger as a commercial routeway has been assessed by several economic planning consultants in recent years and the various recommendations on the future policy to be adopted on river transport by the states within the Niger basin provide a useful basis for surveying the general potential of inland waterways in the developing countries of the tropical world.

THE NIGER WATERWAY SYSTEM

The many problems encountered in maintaining an efficient and economic freight-carrying service on a tropical waterway in the face of strong competition from road and rail haulage interests are clearly illustrated by the Niger system. Physical hazards presented by seasonal variations in water depth, falls, rapids and changing patterns of deposition combine with the economic problems associated with unbalanced traffic patterns and fluctuating crop yields to create difficulties in the organisation of river services which are not usually encountered to the same extent by land-based communications.

Only the lower Niger, from its delta upstream to the Benue confluence at Lokoja (Figure 17.1), offered a direct route for nineteenth century European penetration, navigation through the middle reaches and up to the source in the Guinea highlands being severely restricted by falls and rapids.

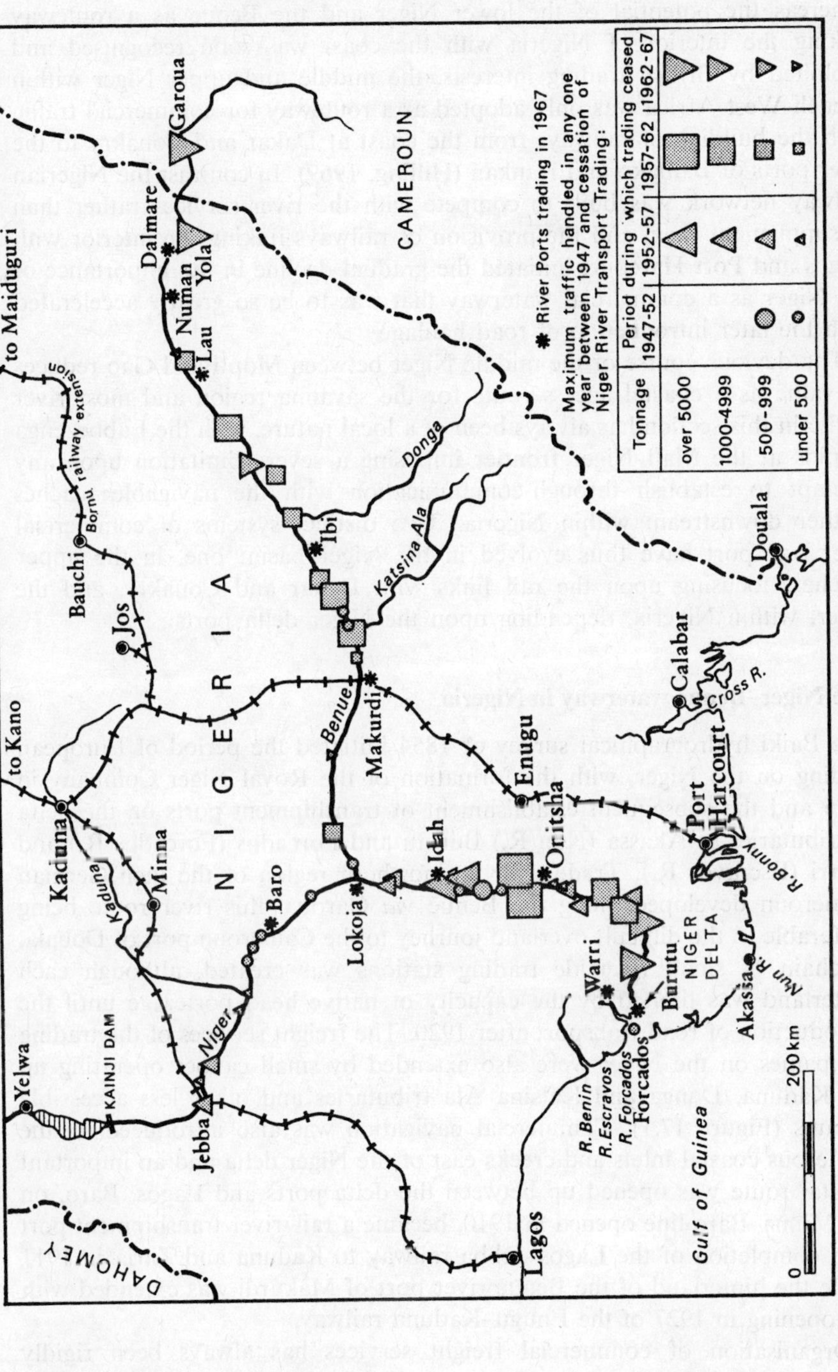

Figure 17.1 The decline of Niger and Benue river trade, 1947–67 (source: Niger River Transport trade records, United Africa Company)

Whereas the potential of the lower Niger and the Benue as a routeway linking the interior of Nigeria with the coast was soon recognised and exploited by British trading interests, the middle and upper Niger within French West Africa was only adopted as a routeway for commercial traffic with the building of railways from the coast at Dakar and Conakry to the river ports of Bamako and Kankan (Hilling, 1969). In contrast the Nigerian railway network was built to compete with the river services rather than to supplement them and the provision of railways linking the interior with Lagos and Port Harcourt initiated the gradual decline in the importance of the Niger as a commercial waterway that was to be so greatly accelerated with the later introduction of road haulage.

The devious course of the middle Niger between Mopti and Gao reduces its value as a coastal access route for the savanna region and most river traffic in this section has always been of a local nature, with the Labbezenga Rapids at the Mali–Niger frontier imposing a severe limitation upon any attempt to establish through-communication with the navigable reaches further downstream within Nigeria. Two distinct systems of commercial river transport have thus evolved in the Niger basin; one, in the upper reaches, focusing upon the rail links with Dakar and Conakry, and the other, within Nigeria, depending upon the Niger delta ports.

The Niger–Benue waterway in Nigeria

The Baiki hydrographical survey of 1854 initiated the period of European trading on the Niger, with the formation of the Royal Niger Company in 1886 and the subsequent establishment of transhipment ports on the delta distributaries at Akassa (Nun R.) Burutu and Forcados (Forcados R.) and Warri (Escravos R.). Trade with the northern region of the then German Cameroun developed along the Benue *via* Garoua, this river route being preferable to the difficult overland journey to the Cameroun port of Douala. A chain of small riverside trading stations was created, although each hinterland was limited by the capacity of native head porterage until the introduction of road transport after 1920. The freight services of the trading companies on the Niger were also extended by small canoes operating up the Kaduna, Donga and Katsina Ala tributaries and other less accessible streams (Figure 17.1). Commercial navigation was also introduced on the numerous coastal inlets and creeks east of the Niger delta and an important coastal route was opened up between the delta ports and Lagos. Baro, on the Minna–Baro line opened in 1910, became a rail–river transhipment port with completion of the Lagos–Jebba railway to Kaduna and Zaria in 1911 while the hinterland of the Benue river port of Makurdi was extended with the opening in 1927 of the Enugu–Kaduna railway.

Organisation of commercial freight services has always been rigidly controlled by the seasonal river régime and the river fleets rarely achieve more than a 30 per cent utilisation factor, with prolonged periods of enforced idleness during low-water periods. High water on the Niger between Lokoja and Gaya, at the Niger state frontier, is maintained from August to late March, the extension into the dry season being due to the delayed contribution of runoff from the upper Niger catchment areas within Mali and Guinea. Navigation by powered commercial vessels is possible throughout this high-water season at Yelwa and for slightly longer periods at Baro and Lokoja. On the Benue, with its more restricted rainfall catchment, the open navigation season is shorter, extending from mid-June to early November at the confluence and decreasing steadily to less than three months at Garoua. Below Lokoja the Niger has a more regular régime, since the effects of the Benue high-water period in June, combined with the sustained high-water levels of the upper Niger between December and May, produce a nine- to ten-month shipping season which below Onitsha becomes all-year navigation (NEDECO, 1960).

Minor rapids, shoals and shifting sandbanks also present hazards to shipping, in the absence of adequate navigational aids, and confine all commercial operations to daylight. The most serious restriction is at the Niger–Benue confluence where a bar prevents full use being made of the Benue navigation. Entry to the latter from the Niger is often delayed, because of insufficient water at the bar, for up to two weeks after the water depths in the Benue itself are adequate for shipping. Progressive silting of the delta distributaries has enforced the abandonment by ocean shipping of the older ports and necessitated expensive channel dredging operations to ensure the survival of the remainder (Udo and Ogundana, 1966). The Escravos channel project, completed in 1964, permits ships of up to 8000 tons to enter Burutu and Warri, which were previously restricted to 2000–3000 ton vessels, but the difficulties of access to the delta ports remain and must be viewed in the light of the much easier approaches to Lagos and Port Harcourt (Ogundana, 1966).

The changing traffic pattern in Nigeria 1945–1970

Prior to the administrative changes of 1969–70, which brought the ports of Warri, Calabar and Burutu under the auspices of the Nigerian Ports Authority, commercial freight services on the Niger–Benue were operated by the United Africa Company (through its subsidiary Niger River Transport (N.R.T.)), Holts' Transport and the Niger–Benue Transport Company. All traffic was allocated on an agreed basis, with the N.R.T. fleet normally carrying about 70 per cent and the other two companies equally sharing the remainder, although at Baro N.R.T. acquired only 50 per cent of the freight transhipped from the railway. Shipping operations were organised to make maximum use of the restricted up-river navigation seasons, with a concentration of carrying capacity on the Benue in summer and the diversion of vessels to the lower Niger and the coastal waterways when the Benue was closed. This deployment ensured the best possible use of the fleet but not of the actual river Niger, which was often deprived of services during its open season in order to concentrate the maximum carrying capacity on the Benue, with its much shorter season. Baro, for example, has a navigable period of about eight months but was usually only served for four (I.B.R.D., 1955). The principal problem encountered by shipping interests on the Benue was the fact that the harvest season of the principal export crops, particularly groundnuts, and the navigation season do not coincide and the fleets were often incapable of despatching all the available cargo downstream during the open period.

The restrictions on the use of the Niger–Benue imposed by physical and economic conditions were reflected in a freight-rate structure in which the differential between rail and river charges was far less than in other regions where the two media compete for traffic. Niger rates had to be high enough to compensate for the long periods during which the fleets were unable to earn revenue and, although similar charges were levied by the major operators, the actual costs varied with the section of waterway concerned, Benue rates being much higher than the average because of the additional navigational problems involved. The road–rail differential was also reduced by the application by the Nigerian Railway Corporation of special reduced rates to the major classes of agricultural freight destined for export. Tariffs of the N.R.T. for 1967–8, the last full trading season of the company, indicate the variations in costs on the Niger and the Benue and may be compared with the appropriate railway tariffs (Table 17.1). The river rates between Baro and the delta ports also reflect the competition offered by the direct railway service to Lagos. All purchasing, collection and transportation of the principal export crops is now controlled by Regional Marketing Boards which, together with the railway undertaking and the Nigerian Ports Authority, are statutory corporations charged with making the most effective use of all transport facilities (Hay, 1970).

Traffic on the Niger–Benue has always been dominated by the downstream movement to the delta ports of groundnuts, cotton seed and lint, benniseed and palm products. At its peak, up-river trade in salt, cement and miscellaneous imported manufactured goods only accounted for 25 per cent of all traffic and this proportion has fallen steadily since the early 1960s. Although summary trade statistics are available for the brief period between the end of the Civil War and state acquisition of the company fleets a more reliable indication of the overall trends in Niger–Benue trade is provided by the detailed traffic returns of N.R.T., the largest single operator, for the period 1947–1967.

Table 17.1 Tariffs of Niger River Transport and the Nigerian Railway Corporation 1967–68

(a) *River Transport* Commodity	Route	Distance (kilometres)	Rate (pence/ton–kilometre)
Groundnuts	Baro – Burutu	634	0·58
Groundnuts	Yola – Burutu	1 365	0·68
Groundnuts	Makurdi – Burutu	776	0·48
Cotton seed	Garoua – Burutu	1 588	0·60
Cotton seed	Numan – Burutu	1 297	0·50
Cotton seed	Lokoja – Burutu	518	0·48
Benniseed	Ibi – Burutu	924	0·53
Palm kernels	Idah – Burutu	416	0·70
(b) *Rail Transport* Commodity	Route	Distance (kilometres)	Rate (pence/ton–kilometre)
Groundnuts	Maiduguri – Lagos	1 785	0·61
Groundnuts	Kano – Pt. Harcourt	1 139	0·67
Groundnuts	Makurdi – Pt. Harcourt	463	0·98
Cotton lint	Maiduguri – Pt. Harcourt	1 441	0·90
Cotton lint	Bukuru – Lagos	1 117	0·93
Cotton lint	Oshogbo – Lagos	293	1·48
Cassiterite	Jos – Lagos	1 118	1·25

Thirty years ago the inland waterways carried an estimated 13 per cent of all freight traffic within Nigeria but by the mid-1950s this share had declined to 9 per cent and today it is negligible (I.B.R.D., 1955). Between 1947 and 1960 the total annual traffic carried by N.R.T. on the Niger–Benue and other waterways often exceeded 180 000 tons but after 1960 a steady contraction of trade began and by 1970 the total had fallen to an estimated 46 000 tons, of which less than 20 000 tons was carried on the Niger–Benue system. The other shippers experienced similar reductions and the total amount of long-distance commercial traffic on the Niger–Benue in 1970 did not exceed 40 000 tons.

This contraction of river trade involved the progressive closure of many of the up-river stations and dramatic declines in the amount of traffic handled at the few ports that survived until the final cessation of services operated by N.R.T. and Holts' Transport. With the exception of Idah, which was served by both the N.R.T. and Holts' fleets, the ports between the delta, Baro and Garoua were associated almost entirely with N.R.T. and the port traffic records of the latter undertaking therefore provide an almost complete picture of the decline of river trade. The distribution of all trading stations handling traffic in the 1947–48 season and the sequence of closures in the subsequent twenty years are summarised in Figure 17.1. At most stations in 1947 up-river traffic represented less than 10 per cent of all tonnage handled and only three ports, Onitsha, Makurdi and Garoua, received more freight from than was despatched to the delta ports. Most of the Niger ports above Lokoja had closed by 1951 and of these only Jebba had handled above 1000 tons annually. Baro survived by virtue of its rail link with the groundnut regions of the north but transhipment traffic here had fallen from prewar levels of over 60 000 tons to 19 000 tons in 1951, with the diversion of export traffic to the direct rail route to Lagos (I.B.R.D., 1955).

Traffic at many of the smaller Benue ports was maintained until the late 1950s, although the total annual trade at each rarely exceeded 2000 tons, but the closure rate then accelerated and most of the surviving ports are now concentrated on the upper reaches between Lau and Garoua. A direct comparison of the traffic handled by river ports in the 1966–67 and 1956–57 seasons also illustrates the dramatic fall in trade (Figure 17.2). In the latter season fifteen up-river stations handled import and export freight with a further eight concerned only with downstream traffic whereas in 1966–67 the comparable totals were two (Baro and Lokoja) and six.

The composition of the down-river freight traffic at individual river stations illustrates the contrast between the groundnut and cotton ports on the Benue in the savanna region and the palm produce centres of the lower Niger (Table 17.2). Lokoja, close to the junction of the savanna and rainforest regions, had a more balanced trade and Baro depended almost entirely upon rail connections for its traffic.

The decline of the Niger–Benue system as a commercial waterway has been due to various factors, some associated with the changing structure of transport within Nigeria and some with the trade policies adopted by the neighbouring independent states of the Cameroun, Chad, Niger and Dahomey. Capture of formerly river-borne freight by road haulage has been the largest single contributory cause although road rates, averaging between 1·5p and 2·5p per ton–kilometre in 1970, are still higher than river tariffs (Table 17.1). In southern Nigeria, however, road transport can now be cheaper than river and rail services for hauls of up to 320 kilometres and this has been largely responsible for the diversion of freight from the ports on the lower Niger to direct road services focusing upon Lagos and Port Harcourt (Ogundana, 1966). In the north also the road transport industry can offer a flexibility of service in both established and new areas of crop production that the river and railway are unable to match. However the negotiated rates for long-distance road haulage of cotton and groundnuts occasionally prove too low to attract adequate carrying capacity during

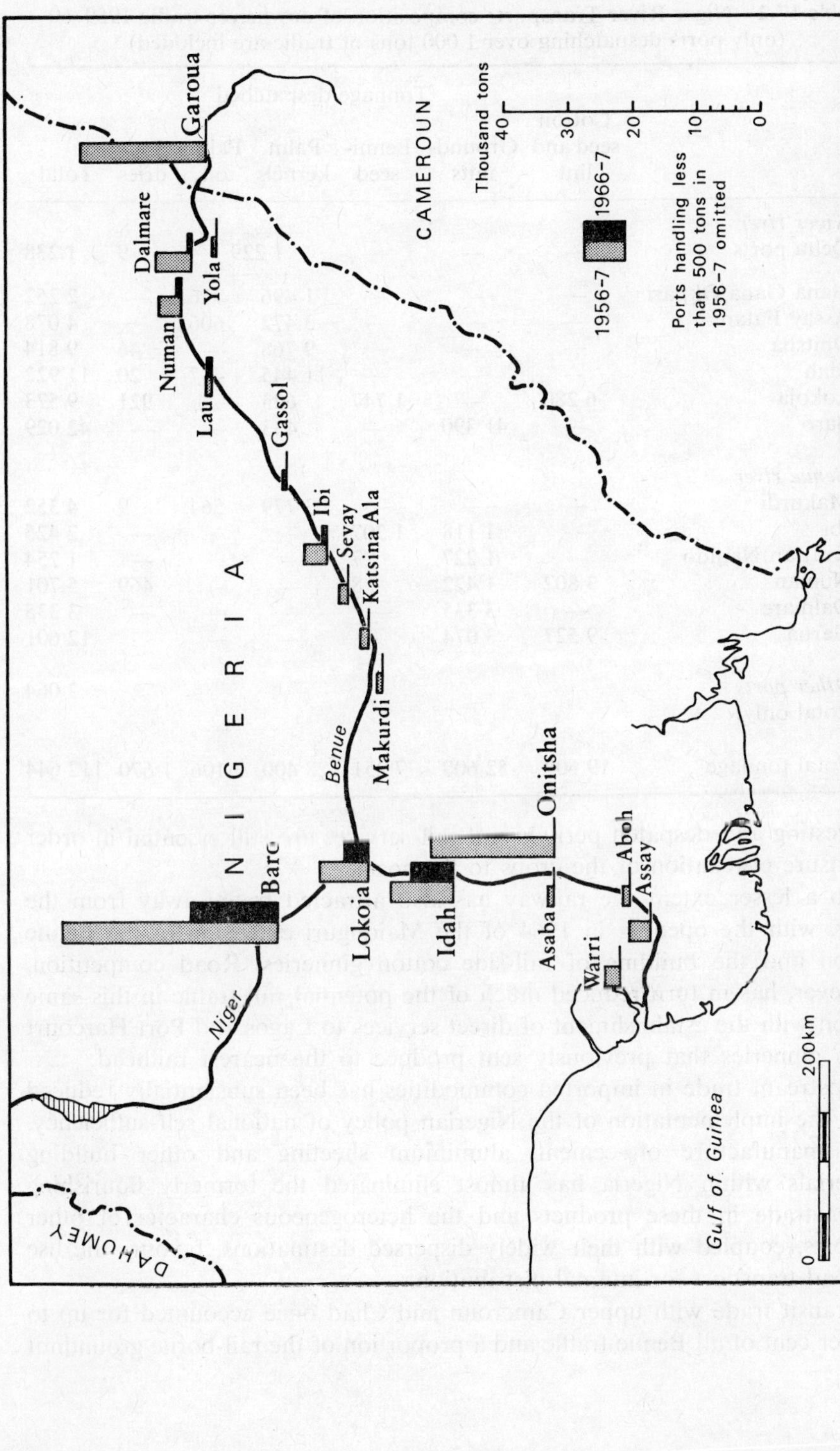

Figure 17.2 Downstream trade of Niger River Transport ports 1956–57 and 1966–67 (source: Niger River Transport trade records, United Africa Company)

Table 17.2 Niger River Transport: composition of downriver traffic 1959–60 (only ports despatching over 1 000 tons of traffic are included)

Port	Tonnage despatched						
	Cotton seed and lint	Ground-nuts	Benni-seed	Palm kernels	Palm oil	Sun-dries	Total
(a) *Niger river*							
Delta ports	—	—	—	1 229		9	1 238
Gana Gana/Okpari	—	—	—	1 496	756	—	2 252
Assay/Fatani	—	—	—	3 472	606	—	4 078
Onitsha	—	—	—	9 768	—	46	9 814
Idah	—	—	—	11 445	457	20	11 922
Lokoja	6 280	—	1 747	623	—	921	9 573
Baro	—	41 390	—	433	—	—	42 029
(b) *Benue river*							
Makurdi	—	—	—	3 779	564	9	4 352
Ibi	—	1 118	1 307	—	—	—	2 425
Krattan Nanido	—	1 227	27	—	—	—	1 254
Numan	3 802	1 422	8	—	—	469	5 701
Dalmare	—	3 335	—	—	—	—	3 335
Garua	9 527	3 074	—	—	—	—	12 601
(c) *Other ports* (total only)							2 064
Total tonnage	19 609	52 608	7 251	28 400	3 106	1 670	112 644

harvesting and despatch periods and rail services are still essential in order to ensure evacuation of the crops to the coast.

To a lesser extent the railway has also attracted traffic away from the river, with the opening in 1964 of the Maiduguri extension in the Benue region and the building of rail-side cotton ginneries. Road competition, however, has in turn reduced much of the potential rail traffic in this same region with the establishment of direct services to Lagos and Port Harcourt from ginneries that previously sent produce to the nearest railhead.

Upstream trade in imported commodities has been substantially reduced with the implementation of the Nigerian policy of national self-sufficiency. The manufacture of cement, aluminium sheeting and other building materials within Nigeria has almost eliminated the formerly flourishing Niger trade in these products and the heterogeneous character of other imports, coupled with their widely dispersed destinations, favours the use of road transport for internal distribution.

Transit trade with upper Cameroun and Chad once accounted for up to 25 per cent of all Benue traffic and a proportion of the rail-borne groundnut and cotton freights exchanged at Baro was also derived in the first instance from the former French colonies now possessing independent status as Niger, Chad and Mali. These states, a part of the French Community, now pursue a policy of using export routes which lie entirely within the Community, with a resultant loss of both rail and river traffic to the Nigerian transport system. Cotton exports from the northern Cameroun are being diverted from the traditional Benue river route to a new road–rail link with Douala, whilst Niger and Mali are abandoning the former export routes through Nigeria in favour of the new roads from Zinder, Maradi and Niamey which provide a direct link, outside Nigerian territory, with the ports of Abidjan (Ivory Coast) and Cotonou (Dahomey). Groundnuts exported from Niger through Cotonou are also offered subsidised rates in order to attract this traffic away from the Nigerian route and an increase in the volume of Niger imports through Cotonou has also contributed to the decline of upstream Niger traffic.

The deficiencies of the delta ports have undoubtedly weakened the effectiveness of the efforts made by Niger shipping interests to resist road and rail competition over the past twenty years. Although cargo handling and storage facilities at Warri and Burutu have been improved, the restricted delta approaches and the superior facilities offered by Lagos and Port Harcourt make the river ports unattractive to ocean shippers serving Nigeria. Down-river traffic arriving at Warri and Burutu could fluctuate from forecast amounts by up to 30 per cent, according to the navigational difficulties encountered and the success of the fleets in attracting freight, whereas Lagos and Port Harcourt could guarantee cargo at any time, generally without the delays and congestion created at the delta ports at the height of the shipping season. Vessels that continued to patronise the delta ports could normally only accept a part-load in order to negotiate the shallow channels, an unpopular policy with shipowners, and only Burutu had a bulk palm oil plant, most of the processing facilities being concentrated at Lagos and Port Harcourt.

RIVER TRANSPORT IN GAMBIA

Apart from the Niger only two other rivers within West Africa have acquired any importance as commercial waterways. The Senegal was adopted as an export route by the French but its value was initially restricted by the silting encountered in the estuary at St. Louis and its full potential was only realised with the opening of the port of Dakar and the Dakar–St. Louis railway, followed by the completion of a direct line from Dakar to the upper Senegal valley near Kayes. The Gambia headwaters are close to those of the Niger in the Guinea highlands, but the lower half

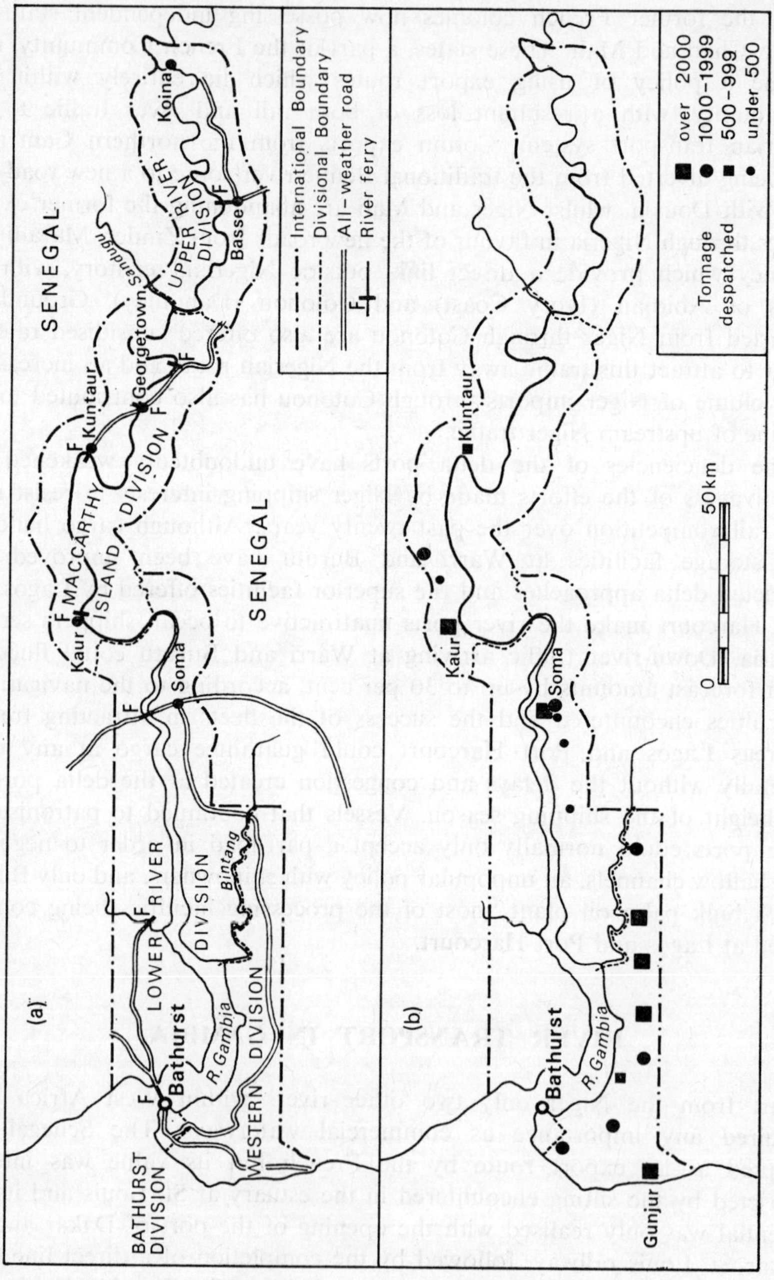
(a)
SENEGAL
SENEGAL
BATHURST DIVISION
Bathurst
LOWER RIVER DIVISION
WESTERN DIVISION
R. Gambia
Bintang
Soma
Kaur
MACCARTHY ISLAND DIVISION
Kuntaur
Georgetown
Sandugu
UPPER RIVER DIVISION
Basse
Koina
International Boundary
Divisional Boundary
All-weather road
River ferry
(b)
Bathurst
R. Gambia
Gunjur
Soma
Kaur
Kuntaur
0
50km
Tonnage despatched
over 2000
1000–1999
500–999
under 500

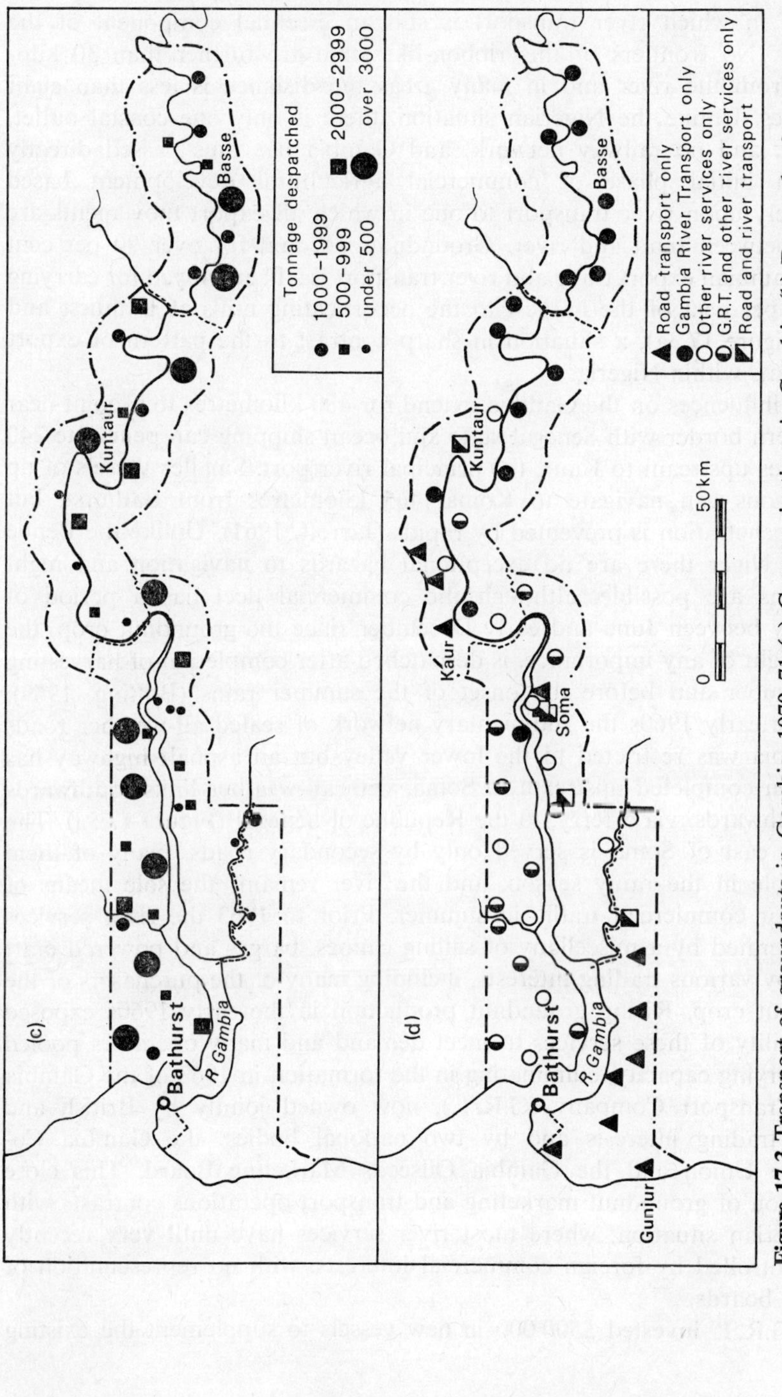

Figure 17.3 Transport of goundnuts in Gambia, 1970–71 (a) Communications in Gambia (b) Despatch of groundnuts by road from buying stations 1970–71 (c) Despatch of goundnuts by river from buying stations 1970–71 (d) Summary map: road and river transport of groundnuts 1970–71 (source: Gambia River Transport and Gambia Produce Marketing Board trade records, United Africa Company)

of its course lies in Gambia, the smallest independent state in West Africa and one in which river transport is still an essential component of the economy. No frontiers of this ribbon-like state are further than 30 kilometres from the river and in many areas the distance is less than eight kilometres. Unlike the Nigerian situation, there is only one coastal outlet, Bathurst, and no railway network, and Gambia has thus passed directly from an initial phase of commercial agricultural development based exclusively upon river transport to one in which all export movements are shared between road and river. Groundnuts account for over 90 per cent of the Gambian export trade and river transport is still employed for carrying over 60 per cent of the harvest to the decorticating mills at Bathurst and Kaur (Figure 17.3a), a situation in sharp contrast to the pattern of export movements within Nigeria.

Tidal influences on the Gambia extend for 480 kilometres to a point near the eastern border with Senegal state and ocean shipping can penetrate 240 kilometres upstream to Kaur, the principal river port. Smaller vessels of up to 200 tons can navigate to Koina, 465 kilometres from Bathurst, but further penetration is prevented by rapids (Jarrett, 1961). Unlike the Benue and the Niger there are no exceptional hazards to navigation and night operations are possible, although the commercial fleet has a period of inactivity between June and early December since the groundnut crop, the only freight of any importance, is despatched after completion of harvesting in December and before the onset of the summer rains (Botting, 1969). Until the early 1960s the rudimentary network of sealed all-weather roads in Gambia was restricted to the lower valley but an asphalt highway has now been completed upstream to Soma, with all-weather links southwards and northwards, via a ferry, to the Republic of Senegal (Figure 17.3a). The territory east of Soma is served only by secondary roads, many of them impassable in the rainy season, and the river remains the sole means of access for commercial traffic in summer. Prior to 1963 the river services were operated by a miscellany of sailing canoes, barges and powered craft owned by various trading interests, including many of the purchasers of the groundnut crop. Rising groundnut production in the early 1960s exposed the inability of these services to meet demand and many operators pooled their carrying capacity, culminating in the formation in 1963 of the Gambia River Transport Company (G.R.T.), now owned jointly by British and French trading interests and by two national bodies, the Gambia Co-operative Union and the Gambia Oilseeds Marketing Board. This close association of groundnut marketing and transport operations contrasts with the Nigerian situation, where most river services have until very recently been controlled by foreign commercial interests, with no representation of produce boards.

The G.R.T. invested £500 000 in new vessels to supplement the existing fleet of powered and sailing craft and the total carrying capacity is now 5470 tons. Freight rates were increased by 30 per cent after a series of deficits but it is difficult to relate rates directly to operating costs since the former are established on a zonal basis by the Oilseeds Marketing Board. Current rates vary from 3·9p per ton–kilometre on hauls of 80 kilometres to 1·9p per ton–kilometre over distances exceeding 200 kilometres. Groundnut exports have increased from 85 000 tons in the 1961–62 season to an annual average of 118 000 tons over the 1966–71 period, with rising output from the lower valley but reductions in the upper, especially the remote Upper River Division (Table 17.3). About 60 per cent of the crop is decorticated at Bathurst, most of the remainder being assembled and processed at Kaur, from whence it is shipped downstream. In recent years the G.R.T. fleet has carried about 70 per cent of all river-borne nuts whilst road haulage has attracted up to 40 per cent of total traffic. Within the Bathurst and Western Divisions, however, where the maximum haul between buying station and Bathurst does not exceed 130 kilometres, over 90 per cent of the crop is now assembled at the mills by road and the small quantities still conveyed by river are handled by minor operators from buying stations on the Bintang tributary (Figure 17.3b and c). The opening of the new route to Soma has now allowed road transport to penetrate into a former preserve of river carriers, and trucks are also used for the assembly of limited quantities of groundnuts at Kaur, but the bulk of the road-borne crop is still derived from buying stations much nearer Bathurst. Groundnut transport in the middle and upper reaches is still dominated by the river and upstream of Kuntaur no road services are employed. Many of these upper river stations despatch over 3000 tons of nuts annually whereas the average tonnage derived from the road-served centres downstream is normally much lower.

INLAND WATERWAYS AND TRANSPORT PLANNING

The problems involved in the planning of transport improvements in the less-developed countries were explored by the United Nations in 1963 (United Nations, 1963) and a subsequent report focused attention on the particular merits and disadvantages of inland waterway transport (United Nations, 1970). Although transport planners do not dispute the advantages offered by water transport for the carriage of low-value bulk commodities there are conflicting views on the extent to which these advantages should be exploited within overall plans for the development of communications. A comparatively modest investment in river services, effectively co-ordinated with road, rail and maritime transport facilities, can often make a valuable contribution to economic advancement in many regions where suitable

Table 17.3 Gambia groundnuts transport

Division	Tonnage 1968–69				Tonnage 1970–71			
	Road	Gambia River Transport	Other river transport	Total	Road	Gambia River Transport	Other river transport	Total
Bathurst and Western	18 684	—	—	18 684	21 729	241	2 124	24 094
Lower River	15 930	20 758	15 632	52 320	4 915	25 563	17 201	47 679
McCarthy Island	11 263	19 339	3 992	34 594	4 907	19 909	2 751	27 567
Upper River	—	15 556	1 294	16 850	—	13 922	—	13 922
Total tonnage	45 877	55 653	20 918	122 448	31 551	59 635	22 079	113 262

water routes are available. However it can be misleading to champion the cause of water transport on the basis of low operating costs alone since the overall costs of improving the associated navigational and cargo-handling facilities may outweigh this advantage and land-based transport might ultimately prove more acceptable (United Nations, 1970).

The adaptability and flexibility of road haulage is particularly suited to the short-term demands for transport in areas of expanding agriculture although planners now recognise the dangers of overestimating the potential of the truck at the expense of established rail and river systems. Pre-investment studies now seek to identify the most efficient mode of transport to meet the demands of the developing states and subsequently to select the most appropriate balance between investment in existing facilities and expenditure on additional roads, railways or waterways (Reid, 1969).

Many tropical rivers would require large-scale channel improvements and flow regulation schemes in order to increase their carrying capacity and the heavy expenditure involved is seldom justified in terms of the additional traffic that would be generated. New areas of agricultural and mineral exploitation are often remote from existing waterways and are thus usually served by road transport or by extensions of railway networks. The use in Venezuela of the lower Orinoco for upstream traffic in coke to the Ciudad Bolivar iron and steel plant and for downstream exports of iron ore provides a rare example of a tropical river playing an integral part in a new industrial development programme and there are few opportunities for similar applications in other regions.

Many attempts have been made to evaluate the future role of the Niger navigation within the transport network of West Africa. The dismemberment of French West Africa and the establishment of the existing pattern of independent states created an urgent need for remodelling the system of communications built up in the colonial era in order to meet the requirements of the new nations, particularly the land locked states of Mali, Upper Volta and Niger (Hilling, 1968). Recommendations for the formation of a Niger river committee with representatives from all states within its basin had been made in 1955 (I.B.R.D., 1955) and the subsequent emergence of the landlocked states focused attention upon the Niger as a potential international routeway providing coastal access for all these new nations. Construction of the Kainji Dam was first recommended by NEDECO as part of an ambitious plan for river regulation and power generation within Nigeria and for the extension of navigation upstream to the Niger state port of Niamey, 1448 kilometres from the delta (NEDECO, 1960). The dam was completed in 1968 and provided with locks and canals negotiable by 2000-ton barges; it has a generating capacity of 320 megawatts and has improved flood control downstream of Jebba to the Benue confluence (Wagland, 1969). Similar dams have been projected for the Kaduna river at Shiroro Gorge, for the Niger at Jebba and for the Benue at Yola and in the Cameroun headwaters, with the object of lengthening the effective navigation season, although reservations have been made on the economic justification for these plans in terms of the additional traffic that would be carried (United Nations, 1962; Ledger, 1963). In the delta the Escravos dredging scheme, already discussed, was completed in 1964 and it was forecast that the navigational improvements derived from this project and from the Kainji Dam could produce a 10 per cent increase in the efficiency of the river fleets by 1972, with a consequent reduction in freight rates (Stanford, 1961).

The failure of these ambitious projects to achieve their aim of improving the efficiency and scope of river haulage has been largely due to the continued capture of the former water-borne traffic by road transport and also to the non-realisation of hopes for agricultural expansion in areas accessible to the river ports. That part in particular of the Nigerian 'middle belt' traversed by the Benue has been regarded as having great potential for irrigated commercial agriculture and thus for yielding a new source of river traffic (I.B.R.D., 1955; Walker, 1959). The optimistic forecasts that 'commercial traffic may well double over the next decade' (Ministry of Trade and Industry, 1963, p. 227) and that 'the Niger and the Benue will be restored to their former place among the main channels of Nigerian commerce' (Walker, 1959, p. 48) have not materialised and the riverine areas of projected development remain largely unchanged. Completion of the overland trans-Cameroun route from Douala to Garoua could eventually eliminate another important source of river trade although it is possible that only the early Cameroun cotton harvest, on whose quality prices are agreed and for which quick despatch to market is essential, will be exported by the faster overland route, leaving the bulk of the crop to be despatched down the Benue at much lower freight rates. It is very doubtful, however, whether even an improved river service could regain any of the lost transit trade with Chad as the planned new road links with the Cameroun and with the Nigerian railhead at Maiduguri will provide direct all-season export routes.

Although the traditional up-river trade in imports has disappeared with the West African policy of self-sufficiency and the use of road haulage it is possible that any large-scale industrial developments along the Niger could best be served by waterway. A proportion of the requirements of the Jebba paper mill (opened in 1968) for imported pulp could be met by Niger transport although Nigerian timber resources will eventually be able to satisfy at least 50 per cent of the demand. Similarly the new cement plant at Garoua will require an estimated 6000 tons of gypsum and fuel oil annually, and both these commodities could be conveyed up-river from a delta port more economically than by the overland route through Douala.

Current railway policy in Nigeria is directed towards maintaining traffic levels in the face of strong competition from road transport, and the retention and rehabilitation of the Minna–Baro branch is essential if traffic on the Niger above Lokoja is to be revived, as transhipment of rail-borne freight at Baro has accounted for up to 40 per cent of all down-river trade. One estimate that annual traffic on this rail link would increase to over 375 000 tons by 1972 has proved widely inaccurate (Stanford, 1961) and, unless the existing branch is modernised or alternative exchange facilities provided at Jebba, this source of river traffic is unlikely to regain its former importance.

The priorities for investment in transport outlined in the Second National Development Plan for Nigeria provide a clear indication of the current official attitude towards the Niger–Benue waterway (Federal Republic of Nigeria, 1970). Despite a total investment of over £12 million in the Kainji Dam and secondary navigational improvements the Niger waterway has not attracted the anticipated traffic and no further large-scale projects are envisaged until another feasibility study of river potential has been commissioned. The £5·7 million allocated for routine maintenance and rehabilitation of waterway facilities represents less than 2·5 per cent of all investment in the transport sector, whereas roads and railways receive 78 per cent during the 1970–74 period of the Plan. Warri, the former headquarters of the Holts' river fleet, is to be rebuilt as a new ocean terminal, but its traditional river-served hinterland is to be replaced by one based on new trunk roads linking the port with Lokoja and Kaduna. Additional berths and new cargo-handling equipment will also be installed and this investment at Warri will accelerate a trend towards the use of road in preference to river services which was already evident in the period before state acquisition when about 50 per cent of exports and 25 per cent of imports were carried by truck.

CONCLUSION

The decline of commercial traffic on the Niger, once a leading transport artery of West Africa, indicates that inland navigation in its present form is very unlikely to retain any importance in the future communication systems of developing nations. River transport however has experienced few technological improvements in recent years beyond the substitution of the traditional towed-barge operation by the more efficient 'push–tow' system; the introduction of new methods of vessel propulsion and freight-handling could provide a basis for a revival of interest in commercial services (United Africa Company, 1961). The introduction of containers on the Niger could stimulate the use of the river for import distribution, although the installation of handling equipment at all the ports would be necessary. In 1961 the NEDECO report concluded that mechanical handling would only be justified where daily traffic totals were over 1000 tons but, up to the late 1960s, the total annual trade at each port rarely exceeded 2000 tons and manual transhipment was adequate in most centres (NEDECO, 1960).

Conventional displacement vessels are inevitably limited in their operations by the régime of the river; their replacement by craft designed to operate at all times irrespective of the water depth could provide a year-round service and eliminate the many disadvantages attached to seasonal navigation. The amphibious air-cushion vehicle, or hovercraft, offers this opportunity and the potential of this type of transport has been noted in many recent surveys of West African communications (Stanford, 1961; United Nations, 1963; United Nations, 1968). Amphibious hovercraft have been operated successfully on tropical rivers and, although their commercial application to date has been confined to passenger and car ferry services, a cargo vessel could be developed. Carriage costs would initially be higher than those of existing river and rail services but the establishment of all-year river services with journey times much shorter than those of conventional vessels could attract trade back to the river and ultimately lead to lower freight costs. If, in the case of West Africa, a hovercraft could successfully navigate both the Niger–Benue and the coastal waterway from the delta to Lagos then the river could once more assume a useful role in the transport of bulk commodities upon which so many of the Niger basin states are dependent. The initial costs of introducing these craft into developing countries would be high but no further investments in conventional river improvement would be necessary and inland navigation could once more make an effective contribution to the transport system.

REFERENCES

Botting, D. (1969) 'Dilemma of Africa's smallest nation', *Geographical Magazine* **41**, 666–669.

Federal Republic of Nigeria (1970) *Second national development plan 1970–74* (Lagos: Federal Ministry of Information).

Hay, A. (1970) 'Crop boards and transport policy in Nigeria', *Journal of Transport Economics and Policy* **4**, 173–177.

Hilling, D. (1968) 'Politics and transportation—the problems of West Africa's land-locked states', *Essays in political geography*, ed. C. A. Fisher (London: Methuen), 253–269.

Hilling, D. (1969) 'The evolution of the major ports of West Africa', *Geographical Journal* **135**, 368–369.

Hodder, B. W. (1973) *Economic development in the tropics* (2nd. ed., London: Methuen).

I.B.R.D. (International Bank for Reconstruction and Development) (1955) *Economic development of Nigeria* (Baltimore: Johns Hopkins).

Jarrett, H. R. (1961) *A geography of Sierra Leone and Gambia* (London: Longmans).
Ledger, D. C. (1963) 'The Niger dams project of Nigeria', *Tijdschrift voor Economische en Sociale Geografie* **54**, 243–246.
Ministry of Trade and Industry (1963) *The industrial potentialities of northern Nigeria* (Kaduna: Ministry of Information).
NEDECO (Netherlands Engineering Consultants) (1960) *The improvement of navigation on the rivers Niger and Benue* (The Hague: NEDECO).
Ogundana, B. (1966) 'Changing port outlets of Nigeria's export commodities', *Nigerian Geographical Journal* **9**, 39–40.
Reid, F. D. T. (1969) 'International highway financing', *International Road Federation African Highway Conference Report* (Washington: IRF).
Stanford Research Institute (1961) *The economic coordination of transport development in Nigeria* (Menlo Park, California: SRI).
Udo, R. K. and B. Ogundana (1966) 'Factors influencing the fortunes of ports in the Niger delta', *Scottish Geographical Magazine* **82**, 169–183.
United Africa Company (1961) 'Importance of transport to tropical Africa', *United Africa Company Statistical and Economic Review* **25**, 6–17.
United Nations Economic Commission for Africa (1962) *Transport problems in relation to economic development in West Africa* (New York: UN).
United Nations Economic and Social Council (1963) *Conference on the application of science and technology for the benefit of less-developed areas* (New York: UN), Vol. 4.
United Nations (1970) *Leningrad symposium on the technico-economic, organisational and administrative aspects of inland waterborne transport, 1968* (New York: UN).
Wagland, P. J. (1969) 'Kainji and the Niger dams project', *Geography* **54**, 459.
Walker, G. (1959) *Traffic and transport in Nigeria* (Colonial Research Studies Series No. 27; London: H.M.S.O.).

ACKNOWLEDGEMENT

The author is deeply indebted to Mr. P. J. Gregor Macgregor, of the United Africa Company, London, for the provision of a wide range of statistical data and other information relating to the operation of UAC river services in Nigeria and Gambia and for unstinted help and guidance during the preparation of this essay.

Chapter 18

The Geography of Development Policy: Reflections on Malaysia

James Osborn

Introduction

While geographers have been pursuing topics concerning areal patterns of human activity and the relationships of man to his environment and the landscape generally, they have failed to analyse systematically the geographical impact and dynamics of one major force in the world today: the *policies* of complex human organizations, most importantly, *governments.* As societies have become more complicated, with higher technologies and greater capacities to affect reality, their governments have assumed greater influence in the world. Government policy is the organised set of decisions and actions of a government. What is the relationship of government policy to *area* (the areal distribution of phenomena and the interaction of phenomena in units of area)? Specifically, what is the *impact* of policy on areal patterns and relationships and the landscape? What is the reciprocal impact of area on policy formulation and implementation? What are the *areal dynamics* of policy action and the relationship of policy actions to other dynamic areal systems, for instance mass or individual behaviour? These question should be answered in a *Geography of Policy*, which is a field of political geography. One kind of policy may be taken as an example. National development is the major problem and preoccupation of almost all countries in the world today; and central government development policy is the major device by which development is (or is hoped to be) induced and channelled in most of these countries. By examining the geography of development policy it is possible both to offer answers to questions like those above, in particular cases, and to draw practical lessons to inform future policy-making. Hence this chapter sets out

---

Dr. Osborn is Senior Research Assistant in the Center for the Study of Democratic Institutions, Santa Barbara, California.

(1) to outline the form and method of the geography of development policy; (2) to present the geography and lessons of Malaysia's experience, giving special attention to the role of cities.

THE GEOGRAPHY OF DEVELOPMENT POLICY

The impact and dynamics of area/development policy relationships may be cast into two sets. The first consists of explicit and implicit locational decisions of a government. These include concern for areal scale of management, as well as pointal, lineal and areal location of activity, and are based on perceptions of areal pattern and its significance in a country. These decisions may be called *area in policy.* Second, there are the actual policy actions and interactions within and among areas, points and lines. They comprise *policy in area.*

Three areal concepts are separable when area in policy is considered. First, there is areal form. The areal forms which development takes may, so far as the operators are concerned, be *points and lines* (as for simple investments), *regions* (as for river basin projects, or semi-autonomous contained political units such as the Borneo States of Malaysia) or not necessarily continuous *areal masses* (for example 'rural areas', or the *loci* of particular racial or economic distress). All have their own special dynamics and prerequisites in the development process.

The second and related areal concept which, with the first, comprises the conceptual area filter of development policy, is that evolved according to the determining basis of the locational decision. Three categories of development expenditure, which are exhaustive and which correspond roughly to the social, local economy and national economy sectors in development policy are: (1) *Population Coverage* (POPC), where the intention is to distribute development fairly within the population and the development activities are, therefore, theoretically concentrated where people concentrate; (2) *Problem Area attention* (PROB), whereby policy may be applied to a location because it supposedly has a problem peculiar to it (such as new land development in a rural mass because there is too much rural poverty or urban housing in certain cities when housing in general is not subsidised for the public); and (3) *Economic-functional* activity (ECFN), the location of which should be determined solely by the nature and specific requirements of each act or investment (e.g. for dams, ports, government buildings).

The third areal concept notable in development is that of scale of areal units. Through this administrative area filter decisions are made as to how large units are to be, at what levels in administrative hierarchies things should work best and what can happen, because of the internal nature and external relations of units of areal organisation (regions, cities, city-regions, etc.), that is functional and dysfunctional in development.

In looking at *policy in area*, i.e. the interaction or impact of national development policy with or on the contents and structure of areas, the analysis of Malaysia will concentrate on middle cities (between 25 000 and 400 000 population). A model of their *Development Policy System* is presented in the form of a flow chart (Figure 18.1). This model is conceived for the purpose of identifying systematically the elements important to policy in area interaction and, as far as possible, their interrelationships. The system contains the elements in a vertical array whereby development policy, and the money, power, and information that are its impulses, are moved through institutions to and through an areal unit of organisation: a middle city. These development policy impulses may have been aimed at points, lines, this area, or a conceptual area or mass of which the particular area is only a part. They are, in any case, then seen to interact with institutions and subsystems of the areal unit and with environments of the local area system that likewise are relevant to centrally defined development.

The method of analysis is twofold: (1) macro-analysis, stressing historical changes in national patterns of development policy expenditure and expectations and the moulding of the landscape by the government in association with private development; and (2) comparative micro-analysis of middle cities. For the macro-analysis, three logical time-periods of development policy were taken: 1950–61, 1961–65 (the Second Malaya Plan) and 1966–68 (the first three, and only researchable, years of the First Malaysia Plan). All areally definable development expenditures were mapped by district and city (Figures 18.2 and 18.3). Appropriate Pearson Product Moment correlations and cluster analyses were made; and orthogonally rotated principal components factor analyses over districts and cities isolated independent patterns among the development expenditure and other variables (see Table 18.1).

THE GEOGRAPHY OF DEVELOPMENT POLICY IN MALAYSIA

Area in development policy

Traditionally, the purpose of British advice to and management of the Malaysian areas was to facilitate private British commercial expansion and exploitation there and to hold these countries as a British Far Eastern trading base. General policy sought to maintain peace and adequate contentment in the traditional native sectors of politics, culture and subsistence agriculture, while allowing Chinese and Indian migrants to work the

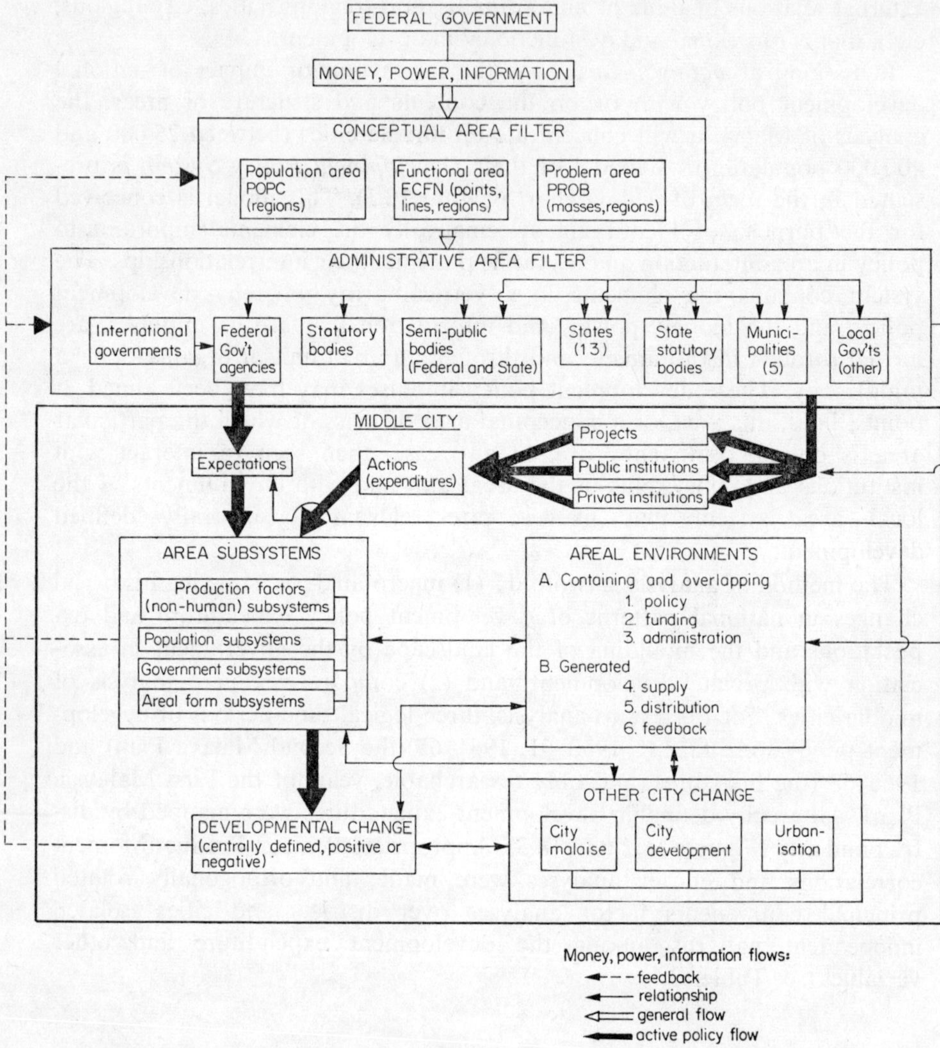

Figure 18.1 Development policy system of a Malaysian middle city

plantation and mining export sectors and populate the cities and their modernising functions. The British employed a development policy which emphasised transport that served the export sectors and enabled the complex administrative system to reach local levels. With these came rudimentary education, health, post and banking and agricultural extension services spreading from the cities and estate areas.

After World War 2 reconstruction proceeded apace and Malaya became

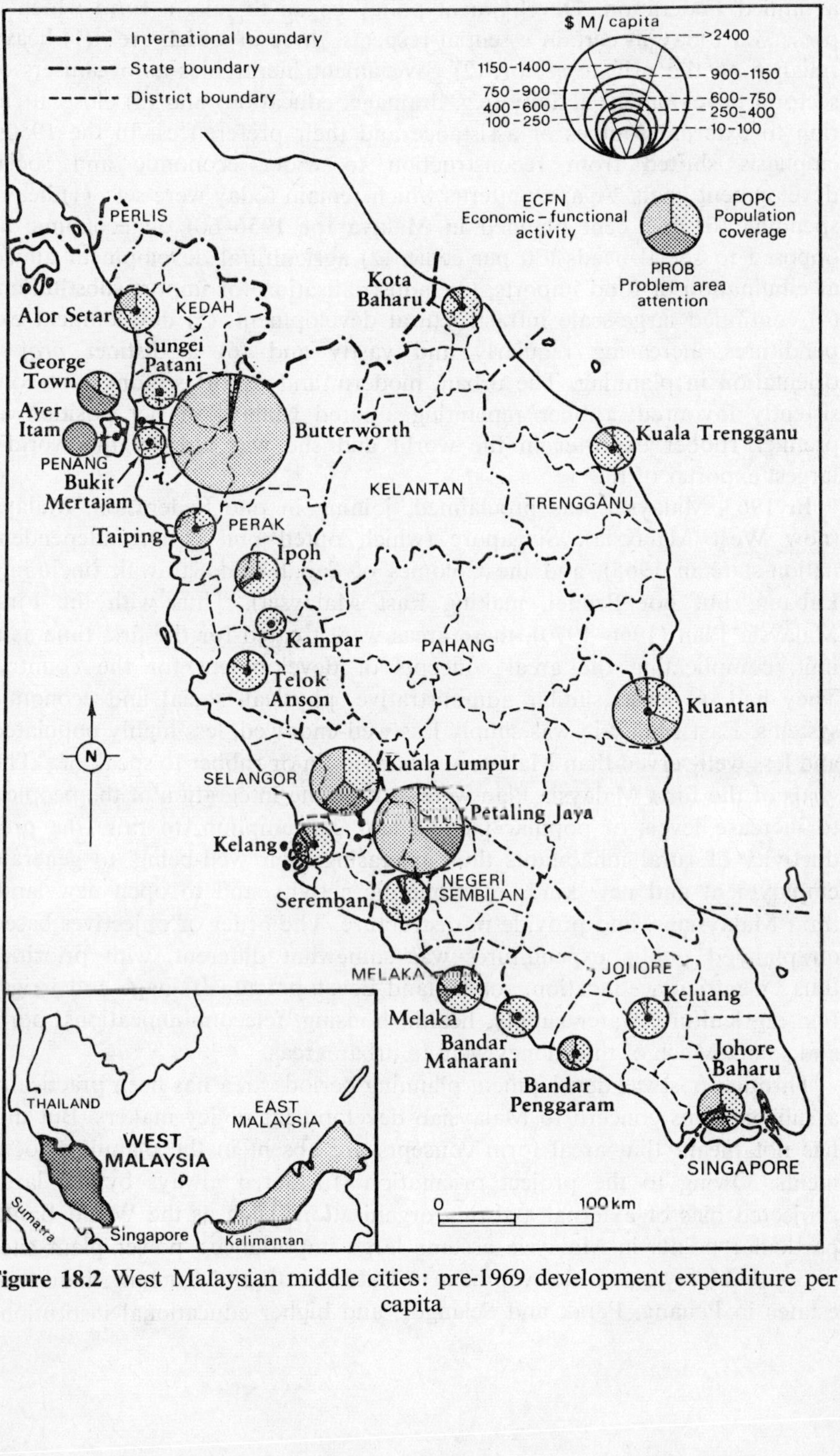

Figure 18.2 West Malaysian middle cities: pre-1969 development expenditure per capita

a unified Federation. Development policy began to take a form which is preserved today in certain essential respects, three of which are: (1) heavy reliance on the private sector, (2) government hierarchies as organisers of sectoral development activity (e.g. drainage, education) and (3) close attention to external sources of assistance and their preferences. In the 1950s, emphasis shifted from reconstruction to wider economic and social development goals. Again, patterns which remain today were set: (1) heavy spending (61 per cent planned in Malaya for 1956–60) on economic as opposed to social needs (30 per cent), (2) agricultural development aimed at elimination of food imports, (3) industrialisation for import substitution, (4) continued large-scale infrastructural development, (5) development expenditures increasing regularly and vastly and (6) a distinct *project* orientation in planning. The urban, modern, and export sectors were consistently favoured; rubber replanting assured Malaya of her position as premier rubber exporter in the world and she was already the world's largest exporter of tin.

In 1963 Malaysia was proclaimed, joining in one Federation, Malaya (now West Malaysia), Singapore (which opted out as an independent nation-state in 1965), and the Colonies of Sabah and Sarawak (including Labaun, but not Brunei, making East Malaysia). Thus with the First Malaysia Plan (1966–1970), these areas were treated for the first time as a unit, complicating the areal concepts of development for the country. They had, at least, similar administrative, physical, social and economic systems. East Malaysia was simply less well-endowed, less highly populated and less well-served than Malaya and had no tin or rubber to speak of. The goals of the First Malaysia Plan were to promote integration of the peoples, to increase levels of popular income and consumption, to raise the productivity of rural inhabitants thus increasing their well-being, to generate employment and new kinds of economic activity and to open new land, train Malaysians, and provide infrastructure. The order of objectives based on planned public expenditures was somewhat different, with priorities thus: electricity, education, roads, land development, drainage and irrigation, agriculture, water supply, health, housing, telecommunications, ports and so on. Much of this money went to urban areas.

Through postwar development planning periods area has been practically a subconscious concern to Malaysian development policy makers. But this has not meant that areal form concepts are absent in the planning documents. Owing to the project orientation (furthered always by the large projected bias of external assisting organisations such as the World Bank) particular points in Malaysia assume large importance: power plant sites in South Malaya, port delevopment in Butterworth and Kuching, industrial estates in Penang, Perak and Selangor, and higher educational institutions in Kuala Lumpur. Likewise electricity distribution, air corridor and telecommunications lines have been specified.

Since it is a large region East Malaysia has been treated somewhat separately as an area with special development needs and one towards which the central government has a special historical-political responsibility. Significantly, the east coast of Malaya, which is as depressed as and has less potential than Sabah, has hardly been mentioned in planning. Real regional concern, not described as such however, has been apparent as directed at particular rural development schemes. Here, for the Muda (Kedah and Perlis) and Kemubu (Kelantan) rice double-cropping water control projects, attention is lavished. Similarly, large new land development projects, like the Jengka Triangle (in Pahang) are conceived as regional in the sense of encompassing large areas and seeking to impose new, functional organisations upon traditional agricultural systems. But this has been in isolation from the larger state–city system and even from the circulation systems into which they have been placed.

In areal form the First Malaysia Plan, like development documents it succeeded, did not seek to organise Malaysia as a whole. It identified and concentrated on basically isolated and functionally limited and specific points, lines and small regions via projects. It was vague on the larger conceptual matters of where to put complex programmatic efforts, such as those for rectifying economic imbalances among groups. Locations for projects were most easily defined when purely economic needs could determine them, with no argument from social planners. Such expenditure (ECFN) constituted 61 per cent, 53 per cent and 49 per cent respectively of the development expenditure in the periods 1950–60, 1961–65 and 1966–68. Expenditure for provision of services to people (POPC) were fairly stable at 27 per cent, 35 per cent and 31 per cent for the three periods. The high point during the Second Malaysia Plan indicates the change at that time away from economic infrastructure but not yet fully into large housing projects and land development schemes. Problem area expenditure (PROB) showed the greatest change over the years: from 11 per cent and 12 per cent of total expenditure in the first two periods to 21 per cent of expenditure in 1966–68.

Thus the First Malaysia Plan period emerges as a time when greater thought and complexity of decision-making were beginning to be required in the location of development activity because more discretion was possible and the decisions were definitely more political. This gradual change in the balance of developmental goals brought a corresponding adjustment in the areal concepts employed. With increasing concern for complex social problems and programmes, especially into the Second Malaysia Plan period (1971–75), pointal and lineal concern has given way to regional and mass concern.

In the context of development administration scale of areal unit of organisation has been a conscious consideration in Malaysian development policy. There are two systems: the hierarchical civil administrative system, in which states are the key units and cities operate under their authority; and the group of autonomous and semi-autonomous special-functional commissions, agencies and boards. Both of these sets are responsible ultimately to the federal government, but they have different areal origins, purposes and impacts.

In the civil administrative system thirteen states accumulate certain revenues (varying widely in amount with the Bornean states generating by far the most) which they spend on development in consultation with national planners. The state government Public Works, Health, Education, Drainage and Irrigation and other departments concerned with development are all staffed by federal service officers who are also the agents of their respective federal ministries. Hence, for line agency activity, there is a large degree of coordination and unity between state and federal goals and processes. The Malaysian system has adapted to the special problems of East Malaysia by leaving in the hands of Sabah and Sarawak much greater autonomy of action and assigned revenue-generating powers. Thus Sabah and Sarawak operate their own land development schemes and health and primary education systems in comparative isolation from federal policies. They have a certain amount of control over immigration into their states and have greater bargaining power in national development policy-making circles.

Local government is assigned to the states. The hierarchy of local government has been remarkable for the inability of most of its constituent units to remain financially viable during the 1960s. At higher levels there has been a tendency for takeover by the state government because of malfeasance or insolvency. The five Municipalities (Kuala Lumpur, George Town, Ipoh, Melaka and Kuching) have by far the most powerful, wealthy and dynamic local governments. Because of their size and organisation they are able to give more to their citizens, including low-cost housing, job training and medical services.

In Malaya, local government is not continuous across the area, nor have government units been standardised or rationalised. Thus great unevenness of local developmental responsibility and resources prevails—made worse by antique assessment rolls and often timid local governments. In the Bornean States, by contrast, Local Councils cover the entire area and District Officers are more important than in Malaya. Whether this is an attribute of comparative underdevelopment or enlightenment is debatable in view of the needs for local government reform in West Malaysia.

Non-hierarchical administrative units of development cover the range of federal agencies which operate outside the state system (such as the Federal

Land Development Authority) and include semi-public, financially autonomous utilities (such as the Penang Port Commission and the National Electricity Board); these authorities receive grants and capital investment from the Federal Government, but operate basically on their self-generated revenues. Private non-profit-making agencies (such as the Borneo Development Corporation, and the Malaya–Borneo Building Society (M.B.B.S.)) are also important and are often associated with the Commonwealth Development Corporation. The newest form of ad hoc non-line development administrative agency is the water resources–food development management organisation. The Muda Agricultural Development Authority is the prototype. The ostensible reason for this innovation (*c.* 1970) was the inadequacy of federal ministries (Drainage and Irrigation, Agriculture, Public Works) and state operators to coordinate their activities sufficiently to bring together the multivarious engineering, human and agronomic factors in a satisfactory sequence. The virtue of the semi-autonomous development agencies is their freedom to mobilise their own resources and to concentrate on specific functions and targets of development.

Thus, scale has a separate meaning for and significance to the civil administrative and the more autonomous government and semi-governmental agencies. In the former, powers are distributed according to old and relatively inflexible forms. There are those areas where there is too little authority (e.g. Pahang, Kelang Valley) and those where too many governments have responsibilities (e.g. Penang before 1966). On the other hand, the public statutory bodies are either areally defined for functional purposes (e.g. the Kuching Water Board, which extends out of town), or they create their own administrative regionalisation of the country and expand or contract according to opportunity. Outside Penang, there has been no areal blending of the two systems or, except in a very vague way, any blending throughout the central government. Future development imperatives will put great pressure on the systems to compete or blend. The singularity and independence of Sabah and Sarawak and their institutions will be an added burden on, though perhaps also a stimulus to, administrative change.

Development policy in area

Malaysia's 111 districts are the basic units of areal analysis. Those of Kuala Lumpur, Johore Baharu, Port Dickson (in Negeri Sembilan) and Province Wellesley Utara (Butterworth) dominate Malaysian manufacturing. Kinta (Ipoh), Kuala Lumpur, Ulu Selangor and Kuantan are by far the largest producers of tin. Estate agricultural production (rubber and palm oil) is concentrated in Keluang and the south coast of Johore and the districts of Selangor and Negeri Sembilan. Riverine districts of Kedah (around Alor Setar), and Kelantan (around Kota Baharu), Perak and mainland Penang produce large amounts of rice. Population is spread through the two or three tiers of districts along the west coast from Johore Baharu to Perlis. Non-Malay population is also localised along the western side of Malaya. In Sarawak the First Division (in the west, around Kuching) is predominant, with Sibu following in all respects save the oil refining and rubber production of Miri. Sabah's two important districts in all respects are Kota Kinabalu and Sandakan, with rubber production spread further along the north-west coast and in Tawau.

These are the areas of traditional inputs into the national economy and the areas that fulfill highest development policy expectations except in terms of employment. Furthermore, these areas generally demonstrate the greatest singularity of blend and receive the greatest absolute amounts of federal development expenditure. A factor analysis of development expenditure and assorted economic-demographic data reveals an unclear distribution of the latter to factors, but shows a breakdown of development expenditure into three factors accounting for 63 per cent of the variance. Factor One contains high factor loadings for expenditures of the to-1961 and the 1961 to 1965 time periods. Factor Two represents 1966 to 1968 expenditure save for ECFN (Economic-functional) which are isolated in Factor Three. Districts high on each of these factors are Kuala Lumpur, Butterworth and Ipoh—clearly the development points of greatest importance in Malaysia. Secondarily (all but high 6668 ECFN scores), Penang Island, Kuala Selangor (Kelang), Ulu Selangor and Kuala Trengganu, in Malaya, and Kuching, Kota Kinabalu and Sibu, in East Malaysia, emerge. It is notable that certain districts with equal or larger total expenditure for 1950–1968 (such as Alor Setar, Batang Padang, Kota Baharu, Johore Baharu and Sandakan) do not take on this historical and overall importance. They are either recently emergent recipients of federal development attention or contain only large special economic-functional (ECFN) projects.

This identification of special development districts in Malaysia is borne out by a series of cluster analyses. These reveal the very great similarity of most of the Malaysian areas insofar as government development activity is concerned. Certain individuals stand out, however in clusters formed on the basis of development expenditure data. (These, like all the cluster analyses, reduce the importance of absolute magnitude differences among the districts by converting the variables to rank scores before clustering according to pattern among them.) The special districts in respect of individuality of character, if not strictly of magnitude, and of development expenditure were: first, Kuching, Kuala Lumpur, Penang and Johore Baharu; and, second, Kinta, Province Wellesley Utara, Kota Setar (containing Alor Setar) and Seremban. Of other districts separated out in cluster analyses of POPC, PROB and ECFN expenditure over the years, Melaka and Kota

Baharu received abnormal social expenditure, and the Kelang Valley and two districts with large hydroelectric facilities (Cameron Highlands and Batang Padang) assumed, of course, great ECFN importance.

Through time there has been an areal spread of development expenditure culminating in a remarkably wide distribution under the First Malaysia Plan. A few new policy points of importance in the smaller middle city regions have arisen. Correlations among the three types of development expenditure to districts during the three periods (1950–60, 1961–65, 1966–68) were only medium high (means of 0·836, 0·807 and 0·758 respectively). But there were discernible trends in the composition of development expenditure as well as in its magnitude. Both POPC and PROB expenditures to large cities (except Kuching) peaked in 1961–65. POPC expenditures generally were proportionally reduced in later years, as economic development policy was pushed outward from the major centres.

Development expenditure for what may very clearly be defined as urban areas and functions (but not counting estate agriculture) from 1950 to 1968 was more than double that for clearly rural activities (including drainage and irrigation). But in the three development periods, the ratios of urban to rural expenditure declined thus: 5·6 : 1, 2 : 1, 1·1 : 1, for similar quantities of development expenditure. If one counted all the 'urbanising' development expenditure in 1966–1968, however, and included communications, transport and administration, but not health or education (whose largest expenditure nevertheless was in urban areas), the ratio for that period is 1·8 : 1. Development policy applied to cities is thus extremely important in Malaysia, as befits a country in which the most important larger areas in development, the districts, are almost all city-centred and operate functionally as city-regions although they are not administered as such.

In mapping the patterns of development expenditure over the years to the 27 middle cities of Malaysia (plus Kuala Lumpur, Malaysia's only proto-'million city', certain cities stand out. The leaders in development policy action have consistently been Kuala Lumpur, Penang and Kuching. As an example of such mapping, Figures 18.2 and 18.3 show total per capita expenditure and the breakdown, from 1950 to 1968, among the three types of conceptual intent. Here, exchanging Butterworth for George Town in Penang, we have the same three most prominent cities; while cluster analyses consistently designate these as special also. Petaling Jaya, the satellite of Kuala Lumpur, and Butterworth emerge as separate clusters over three periods of ECFN expenditure, with George Town and Kuching falling back into the primary cluster. Ipoh clusters with Kuching when all variables are considered. The primary clusters, however, contain more than three-quarters of the cities for all combinations of variables. This shows the basic similarity of most Malaysian middle cities, *vis à vis* the three to six special cases.

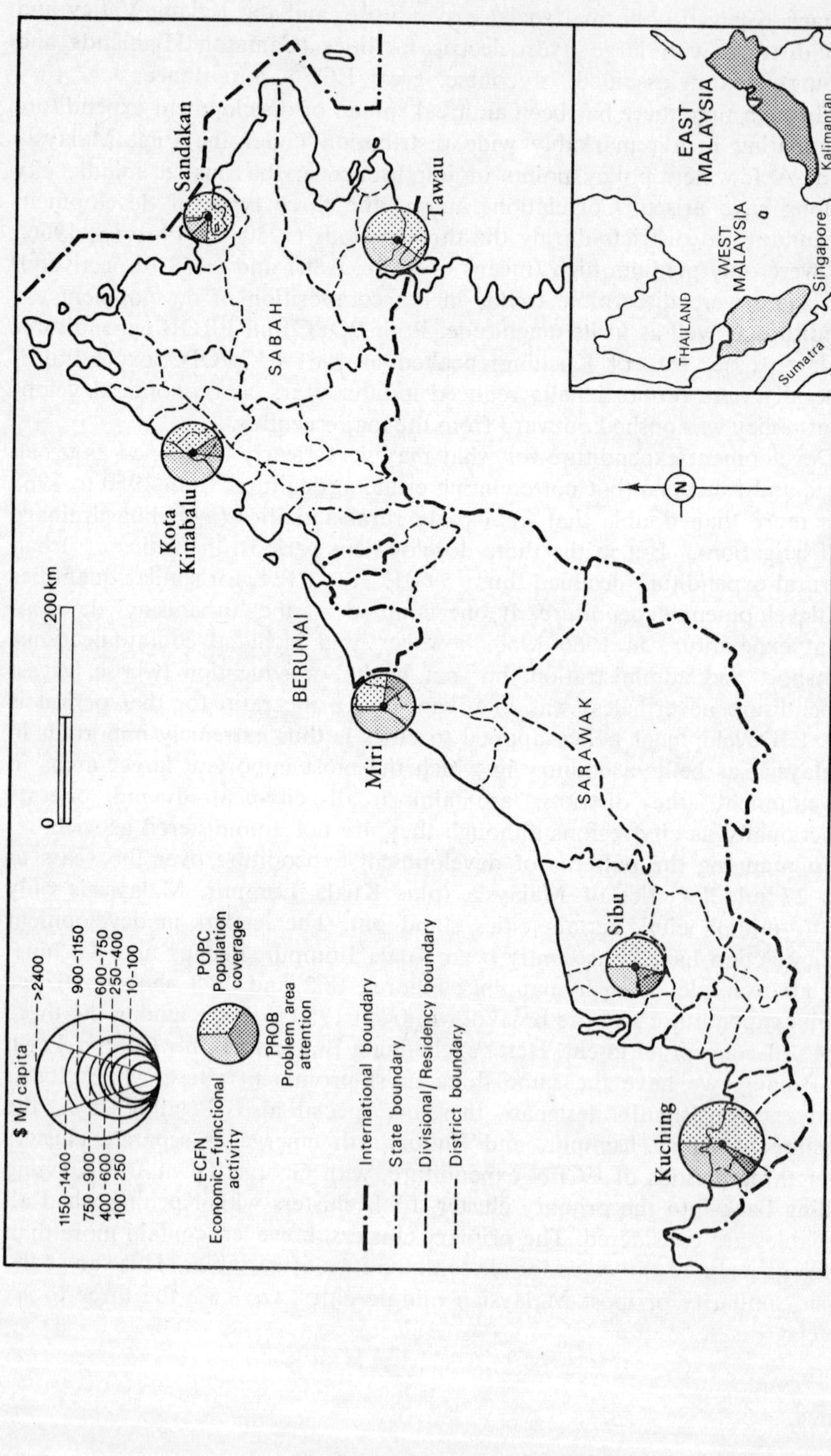

Figure 18.3 East Malaysian middle cities: pre-1969 development expenditure per capita

Table 18.1 Pearson product moment correlation coefficients 1966–68 F.M.P. development expenditures

	6668POPC	6668PROB	6668ECFN	Districts 6668URBN	6668TOTL	6767POPN	65R–ELA
6668POPC	—	0·884	0·504	0·845	0·794	0·749	0·892
6668PROB	0·899	—	0·417	0·864	0·750	0·708	0·968
6668ECFN	0·793	0·822	—	0·680	0·920	0·575	0·982
6668URBN	—	—	—	—	0·865	0·750	0·867
6668TOTL	0·924	0·967	0·373		—	0·738	0·758
6760POPN	0·813	0·906	0·335		0·770	—	0·655
65R–ELA	0·109	0·817	0·298		0·441	0·892	—
65R–EPP	0·406	0·694	0·192		0·439	0·694	0·828
				Cities			

Notes: POPC – Population coverage expenditures
PROB – Problem area expenditures
ECFN – Economic-functional expenditures
TOTL – Total expenditures
6760POPN – Population according to 1960 Censuses in East Malaysia and 1967 estimates in West Malaysia
65R–ELA – 1965 revenues plus expenditures of local authorities
65R–EPP – 1965 revenues plus expenditures of local authorities per capita

A correlation coefficient indicates the degree to which the pattern of a variable's values across a set of observations (in these cases, districts or cities) is similar to the pattern of another variable across the same observation points. In this table a *z*-score form is used, in which positive or negative signs indicate positive or negative correlation, with maximum absolute values of 0·999, and minima of 0·000.

Table 18.2 Middle cities' threshold of dynamism

City	functions[1]	1968 Manuf.	Regd. unem-ploy.	Admin. type[2]	1970 popn.	6165 TOTL proto-type	6668 TOTL	6668 ECFN	65R+ EXPN	65R+ EXPP	47–57 % popn. growth
		Ranks									
Above:											
1. Kuala Lumpur	23456*	1	1	1	1	1	1	2	2	6	14
2. George Town	123456*	7	4	2	2	3	7	6	1	1	26
3. Ipoh	3456	5	2	2	3	11	5	4	3	4	9
4. Johore Baharu	3456	3	3	5	4	5	9	14	6	9	5
5. Kelang	(1)45	6	11	5	5	16	20	—	7	13	6
6. Petaling Jaya	345	2	15	6	6	4	3	3	N.D.	N.D.	1
7. Kuching	13456*	N.D.	18	3	10	7	4	11	4	2	23
8. Butterworth	1245	4	5	4	11	2	2	1	N.D.	N.D.	11
9. Sandakan	1345	N.D.	20	6	21	18	16	16	10	5	10
10. Kota Kinabalu	2456	N.D.	19	6	22	9	8	7	13	8	4
Below:											
11. Melaka	2456	10	7	3	7	6	12	9	5	3	24
12. Seremban	356	12	12	5	8	17	21	20	11	12	16
13. Alor Setar	2456	9	6	5	9	13	15	10	19	22	22
14. Bandar Maharani	35	15	8	5	12	19	23		15	16	18
15. Kota Baharu	256	11	10	5	13	8	19	19	14	18	15
16. Taiping	45	8	9	5	14	14	14	12	9	7	27
17. Bandar Penggaram	5	13	N.D.	5	15	22	18	15	16	14	20
18. Kuala Trangganu	256	21	16	5	16	12	17	17	20	21	19
19. Sibu	235	N.D.	N.D.	6	17	20	13	18	8	N.D.[3]	3
20. Telok Anson	35	18	14	5	18	25	24		18	17	21
21. Kuantan	356	17	17	5	19	15	6	5	22	19	7

22. Keluang	35	16	N.D.	5	20	21	22		17	10	12
23. Niri	235	N.D.	N.D.	6	23	26	10	8	12	N.D.[3]	8
24. Sungei Patani	235	20	13	5	24	23			24	20	13
25. Bukit Mertajam	5	14	N.D.	5	25				N.D.	N.D.	17
26. Kampar	35	19	N.D.	5	26	27			21	15	25
27. Ayer Itam	5	N.D.	N.D.		27	28			N.D.	N.D.	28
28. Tawau	35	N.D.	21	6	28	10	11	13	23	11	2

[1] * = multifunctional (−5); 1 = export; 2 = rice hinterland; 3 = economic primary products hinterland; 4 = 1968 manufacturing greater than $50 million gross sales of own products; 5 = services centre; 6 = state capital

[2] 1 = Federal capital; 2 = City; 3 = Municipality; 4 = Rural District Council; 5 = Town Council; 6 = Town Board

[3] N.D. here because of discrepancies in urban population–urban revenue expenditure definitions

N.B. Blanks mean rank is last, i.e. no absolute score on this variable.

Pearson Product Moment Correlation analysis of all development expenditures to middle cities (see Table 18.1) reveals the 1961–1965 total expenditure pattern to be the prototype of the distribution of all expenditure in the three periods (all but three coefficients of correlation >·800). Comparing this distribution to that of various functional attributes of the middle cities suggests that manufacturing towns have received relatively more expenditure than they otherwise should (Butterworth, Petaling Jaya, Kelang); those commanding primary and agricultural hinterlands have drawn less (Kuching, Ipoh, Alor Setar, Bandar Maharani, Seremban, Sibu, Miri).

The following trends over time are clear: (1) the increasing dominance of Kuala Lumpur–Petaling Jaya, within an overall spread effect of development expenditure to cities; (2) an increasing specialisation of the super-cities: Kuala Lumpur and Kuching to POPC and the government building component of ECFN, Penang to PROB, Butterworth and Ipoh to ECFN; (3) a general decrease (except in Kuala Lumpur and Kuching) in the proportion of POPC expenditure; and (4) a tendency for the larger cities to have higher proportions of PROB expenditures. These trends are reinforced by patterns of policy expectation fulfilment among cities: Kuala Lumpur, Ipoh and the Penang Metropolitan Area had the largest number of M.B.B.S.-supported housing units in process in 1968 and the largest gross value of sales of their own manufactured products.

A large variety of development inputs are directed at middle cities in Malaysia. Certain generalisations emerge from examination of the nine middle city areas–Butterworth, George Town, Ayer Itam (The Penang Metropolitan Area), Alor Setar, Ipoh, Keluang, Kota Baharu, Sandakan and Kuching. Firstly, local governments including localised, semi-autonomous authorities, are key factors in development. With better resources, connections and motivation, they bring more national development activity to their areas (e.g. industrial estates, port development, employment opportunities) and generate their own financial contributions. When stagnant or depressed, local governments are a negative force in city development and for their cities in national development; for instance, they resist state take-over and fail to keep their cities attractive for investment.

Secondly, export and manufacturing functions accompany and perhaps to a degree generate dynamism. Thus a 'threshold of dynamism' is observed among Malaysian cities. Those beyond this threshold (Table 18.2) have generative powers of self-development and at the same time attract large POPC, PROB and ECFN federal expenditure, though in widely differing balances. Their urban malaise, while greater absolutely, like their size, is proportionally less than that of undynamic cities. Most important, perhaps, governments have the impression that they can mould their cities to meet modern challenges.

Thirdly, the more successful elements in the middle city development policy systems are: (1) the welfare ministries, being centralised, hierarchical, standardised, professional and in possession through their state networks of good information, including that bearing on political necessity; (2) the semi-autonomous authorities, owing especially to their independent sources of revenue and business function; (3) opposition governments in cities (Kota Baharu, to a degree; Ipoh); (4) state-controlled city government when the city houses the state government (Alor Setar, George Town; excepting Melaka); (5) local transportation and intra-regional connections and circulation; and (6) the civil services generally.

Fourthly, the least successful elements are: (1) local government financing, except for municipalities; (2) state–city relations, of an integrative and imaginative sort (except in Penang, which is reorganising itself into an integrated city-state, and to a lesser extent in Kuching); (3) private sector–population subsystem relations (to yield employment, lower-cost housing, savings mobilisation); (4) public infrastructural development not managed by authorities; and (5) control of areal development in cities.

Fifthly, one may further generalise on certain Development Policy System elements. *Power* flows poorly. *Information* flows downwards well and, in statistical form, upwards well, but upwards poorly in the case of qualitative judgements. *Projects* usually have the problem of several governments and departments bearing on them all at once, yielding blockages unless they are strictly locally operated (e.g. George Town sewerage versus its water supply—the former federally-financed and inadequate, the latter city-operated and exemplary). The *Production* subsystem of Middle Cities is really two sub-subsystems: the larger-scale western-organised and export (from the city) oriented and the small-scale family business, mostly producing consumer items and services, including professional services. The former is the more amenable to policy and causes fewer internal problems.

CONCLUSION

Development policy

Malaysia has been a prosperous and vigorous country for a long time and her development problems have not been grinding or over-complex. Development policy has, understandably, been consistent over time, yet has had certain discernible trends in its areal content. Development policy has been city-centred in the post-war period, partly because the private sector has so successfully managed the estate system, partly because a paternalistic policy towards the rural Malays left them to their own devices, and partly because Malaysia is so highly urbanised. Nevertheless, development policy has had wide areal coverage in Malaysia with heavy emphasis on infrastructure. But the main trend over time has been one of increasing homogeneity of development policy for districts and cities, in line with general increasing areal homogeneity in Malaysian reality which was accompanied by a tendency for certain areas to become increasingly specialised in their development roles. Thus, increasing similarity among the systems of similar units (for example cities) has resulted; a very few have become more individual and there has been increasing heterogeneity among the sets of areal units of different types. Three types of impact area, among which there is increasing polarisation, are the *cities*, the *export products rural areas* and the *subsistence agricultural areas.*

Kuala Lumpur, with the exit of Singapore from Malaysia, has inexorably become the capital in all respects; its metropolitan area suffers from attendant congestion and poor organisation. There has been a tendency, moreover, to continue putting POPC and even PROB expenditure into old-established large centres and to ignore newer and smaller ones, even when they are growing faster. This seems to inhibit the development of the deprived cities and their regions. Except in magnitude of expenditure the approach to cities in national development has not differentiated systematically between those below and those beyond the 'threshold of Dynamism'.

Cities in development

The functions that middle cities perform in national development in Malaysia, in order of importance in the national view are: (1) production and handling for export; (2) high productivity industrial production and employment; (3) mobilisation of financial, administrative, educational and entrepreneurial resources from national and internal sources; (4) provision of an increasing number of high and middle order services to rural areas as agriculture is modernised; (5) housing development administration; (6) providing physical arrangements for convenient provision of welfare and other social policy incremental inputs; (7) cultural modernisation; and (8) focusing of political-popular-administrative self-help dynamism.

The impact of urban reality on policy, meanwhile, has been simple in Malaysia. One reason why there has been no specifically urban policy until the advent of the Second Malaysia Plan has been that, while Malaysia has become comparatively highly urbanised, gross city malaise and self-development problems have not arisen. Therefore programmes for popular welfare, such as provision of health services, have been national in scope and have closely followed the distribution of population. Cities until 1970 were regarded simply as those places to which people went *from* the problem areas, i.e. the subsistence agriculture sector, and, as such, were seen as places that should be able to take care of themselves. The problems have been that (1) they have not taken care of themselves, in the cases of most undynamic cities, and call out for more support from the state and national governments to spur development; and (2) they have not taken care of their own Malays in the style enjoyed by urban non-Malays. The reality of these points, when combined with the failure of rural development programmes significantly to raise incomes and prospects of Malays, yields now an urban policy of limited but important scope in which cities are treated as working machines for the modernisation of attitudes and lifestyles and thus for the generation of income for everyone.

Further research

The matter of examination and classification of areal policy strategies—of differing geographies of development policy—is an important next step in the evolution of this field of geography. Shortage of data and methodological constraints, however, may be severe. In the study of Malaysia, it has been possible to determine fairly exactly the areal concepts and intents in policy because of the simplicity of the policy for development in Malaysia and the high degree of articulation of it in documents, combined with good reporting of results. None of this can be said, for example, of Indonesia, which has no central complete list of expenditure on national development projects in all provinces for any of the first three years of the five-year development plan. Problems of analysis are worse in individual cities in Indonesia where (1) fewer data are available and (2) the development policy systems are more encumbered by exogeneous systems such as military and village organisation.

Future work in this field might include: development of (1) a workable standardised typology of development policy expenditure and expectations to permit international comparisons, combined with (2) a means of obtaining the necessary data. In fact, the methodology of data gathering in this field is a far larger problem than analysis. While fairly simple machine processing and systems analysis yield profitable results from aggregated data and observations, the gathering of these data and their sifting and arrangement for analysis is very difficult, even in Malaysia (where the largest obtainable amount of mappable development expenditures for any period was 63 per cent for 1966–68). It is remarkable that countries can be so uninformed about their development policy implementation.

It is not so remarkable, however, that countries cannot readily provide areal data of the specific kinds and in the specific forms most useful to research in the geography of development policy; for this is a new endeavour of as yet undemonstrated utility. Further specific research of imporance might include: (1) additional and comparable case studies, particularly from Latin America; and (2) work on the limits imposed by the nature of the larger systems to which they are applied on the possible impact of policies, and work on the limits imposed by there being aspects of the systems which are outside the policy-makers control. In Malaysia an example of such a larger system is the private sector and, in Indonesia, the military; (3) studies of the political factor in explicit and implicit areal policy, particularly the competition of local versus national or regional perceptions and interests, and the roles of particular leaders; and (4) investigation of the applicability of centre-periphery and gravity models in the areal disposition of policy fulfilment (both would seem to be inapplicable in the 'inner' areas of Malaysia but appropriate to the macro-regionalisation of the countries). Such work might be useful for improving policy operation and as a means towards a better understanding of the reality of change. For geography and planning, progress along these lines would be an important and complementary approach in relation to the concept of 'regional planning' as a geographical approach to development. Regions are only one areal form in which development and development policy are disposed, with varying relevance depending upon the level of development of a country and upon the combination of areal concepts employed by policy-makers. Geographers are becoming equipped to apply more subtle and complex concepts and analytical methods and are thereby helping to arrive at a better understanding of the processes and problems of economic development in the less-developed world.

Index